FOREIGN-EXCHANGE-RATE FORECASTING WITH ARTIFICIAL NEURAL NETWORKS

***Recent titles in the* INTERNATIONAL SERIES IN OPERATIONS RESEARCH & MANAGEMENT SCIENCE**

Frederick S. Hillier, Series Editor, *Stanford University*

Gass & Assad/ *AN ANNOTATED TIMELINE OF OPERATIONS RESEARCH: An Informal History*
Greenberg/ *TUTORIALS ON EMERGING METHODOLOGIES AND APPLICATIONS IN OPERATIONS RESEARCH*
Weber/ *UNCERTAINTY IN THE ELECTRIC POWER INDUSTRY: Methods and Models for Decision Support*
Figueira, Greco & Ehrgott/ *MULTIPLE CRITERIA DECISION ANALYSIS: State of the Art Surveys*
Reveliotis/ *REAL-TIME MANAGEMENT OF RESOURCE ALLOCATIONS SYSTEMS: A Discrete Event Systems Approach*
Kall & Mayer/ *STOCHASTIC LINEAR PROGRAMMING: Models, Theory, and Computation*
Sethi, Yan & Zhang/ *INVENTORY AND SUPPLY CHAIN MANAGEMENT WITH FORECAST UPDATES*
Cox/ *QUANTITATIVE HEALTH RISK ANALYSIS METHODS: Modeling the Human Health Impacts of Antibiotics Used in Food Animals*
Ching & Ng/ *MARKOV CHAINS: Models, Algorithms and Applications*
Li & Sun/ *NONLINEAR INTEGER PROGRAMMING*
Kaliszewski/ *SOFT COMPUTING FOR COMPLEX MULTIPLE CRITERIA DECISION MAKING*
Bouyssou et al/ *EVALUATION AND DECISION MODELS WITH MULTIPLE CRITERIA: Stepping stones for the analyst*
Blecker & Friedrich/ *MASS CUSTOMIZATION: Challenges and Solutions*
Appa, Pitsoulis & Williams/ *HANDBOOK ON MODELLING FOR DISCRETE OPTIMIZATION*
Herrmann/ *HANDBOOK OF PRODUCTION SCHEDULING*
Axsäter/ *INVENTORY CONTROL, 2nd Ed.*
Hall/ *PATIENT FLOW: Reducing Delay in Healthcare Delivery*
Józefowska & Węglarz/ *PERSPECTIVES IN MODERN PROJECT SCHEDULING*
Tian & Zhang/ *VACATION QUEUEING MODELS: Theory and Applications*
Yan, Yin & Zhang/ *STOCHASTIC PROCESSES, OPTIMIZATION, AND CONTROL THEORY APPLICATIONS IN FINANCIAL ENGINEERING, QUEUEING NETWORKS, AND MANUFACTURING SYSTEMS*
Saaty & Vargas/ *DECISION MAKING WITH THE ANALYTIC NETWORK PROCESS: Economic, Political, Social & Technological Applications w. Benefits, Opportunities, Costs & Risks*
Yu/ *TECHNOLOGY PORTFOLIO PLANNING AND MANAGEMENT: Practical Concepts and Tools*
Kandiller/ *PRINCIPLES OF MATHEMATICS IN OPERATIONS RESEARCH*
Lee & Lee/ *BUILDING SUPPLY CHAIN EXCELLENCE IN EMERGING ECONOMIES*
Weintraub/ *MANAGEMENT OF NATURAL RESOURCES: A Handbook of Operations Research Models, Algorithms, and Implementations*
Hooker/ *INTEGRATED METHODS FOR OPTIMIZATION*
Dawande et al/ *THROUGHPUT OPTIMIZATION IN ROBOTIC CELLS*
Friesz/ *NETWORK SCIENCE, NONLINEAR SCIENCE AND DYNAMIC GAME THEORY APPLIED TO THE STUDY OF INFRASTRUCTURE SYSTEMS*
Cai, Sha & Wong/ *TIME-VARYING NETWORK OPTIMIZATION*
Mamon & Elliott/ *HIDDEN MARKOV MODELS IN FINANCE*
del Castillo/ *PROCESS OPTIMIZATION: A Statistical Approach*
Józefowska/*JUST-IN-TIME SCHEDULING: Models & Algorithms for Computer & Manufacturing Systems*

**** A list of the early publications in the series is at the end of the book ****

FOREIGN-EXCHANGE-RATE FORECASTING WITH ARTIFICIAL NEURAL NETWORKS

Lean YU, Shouyang WANG and Kin Keung LAI

Springer

Lean Yu
Chinese Academy of Sciences
Beijing, China

Shouyang Wang
Chinese Academy of Sciences
Beijing, China

King Keung Lai
City University of Hong Kong
Kowloon, Hong Kong

Series Editor:
Fred Hillier
Stanford University
Stanford, CA, USA

Library of Congress Control Number: 2007926588

ISBN-13: 978-0-387-71719-7 e-ISBN-13: 978-0-387-71720-3

Printed on acid-free paper.

9 8 7 6 5 4 3 2 1

springer.com

Table of Contents

Preface

The foreign exchange market is one of the most complex dynamic markets with the characteristics of high volatility, nonlinearity and irregularity. Since the Bretton Woods System collapsed in 1970s, the fluctuations in the foreign exchange market are more volatile than ever. Furthermore, some important factors, such as economic growth, trade development, interest rates and inflation rates, have significant impacts on the exchange rate fluctuation. Meantime, these characteristics also make it extremely difficult to predict foreign exchange rates. Therefore, exchange rates forecasting has become a very important and challenge research issue for both academic and industrial communities.

In this monograph, the authors try to apply artificial neural networks (ANNs) to exchange rates forecasting. Selection of the ANN approach for exchange rates forecasting is because of ANNs' unique features and powerful pattern recognition capability. Unlike most of the traditional model-based forecasting techniques, ANNs are a class of data-driven, self-adaptive, and nonlinear methods that do not require specific assumptions on the underlying data generating process. These features are particularly appealing for practical forecasting situations where data are abundant or easily available, even though the theoretical model or the underlying relationship is unknown. Furthermore, ANNs have been successfully applied to a wide range of forecasting problems in almost all areas of business, industry and engineering. In addition, ANNs have been proved to be a universal functional approximator that can capture any type of complex relationships. Since the number of possible nonlinear relationships in foreign exchange rates data is typically large due to the high variability, ANNs have the advantage in approximating them well.

The main motivation of this monograph is to provide academic researchers and business practitioners recent developments in forecasting foreign exchange rates with ANNs. Therefore, some of the most important progress in foreign exchange rates forecasting with neural networks are first surveyed and then a few fully novel neural network models for exchange rates forecasting are presented. This monograph consists of six parts which are briefly described as follows.

Part I presents a survey on ANNs in foreign exchange rates forecasting. Particularly, this survey discusses the factors of affecting foreign exchange rates predictability with ANNs. Through a literature review and analysis, some implications and research issues are presented. According to the results and implications of this survey, the sequel parts will discuss those research issues respectively and provide the corresponding solutions.

In Part II, we provide a data preparation scheme for neural network data analysis. Some basic learning principles of ANNs are presented before an integrated data preparation scheme is proposed to remedy the literature gap.

In terms of the non-optimal choice of the learning rate and momentum factor in neural network learning algorithms in the existing literature, Part III constructs three single neural network models through deriving optimally adaptive learning rates and momentum factors. In the first single neural network model, optimal instantaneous learning rates and momentum factors are derived from the back-propagation (BP) algorithm. Using adaptive forgetting factors, an online BP learning model with optimally adaptive learning rates is developed to realize the online prediction of foreign exchange rates. In the third single neural network model, adaptive smoothing techniques are used to determine the momentum factor of neural network algorithm. Meantime, the proposed neural network model with adaptive smoothing momentum factors is applied to exchange rate index prediction.

In accordance with the analysis in the survey, the hybrid and ensemble models usually achieve better prediction performance than that of individual ANN model, Part IV and Part V present three hybrid neural network models and three ensemble neural network models, respectively.

In the first hybrid neural network model of Part IV, neural network and exponential smoothing model are hybridized into a synergetic model for foreign exchange rate prediction. Subsequently, the generalized linear autoregression (GLAR) and neural network models are fused by another neural network model in a nonlinear way. This model is applied to three typical foreign exchange rate forecasting problems and obtained good prediction performance. In the third model, a hybrid intelligent data mining approach integrating a novel ANN model — support vector machine (SVM) with genetic algorithm (GA) for exploring foreign exchange market movement tendency. In this model, a standard GA is first used to search through the possible combination of input features. The input features selected by GA are used to train SVM. The trained SVM is then used to predict exchange rate movement direction.

In the three ensemble neural network models of Part V, the first model presents a multistage ensemble framework to formulate an ensemble neural network model. In these stages of formulating ensemble models, some crucial issues are addressed. The second ensemble model introduces a meta-learning strategy to construct an exchange rate ensemble forecasting model. In some sense, an ensemble model is actually a meta-model. In the third ensemble model, a confidence-based neural network ensemble model is used to predict the exchange rate movement direction. In the last chapter of Part V, we propose a double-phase-processing procedure to solve the following two dilemmas, i.e., (1) whether should we select an appropriate modeling approach for the prediction purpose or should combine these different individual approaches into an ensemble forecast for the different/dissimilar models? (2) whether should we select the best candidate model for forecasting or to mix the various candidate models with different parameters into a new forecast for the same/similar modeling approaches?

Depended upon the previous methodology framework, an intelligent foreign exchange rates forecasting support system is developed by using client/server model and popular web technologies in Part VI. The description of this intelligent system is composed of two chapters. First of all, system conceptual framework, modeling techniques and system implementations are illustrated in details. Then an empirical and comprehensive assessment is performed. Empirical and comprehensive assessment results reveal that the intelligent exchange rate forecasting support system is one of the best forecasting systems by evaluating the performance of system implementation and comparing with the existing similar systems.

We would like to thank many friends and colleagues for their help and support in preparing this monograph. First, we thank Professor Wayne W. Huang of Ohio University, Professor Y. Nakamori of Japan Advanced Institute of Science and Technology and Professor Wei Huang of Huazhong University of Science and Technology for their contributions to this book. Three chapters are based on the results that we achieved jointly with them. We would like to thank several other scientists for their helpful suggestions and valuable comments on our research in this area, among them are Professor Kate A. Simth of Monash University in Australia, Professor Wuyi Yue of Konan University in Japan and Professor M. Makowski of International Institute of Applied Systems Analysis in Austria.

Finally, we would like to thank the National Natural Science Foundation of China (NSFC), Chinese Academy of Sciences (CAS), Academy of Mathematics and System Science (AMSS) of CAS, Hong Kong Research Granting Committee (HKRGC), City University of Hong Kong and University of Tsukuba in Japan for their financial support to our research in this promising area.

Lean YU
Institute of Systems Science
Academy of Mathematics and Systems Science
Chinese Academy of Sciences
Beijing, 100080, China

and

Department of Management Sciences
City University of Hong Kong
83 Tat Chee Avenue, Kowloon, Hong Kong
Email: yulean@amss.ac.cn

Shouyang WANG
Institute of Systems Science
Academy of Mathematics and Systems Science
Chinese Academy of Sciences
Beijing, 100080, China

and

Graduate School of Systems and Information Engineering
University of Tsukuba
Tsukuba, Ibaraki 305-8573, Japan
Email: sywang@amss.ac.cn

Kin Keung LAI
Department of Management Sciences
City University of Hong Kong
83 Tat Chee Avenue, Kowloon, Hong Kong
Email: mskklai@cityu.edu.hk

Biographies of Three Authors of the Book

Lean Yu received his Ph.D. degree in Management Sciences and Engineering from Institute of Systems Science, Academy of Mathematics and Systems Science, Chinese Academy of Sciences. He is currently a research fellow at Department of Management Sciences of City University of Hong Kong and an assistant professor of Institute of Systems Science, Academy of Mathematics and Systems Sciences. In the past few years, he received many awards and honors, such as "President Prize of Chinese Academy of Sciences" awarded by the Chinese Academy of Sciences, "First Class Prize for Beijing Science and Technology Progress" awarded by the Beijing Municipal Government and "First Class Prize for Applications of Operations Research in China" awarded by the Operations Research Society of China. He has published 15 papers in journals including IEEE Transactions on Knowledge and Data Engineering, European Journal of Operational Research, International Journal of Intelligent Systems, Computers and Operations Research, International Journal of Information Technology and Decision Making, and International Journal of Knowledge and Systems Sciences. He was one of the guest editors of a special issue on risk management of International Journal of Information Technology and Decision Making. His research interests include artificial neural networks, decision support systems, knowledge management and financial forecasting.

Shouyang Wang received his Ph.D. degree in Operations Research from Institute of Systems Science, Chinese Academy of Sciences in 1986. He is currently a Bairen distinguished professor of Management Science at Academy of Mathematics and Systems Science of Chinese Academy of Sciences and the Lotus distinguished professor management science of Hunan University at Changsha. He is also an adjunct professor of over 20 universities in the world. He is the editor-in-chief, an area editor or a co-editor of 12 journals including Information and Management, International Journal of Information Technology and Decision Making, Pacific Journal of Optimization, and Journal of Management Systems. He was/is a guest editor for a special issue/volume of over 10 journals including European Journal of Operational Research, Optimization, Engineering and Optimization, Annals of Operations

Research, Dynamics of Continuous, Discrete and Impulsive Systems, International Journal of Information Technology and Decision Making, IIE Transactions, and Journal of Industrial and Management Optimization. He has published 10 monographs and over 150 journal papers. His current research interests include financial engineering, forecasting, knowledge management and decision analysis.

Kin Keung Lai received his Ph. D. degree at Michigan State University in USA in 1977 and is currently a Chair Professor of Management Science at City University of Hong Kong. Prior to his current post, he was a Senior Operational Research Analyst at Cathay Pacific Airways and an Area Manager for Marketing Information Systems at Union Carbide Eastern. He is the president of the Asia-Pacific Industrial Engineering and Management Society, the general secretary of the Hong Kong Operational Research Society and a council member of the International Federation of Operations Research Societies. He is also a co-editor of 8 journals including Journal of the Operational Research Society and the managing editor of book series "Lecture Notes in Decision Science". He has published two monographs and over 100 journal papers. His main research interests include supply chain and operations management, forecasting, computational intelligence and risk analysis.

List of Figures

List of Tables

Part I: Forecasting Foreign Exchange Rates with Artificial Neural Networks: An Analytical Survey

1 Are Foreign Exchange Rates Predictable? — A Literature Review from Artificial Neural Networks Perspective

1.1 Introduction

After more than two decades of research since Meese and Rogoff's seminal work on exchange rates predictability (see Meese and Rogoff, 1983a, 1983b), the goal of exploiting foreign exchange rates forecasting model to beat naïve random walk forecasts remains as elusive as ever (Taylor, 1995) due to the fact that evidence supporting or refuting the exchange rate predictability seems plausible. For example, Bekaert and Hodrick (1992), Fong and Ouliaris (1995), LeBaron (1999), Levich and Thomas (1993), Liu and He (1991), McCurdy and Morgan (1988), Baillie and Bollerslev (1989), Sweeney (1986) and Soofi et al. (2006) found evidence contrary to the martingale (random walk or pure unit-root) hypothesis for nominal or real exchange rates, indicating that exchange rates are predictable. While Diebold and Nason (1990), Fong, Koh and Ouliaris (1997), Hsieh (1988, 1989, 1993), McCurdy and Morgan (1987), and Meese and Rogoff (1983a, 1983b) found little evidence against the martingale (random walk or pure unit-root) hypothesis for nominal or real exchange rates, implying that predictability of exchange rates is impossible. One simple and possible explanation is that traditional exchange rate forecasting models are inadequate. Due to the fact that exchange rate forecasting is of practical as well as theoretical importance, a large number of methods and techniques (including linear and nonlinear) were introduced to beat random walk model in foreign exchange markets. With increasing development of artificial neural networks (ANNs), researchers and investors are hoping that the foreign exchange market mysteries can be unraveled with neural network models. The main reason of selecting ANNs as an exchange rate forecasting tool is that several distinguishing features of ANNs make them valuable and attractive in forecasting. First of all, in contrast to many model-based forecasting methods, ANNs are data-driven self-adaptive methods in that there are few restrictive assumptions involved in these models for problems under study. This unique feature is highly desirable in several financial forecasting situations, where data are generally abundant but the underlying data generating mechanism is often unknown (Qi and Zhang, 2001). Second, ANNs can generalize. Third, ANNs are universal functional approximators (Hornik et al., 1989). Finally, ANNs are a class of nonlinear model (Zhang et al., 1998). Since Lapedes and Farber (1987) first proposed using multi-layer feedforward neural networks (MLFNN) for nonlinear signal prediction, much research using ANNs have justified their use for nonlinear time series forecasting (Shin and Han, 2000).

However, no one technique has been successful enough to consistently beat other methods in predicting foreign exchange market in any situations. Therefore, it is difficult to say that ANNs uniformly perform better than other methods. Some articles show that ANNs perform well in foreign exchange rates forecasting, while others give negative conclusions. Even in the same article, conflicting results are often presented. For example, Hann and Steurer (1996) found that if monthly data are used, ANNs do not show much improvement over some of the linear models; but for weekly data, ANNs are much better than both monetary and random walk models in forecasting the exchange rate for Deutsche mark (DEM) against U.S. dollar (USD). Hence, the main purpose of this study is to investigate whether foreign exchange rates are predictable by ANNs. In concrete terms, we examine the following five issues.

(a) Why there are mixed results in the literature?
(b) In what situations foreign exchange rates are predictable by ANNs?
(c) In what situations exchange rates are unpredictable by ANNs?
(d) What are the factors that affect performance of ANNs?
(f) What can be done to improve the performance of ANNs?

Due to unique features of the ANNs, their financial applications of ANNs have been a research stream. Accordingly, these applications are reviewed by some researchers. Typically, Wong et al. (1995, 2000) have presented a bibliography of neural network business applications between 1988 and 1998. Wong and Selvi (1998) gave a literature review and analysis of neural networks' applications in finance during the period 1990-1996. Likewise, Fadlalla and Lin (2001) also presented an analysis of application of neural networks in finance. In these three reviews, there are few neural networks' applications for foreign exchange rates. Recently, Huang et al. (2004a) and Yu et al. (2005e) provided a review of foreign exchange rates forecasting with ANNs and described several important design features of ANNs relevant to foreign exchange rates forecasting applications. Their work is, however, different from the current study, presented in this chapter, because the main purposes of this chapter are to investigate whether foreign exchange rates are predictable and what can be done to improve the performance of ANNs (i.e., future research topics).

In view of these stated objectives of this study, we present a general analytical framework of the survey in Fig. 1.1. As can be seen from Fig. 1.1, the analytical process is as follows. First of all, we collect the literature about exchange rates forecasting with ANN approach. Then some related articles are classified into three types in terms of conclusions of articles: (i) ANNs perform better than other models; (ii) ANNs perform worse than other models; and (iii) ANNs give a mixed results, i.e., ANNs perform better than other methods in some situations and worse under some other situations. Subsequently, several main factors that affect the performance of the ANNs, such as prediction horizon, data frequency, train set size, network type, control strategy and training algorithm, are investigated, for each article, to further analyze the reasons behind the mixed results. Furthermore, some general situations of foreign exchange rates predictability are introduced and summarized. Finally, some future research directions for ANNs in exchange rate prediction are given.

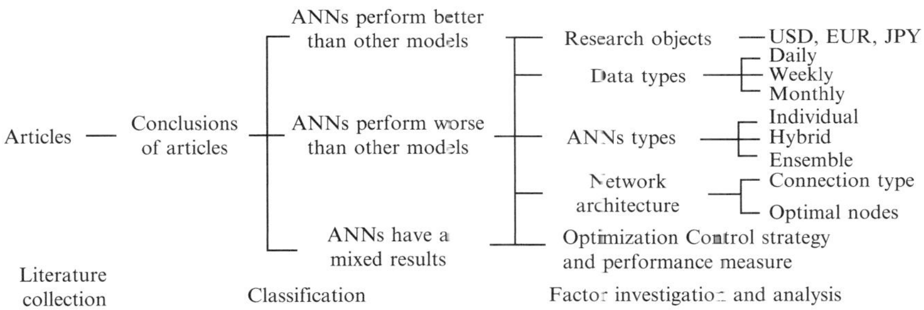

Fig. 1.1. A general analytical framework

The remainder of the chapter is organized as follows. In next section, we explain how articles were selected. Section 1.3 examines and analyzes 45 articles in detail and investigates some main factors that affect the performance of the ANNs. Subsequently, some implications and future research directions are also pointed out in Section 1.4. Finally, Section 1.5 concludes the paper.

1.2 Literature Collection

Since we are interested in investigating some main factors that affect the performance of the ANNs, the criteria of literature selection for this survey is that they should have detailed discussions on the development process of neural networks for exchange rates forecasting. The literature collection process was carried out in two steps. First, ten databases (Science Citation Index, Social Science Citation Index, ScienceDirect, Wiley InterScience, IEEE Xplore, JSTOR, Kluwer online, ProQuest Database, Springerlink, and Academic Search Premier) are searched with the keywords "(artificial) neural network(s) (in) exchange rates forecasting (prediction)" for the period 1971-2004. Searching the ten databases is the most important step in the literature collection process since they include more than 2000 different business-related international journals. In those databases, we were able to retrieve about 300 abstracts that answered to the keywords for the specified period.

Second, a reference search of textbooks on neural networks and exchange rates forecasting (prediction) was conducted. We considered a total of fourteen textbooks: Azoff (1994), Masters (1995), Beltratli et al. (1996), Gately (1996), Kacapyr (1996), Trippi and Turban (1996), Kindon (1997), Lisboa et al. (2000), Kovalerchuk and Vityaev (2000), Graf (2002), Shadbolt and Taylor (2002), Soofi and Cao (2002), Smith and Gupta (2003), and Zhang (2003). However, the related articles in Zhang and Smith & Gupta's book are the same as the journal articles searched in the ten databases, while other books are about applications of ANNs to other financial research, such as stock, interest rate and bank failure prediction. Therefore, these books are excluded in the investigations.

Thus, we investigated only the journal articles. Some of the retrieved articles did not present detailed discussions on the development process of neural networks for exchange rates forecasting. Consequently, we review only 45 articles with detailed forecasting process in the survey (see Tables 1.1-1.4). Fig. 1.2 shows the distribution of the articles by the year of publication. Although our research covers the period 1971-2004, we find no applications about neural networks in exchange rates forecasting published earlier than 1993.

We can also find that the number of articles published in 1998, 2000 and 2002 are fewer than their respective preceding and following years, i.e., articles in 1998 are fewer than those in 1997 and 1999. Similarly, articles in 2000 are fewer than those in 1999 and 2001 and articles in 2002 are fewer than those in 2001 and 2003. There are several articles related to ANNs and exchange rates forecasting, but some articles do not satisfy the above selection criteria and therefore they are excluded from this survey.

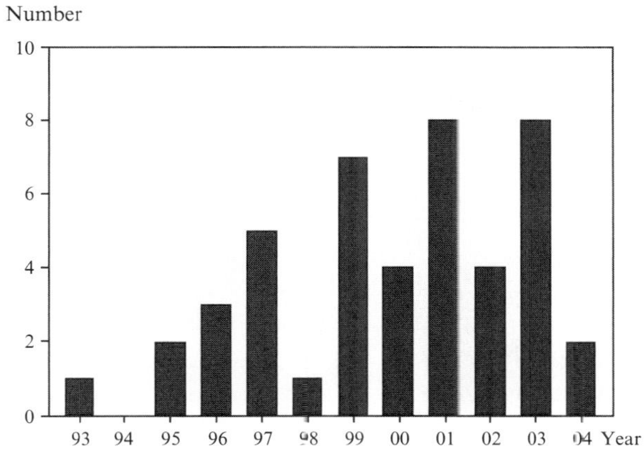

Fig. 1.2. Distribution of articles by year

1.3 Analytical Results and Factor Investigation

1.3.1 Basic Classifications and Factors Summarization

As earlier noted, a final total of 45 articles are selected for further analysis in this survey. In accordance with the general analytical framework, the articles are classified into three categories by the forecasting performance of the ANNs, as follows:

(I) ANNs perform better than other models;
(II) ANNs perform worse than other methods;
(III) ANNs perform better than other methods in some situations, while worse in other situations.

Table 1.1 shows classification of 45 articles by prediction performance. It indicates that the ANNs are not uniformly better than other methods in all situations, even though ANNs are a class of advanced artificial intelligence (AI) technique. Even within the same article, mixed conclusions are often presented, as shown in third category.

Table 1.1. Classification of 45 articles by performance

Category	Articles	Ratio (%)
I	27	60.00
II	2	4.44
III	16	35.56

To find out the reasons for the inconsistent conclusions, we decompose and investigate the factors that affect the performance of the ANNs, from the following aspects:

(a) Prediction horizon, including short-term (1-3 steps), medium-term (4-8 steps) and long-term (more than 8 steps) forecasting;
(b) Data frequency (daily, weekly, monthly and quarterly);
(c) Training set size;
(d) Network type (e.g. MLFNN, RNN, hybrid etc.);
(e) Control strategy (recurrent, feedforward etc.);
(f) Training algorithm;
(g) Transfer function;
(h) Performance measure.

Accordingly, the factors are summarized in Tables 1.2-1.4. For convenience, we give classification order of the references in terms of the above categories: articles of nos. 1-27 are the first category, articles of nos. 28-29 are the second category and those of nos. 30-45 are the third category.

1.3.2 Factor Analysis

In Tables 1.2-1.4, some basic information about these articles, including background information (authors, year and classification), research objects and forecasting horizon, information about the data used (data type, time range, number of training samples, number of testing samples), information about network architecture (connection type, model type, number of nodes per layer, comparable methods), and information about optimization strategies used (control strategy, training algorithm, transfer function, performance measures), is presented. In order to mine useful information and present a reasonable explanation for the issues we are concerned with, it is necessary to further investigate the factors and summarize the comparisons. It should be noted that some of the values shown in Tables 1.2-1.4 are not explicit in the articles and some of them are inferred from the information provided.

1.3.2.1 Research objects and data type

In all examined articles, our survey finds that 43 applications (95.56%) are related to several internationally traded currencies, such as Canadian dollars (CAD), Australian dollars (AUD), German marks (DEM), Swiss francs (CHF), Japanese yen (JPY), and British pounds (GBP). Only two articles (Shin and Han (2000), Wu (1995)) focus on Korean wons (KRW).

Table 1.2. Details of the articles reviewed (Basic information and research objects)*

Basic information and classification				Research objects and prediction horizon		Comparable methods
No	Author(s)	Year	Classification	Object(s)	Prediction horizon	Benchmark models
1	Bolland & Connor	1997	Better	USD/DEM	One step	AR, Kalman filter
2	Chen & Leung	2004	Better	(GBP, CAD, JPY)/USD	1 month	RW, MTF, GMM, BVAR, Error correction
3	Chun & Kim	2003	Better	JPY/USD	1 day	CBR, PCA, FA
4	El Shazly & El Shazly	1997	Better	(GBP, DEM, JPY)/USD	4 weeks	?
5	El Shazly & El Shazly	1999	Better	(GBP, DEM, JPY, CHF)/USD	1 quarter	GA
6	Giles et al.	2001	Better	(DEM, GBP, CAD, JPY, CHF)/USD	1-7 day(s)	RW
7	Jamal & Sundar	1997	Better	USD/DEM, USD/FRF	1 month	Regression model
8	Jasic & Wood	2003	Better	USD/(DEM,JPY,CHF,GBP)	1 day	RW, Linear model
9	Kaashock & Van Dijk	2002	Better	(GBP, NLG, FRF, DEM, JPY)/USD	1 month	ARIMA
10	Kodogiannis & Lolis	2002	Better	USD/GBP	1, 4, 8 day(s)	BP
11	Kumar et al.	2003	Better	CAD/USD	1 month	Regression
12	Leung et al.	2000	Better	(GBP, CAD, JPY)/USD	1 month	RW, ARIMA, MTF, MLFNN
13	Li et al.	1999	Better	USD/SGD	1 month	MLFNN
14	Nag & Mitra	2002	Better	DEM/USD, JPY/USD, USD/GBP	1 day	GARCH, ARCH, FGNN
15	Parhizgari& De Boyrie	1997	Better	(GBP, CAD, DEM, JPY, CHF)/USD	1, 3, 6 day(s)	RW, Locally weighted regression
16	Poddig & Rehkugler	1996	Better	USD/DEM, JPY/DEM, USD/JPY	6 month	RW, Regression, SLP
17	Qi & Zhang	2001	Better	GBP/USD	1 week	RW, AR
18	Refenes et al.	1993	Better	DEM/USD	Varied	AR, ES
19	Rivas et al.	2003	Better	GBP/USD	1 week	?
20	Shin & Han	2000	Better	KRW/USD	1 day	RW
21	Tenti	1996	Better	DEM/USD	2 days	BP

22	Vojinovic et al.	2001	Better	USD/NZD	1, 3, 5 day(s)	AR, RW
23	Walczak	2001	Better	(GBP, DEM, JPY)/USD	?	?
24	Wu	1995	Better	TWD/USD	1, 6 month(s)	ARIMA
25	Yao & Tan	2000	Better	USD/(JPY, GBP, DEM, CHF, AUD)	1 week	RW
26	Yu et al.	2004	Better	USD/(DEM, GBP, JPY)	1 Month	GLAR, Single MLFNN
27	Zhang	2003	Better	GBP/USD	1, 6, 12 week(s)	ARIMA, Single MLFNN
28	Qi & Wu	2003	Worse	(JPY, DEM, CAD, GBP)/USD	1, 6, 12 month(s)	RW, Linear regression
29	Gencay	1999	Worse	(GBP, DEM, FRF, JPY, CHF)/USD	1, 5, 10 day(s)	RW, GARCH, GARCH, KNN
30	Davis et al.	2001	Mixed	CAD/USD	1-15 day(s)	RW
31	Dempster et al.	2001	Mixed	GBP/USD	1 month	RW, Logit, RL, GA, LP, Heuristic
32	Dunis & Huang	2002	Mixed	GBP/USD, USD/JPY	1 day	GARCH, Regression
33	Franses & Van Homelen	1998	Mixed	(USD, CAD, GBP, JPY)/NLG	1 day	GARCH, Bilinear
34	Hann & Steurer	1996	Mixed	DEM/USD	4 weeks	RW, Linear model
35	Hong & Lee	2003	Mixed	(CAD, DEM, GBP, JPY, FRF)/USD	1 week	RW, AR, FC, PN
36	Hu & Tsoukalas	1999	Mixed	12 currencies of EU	1 day	MAV, GARCH, EGARCH, IGARCH
37	Hu et al.	1999	Mixed	GBP/USD	(1, 6, 12) month(s)	RW
38	Huang et al.	2003	Mixed	USD/GBP, USD/JPY	(1, 3, 5, 10, 30) day(s)	RW
39	Kuan & Liu	1995	Mixed	(GBP, CAD, DEM, JPY, CHF)/USD	1 day	RW
40	Lisi & Schiavo	1999	Mixed	(FRF, DEM, LIT, GBP)/USD	1 month	RW, Chaotic model
41	Medeiros et al.	2001	Mixed	14 exchange rates	1-4 month(s)	RW, AR, NCSTAR
42	Taylor	2000	Mixed	(DEM, JPY)/USD	1 day	GARCH, Line quantile regression
43	Walczak & Cerpa	1999	Mixed	(GBP, JPY, DEM)/USD	?	BP
44	Yao et al.	1997	Mixed	CHF/USD	1 day	RW
45	Zhang & Berardi	2001	Mixed	GBP/USD	1 day	KTB

*AR: auto-regression; ?: not specified; RW: random walk; MTF: multivariate transfer function; GMM: generalized method of moments; BVAR: Bayesian vector auto-regression; CBR: case based reasoning; RW: random walk; GARCH: generalized autoregressive conditional heteroskedasticity; BP: back-propagation; ARIMA: autoregressive integrated moving average; ES: exponential smoothing; GLAR: generalized linear auto-regression; KTB: keep-the-best.

Table 1.3. Details of the articles reviewed (Network architecture and data types)*

	Network type and its architecture			Data type and its specification			
No	Network type	I/O nodes	Hidden layer: nodes	Data type	Time range	Training size	Testing size
1	Hybrid RNN	?/1	1:4	Tick data	1993:03-1995:04	?	?
2	GRNN	?/?	2:?	Monthly	1980:01-2001:12	144	60
3	MLFNN	6/1	1:4	Daily	93:01:25-94:09:30	428	100
4	MLFNN	4/1	1:10	Weekly	88:01:08-94:04:08	289	32
5	MLFNN+GA	5/1	?	Quarterly	1977:01-1996:04	72	8
6	SOM, RNN	?/2	5	Daily	73:09:03-87:05:18	1210-1959	750
7	MLFNN	?/?	?	Monthly	1984:01-1993:12	?	?
8	MLFNN	(6, 7)/1	1:4	Daily	86:01:02-99:11:11	2606	1010
9	MLFNN	?/?	?	Monthly	1957:01-1998:03	?	?
10	RNN, RBF, ELM, AFLS	(24, 4)/(6, 1)	2:(34, 16), (8, 18, 24)	Daily	1997:12-2000:03	800	200
11	RNN	?/?	?	Monthly	1974:01-2001:11	300	35
12	GRNN	(1, 2, 12)/1	?	Monthly	1974:01-1995:07	129	65
13	Fuzzy NN	3/3	2:9, 27	Monthly	1990:01-1995:09	50	19
14	GANN	?/1	2:(2-16)	Daily	1992:01-1998:05	?	?
15	MLFNN	(2-18)/1	1:10	Daily	85:01:02-94:06:30	1989-2078	248-251
16	MLFNN, RNN	?/?	?	Monthly	1978:01-1994:05	143	24
17	MLFNN	(1-5)/1	1:1-5	Weekly	1976:01-1993:12	1976:01-1989:12	1990:01-1993:12
18	MLFNN	9/3	2:12, 6	Daily	1988:01-1989:01	200	60
19	Evolved RBF	?/?	?	Weekly	79:12:31-83:12:26	?	?
20	GANN	(4, 12)/1	1:(4, 12)	Daily	90:01:10-97:06:25	90:01:10-95:08:15	95:08:07-97:06:25
21	RNN	18/1	1:5	Daily	1990:01-1994:12	424	100
22	RBF	?/?	?	Daily	1997:01-2001:12	100, 600, 1300	?
23	MLFNN	(1, 2, 3, 5)/?	?	Daily	73:03:01-95:06:30	73:03:01-94:12:30	95:01:01-95:06:30

24	MLFNN	8/1	1:25	Monthly	1979:01-1992:12	162	6
25	MLFNN	(5, 6)/1	1: Varied	Weekly	84:05:18-95:07:07	1984:05-1993:10	1993:11-1995:07
26	Hybrid MLFNN	4/1	1:4	Monthly	1971:01-2003:12	360	36
27	Hybrid MLFNN	7/1	1:(5,6)	Weekly	1980:01-1993:12	?	?
28	MLFNN	?/?	?	Monthly	1973:03-1997:07	?	?
29	MLFNN	?/?	?	Daily	73:01:02-92:07:07	?	?
30	BP, MOD, RBF, LVQ, ARTMAP, GRL	1/1 (BP, RBF, GRL); 2 (LVQ ARTMAP);4 (MOD)	2:5, 2 (BP, RBF); 2:9, 4 (MOD); 1:6 (LVQ); 2:4,5 (ARTMAP); 1:5(GRL)	Daily	92:01:02-94:12:15	544	200
31	BP	?/?	?	Monthly	1994:01-1998:01	?	?
32	RNN &Combined	44/1	1:1 (2:10, 5)	Daily	93:12:31-00:05:09	1329	280
33	MLFNN	?/?	?	Daily	1986-1992	Three years data	One years data
34	MLFNN	?/?	?	Weekly/Monthly	86:01:27-94:10:10	368	87
35	MLFNN	?/?	?	Weekly	75:01:01-89:12:31	?	?
36	MLFNN	4/1	1:4	Daily	79:03:13-94:12:30	79:03:13-90:04:04	93:06:03-94:12:30
37	MLFNN	10/1	1:(4, 8, 12, 16, 20)	Weekly	1976:01-1993:12	?	?
38	MLFNN	(3, 5, 7, 9)/1	1:4	Daily	97:01:01-02:09:06	2000	70
39	MLFNN, RNN	(1-6)/?	1:2-6	Daily	80:03:01-85:0128	50, 100, 150	1194, 1144, 1094
40	MLFNN	(2, 6, 10, 20)/1	1:(1-5, 7, 10)	Monthly	1973:01-1995:10	175	48
41	Bayesian MLFNN	?/?	?	Monthly	1971:01-2000:07	Varied	Varied
42	QRNN	?/?	?	Daily	88:07:04-96:07:05	1014	1000
43	ART, RBF	?/?	?	?	?	?	?
44	MLFNN	6/1	1: Varied	Daily	1983:03-1995:11	Varied	Varied
45	NN ensemble	(1-5)/1	1:(2 ,4, 8)	Daily	1976-1994	782	92

*RNN: recurrent neural network; ?: not specified; GRNN: generalized regression neural network; MLFNN: multilayer feedforward neural networks; PCA: principle component analysis; FA: factor analysis; MOD: modular; RBF: radial basis function; LVQ: learning vector quantization; ARTMAP: adaptive resonance theory map; GRL: genetic reinforcement learning; RL: reinforce learning; GA: genetic algorithm; LP: linear programming; GA: genetic algorithm; SOM: self-organizing map; PN: polynomial; FC: functional coefficient; NCSTAR: Neuro-coefficient smooth transition autoregressive model; GANN: genetic algorithm neural networks; QRNN: quantile regression neural network; GRG: generalized reduced gradient.

Table 1.4. Details of the articles reviewed (Network strategy and optimization algorithm)*

	Network strategy		Network algorithm and transfer function			Performance measure
No	Model type	Control strategy	Training algorithm	Learning rate	Transfer function	Indicator
1	Hybrid	Recurrent	Estimation maximization	?	Logistic	MSE, MAD
2	Individual	Feedforward	Modified backpropagation	?	Gaussian	RMSE, U_{stat}, R^2, Profit
3	Individual	Feedforward	Backpropagation	Fixed value	Logistic	MAPE
4	Individual	Feedforward	Backpropagation	Fixed value	Logistic	TAFE, MAFE, Correct
5	Hybrid	Feedforward	Genetic algorithm (GA)	Fixed value	Logistic	TAFE, MAFE, Correct
6	Individual	Recurrent	?	?	Gaussian	Error rate
7	Individual	Feedforward	Backpropagation	Fixed value	Logistic	Percent error
8	Individual	Feedforward	Conjugate gradient	Fixed value	Logistic	RMSE, NMSE, S_{stat}, D_{stat}
9	Individual	Feedforward	Backpropagation	Fixed value	Logistic	RMSE
10	hybrid	Hybrid	Gradient descent	?	Logistic/RBF	RMSE, SDE
11	Individual	Recurrent	?	?	Hyperbolic Tangent	RMSE
12	Individual	Feedforward	?	?	Gaussian	RMSE, MAE, U_{stat}
13	Hybrid	Feedforward	Self-organized learning	?	Gaussian	MSE, MAPE, U_{stat}
14	Hybrid	Feedforward	GA	?	Hyperbolic Tangent	AAE, MAPE, MSE
15	Individual	Feedforward	Backpropagation	Fixed value	Logistic	MSE, RMSE
16	Individual	Hybrid	?	Fixed value	?	MSE, hit rate, return
17	Individual	Feedforward	Backpropagation	Fixed value	Logistic	RMSE, MAE, DA, Sign
18	Individual	Feedforward	Backpropagation	Fixed value	Squashing function	Returns
19	Hybrid	Feedforward	GA, Hill climbing	?	Radial basis function	MSE
20	Hybrid	Feedforward	GA, Hill climbing	?	?	RMSE
21	Individual	Recurrent	Backpropagation	?	Logistic	NMSE, ROE, ROC
22	Individual	Feedforward	?	?	Radial basis function	RMSE, Direction

23	Individual	Feedforward	Backpropagation	Fixed value	Logistic	Forecast accuracy
24	Individual	Feedforward	Backpropagation	Fixed value	Logistic	RMSE, MAPE, MAE
25	Individual	Feedforward	Backpropagation	Fixed value	Logistic	NMSE, D_{stat}, return
26	Hybrid	Feedforward	Levenberg-Marquardt	Fixed value	Logistic	NMSE, D_{stat}, return
27	Hybrid	Feedforward	GRG2	Fixed value	Logistic	MSE, MAD
28	Individual	Feedforward	Levenberg-Marquardt	Fixed value	Logistic	RMSE, Direction accuracy
29	Individual	Feedforward	Backpropagation	Fixed value	Logistic	MSPE, Sign
30	Individual	Hybrid	Hybrid	?	Varied	Sign
31	Individual	Feedforward	GA, RL, LP, Heuristic	Fixed value	?	Monthly returns
32	Ensemble	Recurrent	?	?	Logistic	RMSE, MAE, U, CDC
33	Individual	Feedforward	Backpropagation	Fixed value	Logistic	Directional accuracy
34	Individual	Feedforward	Backpropagation	Fixed value	Logistic	Return, hit rate, U_{stat}
35	Individual	Feedforward	Backpropagation	Fixed value	Logistic	MSPE, sign
36	Individual	Feedforward	Backpropagation	Fixed value	Logistic	RMSE
37	Individual	Feedforward	GRG2	Fixed value	Logistic	RMSE
38	Individual	Feedforward	Backpropagation	Fixed value	Logistic	RMSE
39	Individual	Hybrid	Backpropagation /Newton	Fixed value	Logistic	RMSPE, Sign
40	Individual	Feedforward	Backpropagation	Fixed value	Logistic	NMSE
41	Individual	Feedforward	Levenberg-Marquardt	?	Logistic	RMSE, MAE, MAD, Sign
42	Individual	Feedforward	Backpropagation	?	Logistic	MSE
43	Individual	Feedforward	Backpropagation	?	Logistic	Forecast accuracy
44	Individual	Feedforward	Backpropagation	Fixed value	Logistic	NMSE, Return
45	Ensemble	Feedforward	Scaled Conjugate gradient	?	Logistic	MSE, MAE

*?: not specified; MSE: mean squared error; MAE: mean absolute error; MAD: mean absolute deviation; RMSE: root mean squared error; MSPE: mean squared prediction error; MAPE: mean absolute percentage error; TAFE: total absolute forecasting error; MAFE: mean absolute forecasting error; AAE: Average absolute error; NMSE: normalized mean squared error. The "hybrid" of the second column represents the mix of recurrent and feedforward strategy.

and Taiwanese dollars (TWD) against U.S. dollars. Some other currencies are included in these applications, as shown in Table 1.2. The data type of these research objects include daily, weekly, monthly and quarterly. Of the 45 journal articles, 21 articles (46.67%) used daily data, 14 articles (31.11%) used monthly data, and 8 articles (17.78%) used weekly data, while only one article (El Shazly and El Shazly, 1999) used quarterly data and one article (Bolland and Conner, 1997) used high frequency tick data. Furthermore, of the 27 articles in category I, 11 articles used daily data, 5 articles used weekly data and 9 articles used monthly data. Besides, one article used quarterly data and another used tick data. Likewise, of the articles in category III, 9 articles used daily data and 3 articles used weekly data and 4 articles used monthly data. While in the two articles of category II, one is daily data and the other is monthly data. Therefore, it is hard to say which data type is easy to predict in terms of previous analysis.

In view of the results in Table 1 2, several conclusions are summarized in the sequel. First of all, different pairs of currencies often lead to different forecasting results. Second, different data types often result in different forecasting performance, which is the same as the first conclusion. The two conclusions from Table 1.2 are quite evident. Third, for the same research objects, different data types may generate different conclusions. For example, the article of Parhizgari and De Boyrie (1997) showed that the ANNs performed better than other methods for GBP, CAD, DEM and JPY, while in the article of Qi and Wu (2003), they found that the ANNs were worse than other methods for the same research objects, i.e., currencies. Even within the same article, this conclusion still holds good. For example, in the article of Hann and Steurer (1996), if monthly data are used, ANNs do not show much improvement over linear models; but for weekly data, ANNs are much better than both the monetary and random walk models in DEM/USD forecasting. Finally, even for the same data type, same search objects may lead to conflicting conclusions. For example, with daily data, Giles et al. (2001) pointed out that the ANNs perform better than other approaches for GBP, DEM, JPY and CHF, but Gencay (1999) found that the ANNs perform worse than other approaches for the same objects (currencies). Therefore, it is difficult to say which currency is predictable or unpredictable to some extent.

Therefore, different research objects and different data types often result in different forecasting performance with ANNs, according to the literature review. This is one of the reasons leading to inconsistent results, as revealed by Hu and Tsoukalas (1999) and Medeiros et al. (2001). It is, however, hard to infer exchange rates predictability from research objects and data types with the use of ANNs, as indicated in the analysis. For interpretation, further factor investigation and analysis about ANN itself is required.

1.3.2.2 Forecasting horizons

From the examined articles, it is found that different forecasting horizons often lead to different forecasting performance, and they sometimes lead to mixed results even in the same article, which is the same as the research objects. As such, different forecasting horizons are one of the reasons that affect the performance of the ANNs. As Huang et al. (2003) noted, for the forecasting horizons of 1, 3, 5 days, the ANNs perform better than the random walk, while for forecasting horizons of 10 and 30 days, the general performance of ANNs is worse than the random walk model. Thus, determining an appropriate forecasting horizon is necessary. In the literature review, Chun and Kim (2003) proposed using Lyapunov exponent to determine the most suitable predictive horizons from the information loss perspective. For the most suitable forecasting horizon, the ANNs only can generate valid forecasts. Generally, for short and medium term forecasting, ANNs can give effective forecasts. On the contrary, for long-term forecasting, foreign exchange rates are not predictable with the use of ANNs, according to the method of Chun and Kim (2003). It is worth noting that we define "short-term" as "1-3 step(s)", "medium-term" as "4-8 steps", and "long-term" as "more than 8 steps". Here "step" denotes data frequency, such as "days", "weeks", "months" or "quarters". Of the 45 journal articles, 39 articles used short-term forecasting, 15 articles used medium-term prediction and only 6 articles used long-term forecasting for foreign exchange rate. Furthermore, of the 27 articles in category I, all articles, except Zhang (2003), used short-term or medium-term forecasting, while the two articles in category II used long-term forecasting besides short-term and middle-term forecasting. In the same way, of the 16 articles in category III, three articles used long-term forecasting. Thus, we can infer that long-term prediction horizons generally lead to negative or mixed forecasting results in foreign exchange rates forecasting with ANNs, according to the literature analysis.

1.3.2.3 Data division and training set size

In applications of ANN in foreign exchange rates forecasting, data partition is necessary. Furthermore, the data split can have a significant impact on the results obtained (Dawson and Wilby, 1998; Maier and Dandy, 1996). In the literature review, data division was carried out in most articles. Although at least two data sets were used, the validation data were used as part of the training process in many instances either by using a trial-and-error procedure to optimize model inputs and some network parameters, or by using cross-validation (CV) to verify the effectiveness of the ANNs (Hu et al. (1999) found that the CV can effectively improve the robustness of the ANNs). Generally, data division is carried out on an arbitrary basis and the statistical properties of the respective data sets were seldom considered. In the first category of the articles we examined, three data sets (i.e., training dataset, validation dataset and testing dataset) are used and good results are obtained.

In the same way, training set size is also an important factor that affects the performance of ANNs. As Walczak (2001) revealed, neural networks, given an appropriate amount of historical knowledge, can predict future currency exchange rates with 60 percent accuracy, while neural networks trained on a smaller or larger training set have a lower forecasting accuracy. His experimental results indicated that for financial time series, two years of training data is frequently all that is required to produce optimal forecasting accuracy. Besides, Huang et al. (2004b) proposed using change-point detection method to seek training sets for neural networks in exchange rates forecasting. Of the 27 articles in category I, 22 articles used data split and had corresponding training set size, while of the two articles in category II, data division is not specified. Of the 16 articles in category III, 11 articles divided the data into at least two data sets and only 5 articles did not provide information related to data division. Thus, appropriate data division and training set size is helpful to predict foreign exchange rates. That is, splitting data into three datasets and use of adequate training data can effectively predict foreign exchange rates by ANNs.

1.3.2.4 Network types and model types

In this survey, multi-layer feedforward neural networks (MLFNN) were used in 30 articles (66.67%) and some other types of networks were used in the remaining articles, such as recurrent neural network (RNN), general regression neural network (GRNN), and radial basis function network (RBFN). In addition, of the 27 articles in category I, 15 articles used multi-layer feedforward neural networks with one hidden layer. This implies that multi-layer feedforward neural networks with single hidden layer and enough nodes can predict the exchange rates well. In addition, Hornik et al. (1989) and White (1990) found that a single hidden layer architecture with an arbitrarily large quantity of hidden nodes in the single hidden layer is capable of modeling any categorization mapping. In a sense, MLFNN is adequate for predicting foreign exchange rates.

In 14 articles, the ANN model type is hybrid or ensemble, while others is individual. Of the 14 articles, 9 articles are the first category. Actually, the hybridization of neural networks with other technologies, such as expert systems, fuzzy logics and genetic algorithms (GA) can improve the applicability of neural networks in addressing various types of finance problems. Although each technique has its own strengths and weaknesses, these technologies are complementary. Weakness of one technology can be overcome by strengths of another by achieving a synergistic effect. Such an effect can create results that are more efficient and effective than the simple sum of their parts. In the survey, five of the 14 articles are integrated with genetic algorithms and two of them are combined with fuzzy logics. In addition, three of the 14 articles used an ensemble strategy for foreign exchange rates prediction and reported some promising and potential results. Furthermore, all hybrid or ensemble models are superior to the individual models in the examined articles. This indicates that the hybridization and ensemble of neural networks is an effective way to improve the performance of exchange rates prediction.

1.3.2.5 Control strategy

In fact, the control strategy describes the way information flows, and the neural network uses the control strategy to learn and to improve its performance. Table 1.5 shows the number of articles employing different control strategy. 36 articles employed feedforward strategy; 5 articles employed recurrent or feedback strategy; and 4 articles employed hybrid strategy, as shown in Table 1.5. It indicates that the feedforward strategy is the most widely used, but recurrent and hybrid strategies are also potential candidates.

As can be seen from Table 1.5, 21 of 27 articles in category I used feedforward control strategy. It indicates that feedforward control strategy is suitable for foreign exchange rates forecasting with ANNs.

Table 1.5. The number of articles employing different control strategies

Control strategy	Category I	Category II	Category III	Sum
Feedforward	21	2	13	36
Recurrent or Feedback	4	0	1	5
Hybrid	2	0	2	4
Sum	27	2	16	45

1.3.2.6 Training algorithm

In this survey, the back-propagation (BP) algorithm was used to search optimal connection weights in 23 of the 45 articles reviewed, as shown in Table 1.4. In 3 articles, Levenberg-Marquardt algorithm was used to speed up the convergence. In addition, other algorithms, such as GA, GRG2 and conjugate gradient are also used. As is well known, BP is by far the most widely used algorithm for optimizing feedforward neural networks; it is based on the method of steepest descent. However, BP is prone to be trapped into local optimum and, therefore GA and simulated anneals are used to improve this defect; but they are time-consuming and offer weak convergence. Therefore, Levenberg-Marquardt algorithm, GRG algorithm and conjugate gradient algorithm are developed to overcome the drawbacks. However, in the survey, both BP and other algorithms perform better than those of other models except Gencay (1999) and Qi and Wu (2003).

It should be mentioned that the selection of optimal learning algorithm is an open problem (Walczak and Cerpa (1999)) and ANNs' design must use the constraints of the training data set for determining the learning method. If a reasonably large quantity of relatively noise-free training examples is available, then BP algorithm can provide effective forecasting results.

In the 45 articles, 24 articles (53.33%) use BP learning algorithm. Moreover, 12 articles in category I also used BP algorithm and obtained good performance. This shows that the BP algorithm is a widely used and good training algorithm for foreign exchange rates forecasting. In addition, 5 articles adopt GA algorithm, while others utilize other training algorithm, such as LM, GRG2, conjugate gradient and self-organized learning algorithms.

In addition, an important factor related to learning algorithms is the learning rate (i.e., step size), which actually controls the learning speed and the convergence property. In the reviewed 45 articles, 39 articles (86.67%) used a fixed learning rate with different values; the remainder does not specify the learning rates. Usually, a fixed learning rate does not lead to an optimal convergence rate. Therefore, deriving an optimally adaptive learning rate is very important for ANN applications (Yu et al., 2005a).

1.3.2.7 Transfer function

The transfer functions that are most commonly used are sigmoidal-type functions, such as logistic and hyperbolic tangent functions. However, other transfer functions may be used as long as they are differentiable. In an empirical article, Moody and Yarvin (1992) compared the performance of logistic, polynomial, rational function and Fourier series transfer functions on datasets containing varying degrees of noise and nonlinearities. They found that the non-sigmoidal-type transfer functions performed best when the data were noiseless and contained highly nonlinear relationships. While performance of the polynomial activation function was inferior to sigmoidal-type transfer functions when the data were noisy and contained mildly nonlinear relationships. Kalman and Kwasney (1992) argue that the hyperbolic tangent transfer function should be used. Another option is to use a radial basis transfer function. Generally, the same transfer function could be used in all layers. However, it is advantageous to use sigmoidal-type transfer functions in the hidden layers and linear transfer functions in the output layer when it is necessary to extrapolate beyond the range of the training data (Kaastra and Boyd, 1995).

In our examined articles, 34 articles (75.56%) use sigmoidal-type transfer function (logistic or hyperbolic tangent) in the hidden layer. Because the foreign exchange markets have the features of high noise, large volatility and complexity, it is advisable to use the sigmoidal-type transfer function. Furthermore, of the 27 articles in category I, 18 articles used logistic or hyperbolic tangent function as transfer function, two articles of category II and 14 of 16 articles in category III also used logistic function as transfer function for predicting exchange rates. This implies that sigmoidal-type transfer function is appropriate for exchange rates prediction.

1.3.2.8 Performance comparison between ANNs and other models

In measurement of performance of foreign exchange rates forecasting, level measurement and directional accuracy are two important criteria. In this review, almost all articles adopt these two performance measures.

Based on the two classes of performance criteria, performance comparisons are performed between neural network and other comparable methods in exchange rates forecasting. 40 of 45 articles compared the performance of the neural networks with performance of statistical models, mainly linear, multiple regression analysis (see Table 1.2). In 27 of 40 articles, the authors concluded that ANNs perform better than other models, while two articles (Gencay, 1999; Qi and Wu, 2003) show that ANNs perform worse than some other models. To some extent, this implies that ANNs is also a promising approach, although ANNs produce many mixed results.

1.3.2.9 General situations of exchange rate predictability by ANNs

In view of the above investigation and analysis, we can summarize the general situation in which foreign exchange rates are predictable with ANNs, as shown in Table 1.6.

As can be seen from previous analysis and Table 1.6, foreign exchange rates are predictable in the following situations: when forecasting horizon is short-term or medium-term, sample data must be divided into at least two data sets (i.e., training set and testing set, and validation set if necessary). These data sets must have an appropriate size individually, especially for training set, network control strategy is feed-forward strategy, training algorithm is back-propagation algorithm, transfer functions in hidden layer and in output layer are sigmoidal type function and linear function respectively, and network type is multi-layer feedforward neural network (MLFNN). On the contrary, if prediction horizon is long-term, sample data cannot be divided into (at least) two data sets and training set size is either too small or too large, then foreign exchange rates are not predictable. It is worth noting that this is only a sufficient condition to guarantee exchange rates predictability with ANNs. That is to say, if the sufficient situations are satisfied, foreign exchange rates are predictable with ANNs. Of course, this does not necessarily mean that other-types of ANNs do not forecast foreign exchange rates; or foreign exchange rates are unpredictable by other-types of ANNs.

Table 1.6. The general situations of foreign exchange rates predictability by ANNs

Factors	General situations
Forecasting horizon	Short and medium-term
Data division	Three data sets, at least two data sets
Training set size	Appropriate size, e.g., two years data
Control strategy	Feedforward strategy
Training algorithm	Back-propagation algorithm
Transfer function in hidden layer	Sigmoidal-type function
Transfer function in output layer	Linear function
Network type	Multi-layer feedforward neural network

1.4 Implications and Research Topics

In the last section, some general situations are given to guarantee predictability in foreign exchange rate forecasting with ANNs. But in the process of our investigations and analyses, some implicit information also indicates some interesting research topics. First of all, in all studies listed in this survey, authors seldom mention data preparation, such as data collection, variable selection and data cleaning, for foreign exchange rates forecasting. Actually, data preparation has a huge impact on the success of foreign exchange rate prediction, as discussed by Yu et al. (2006a).

Second, in almost all listed studies about foreign exchange rate prediction with ANNs, ANN models do not use optimal learning rates. In these studies, the learning rates are set to a fixed value during training. It is, however, crucial to determine a proper fixed learning rate for ANN applications. If the learning rate is too large, learning may occur quickly, but ANN may also become unstable and may even learn at all. If the learning rate is too small, it may lead to a long learning time and a slow convergence. Also, a suitable fixed learning is usually problem-depended and it varies with different neural network structures for different problem applications (Sha and Bajic, 2002; Yu et al., 2005a, 2006b, 2006c). Therefore, a suitable learning rate is important for any learning algorithm. However, in the examined 45 articles, almost all authors do not use optimal learning rates, which might be a subject for further research to improve the forecasting performance.

Third, as earlier noted, hybrid and ensemble strategies usually achieve better prediction performance than individual ANN models, implying that the hybrid and the ensemble ANNs will also be promising research topics for foreign exchange rates prediction with ANNs. Lai et al. (2006a, 2006b, 2006c) and Yu et al. (2005c, 2005f, 2005h, 2006d, 2006f) have done some innovative work in these research topics. Interested readers can be referred to Lai et al. (2006a, 2006b, 2006c) and Yu et al. (2005c, 2005f, 2005h, 2006d, 2006f) for more details.

Relying on the above three implications, this book will extend the research from following five aspects:

(1) Proposing an integrated data preparation scheme for neural network data analysis, including foreign exchange rates prediction.

(2) Deriving an optimally adaptive learning rate for individual neural network model in foreign exchange rates prediction.

(3) Constructing hybrid neural network models for foreign exchange rates forecasting.

(4) Combining different neural network models into an ensemble model for foreign exchange rates prediction.

(5) Developing an intelligent decision support system (DSS) for foreign exchange rates forecasting and trading decision.

1.5 Conclusions

This chapter provides an analysis of verifying exchange rates predictability from neural networks perspective. We have described the literature collection process, examined 45 articles with details about predicting exchange rates with ANNs, and reported the analytical results. In terms of factor decomposition and investigation, some general situations of exchange rate predictability are summarized by literature analysis and inference. Meanwhile, we also pointed out several future research topics that have some implications for exchange rates forecasting.

To summarize, we can have the following conclusions. First of all, within the constraints, ANNs can effectively forecast foreign exchange rates, although there are some negative or mixed results reported in some literature. Second, from the analysis of ANNs in exchange rate forecasting, we can confirm that the performance of exchange rates forecasting can be improved by adjusting some factors that affect the performance of ANNs. As advancements are made in AI technology and computer-based systems, there should be new opportunities to apply neural network technology for exchange rates forecasting. This would encourage or motivate academics and practitioners to collaborate in further exploration of the ANNs applications.

It is necessary to be cautious in explaining the results of this survey since the main finding are based on literature collected only from journal articles. The results therefore do not include all actual business applications. Although a large number of actual systems may be in use, the sponsoring companies may not wish to divulge information on successful applications. Furthermore, we have examined only academic journal articles, conference proceedings and doctoral dissertations are excluded, as we assume that high-quality research is eventually published in journals. Also, some related journals might not have been included in this survey since they may not have been within the scope of our computer searches.

Part II: Basic Learning Principles of Artificial Neural Networks and Data Preparation

2 Basic Learning Principles of Artificial Neural Networks

2.1 Introduction

Artificial neural networks (ANNs), as an emerging discipline, studies or emulates the information processing capabilities of neurons of the human brain. It uses a distributed representation of the information stored in the network, and thus resulting in robustness against damage and corresponding fault tolerance (Shadbolt and Taylor, 2002). Usually, a neural network model takes an input vector $\boldsymbol{X}$ and produces output vector $\boldsymbol{Y}$. The relationship between $\boldsymbol{X}$ and $\boldsymbol{Y}$ is determined by the network architecture. There are many forms of network architecture inspired by the neural architecture of the human brain. The neural network generally consists of one input layer, one output layer, and one or more hidden layers, as illustrated in Fig. 2.1.

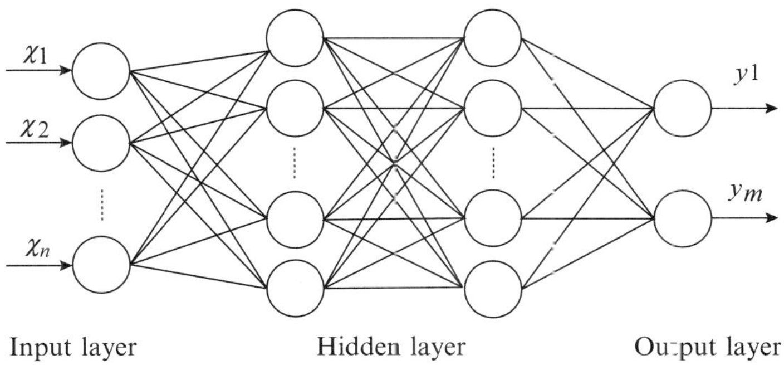

Fig. 2.1. The basic architecture of neural network

In the neural network model, it is widely accepted that a three-layer back propagation neural network (BPNN) with an identity transfer function in the output unit and logistic functions in the middle-layer units can approximate any continuous function arbitrarily well given a sufficient amount of middle-layer units (White, 1990). Furthermore, in the practical applications, about 70 percent of all problems are usually trained on a three-layer back-propagation network, as revealed by Chapter 1. The back-propagation learning algorithm, designed to train a feed-forward network, is an effective learning technique used to exploit the regularities and exceptions in the training sample.

A major advantage of neural networks is their ability to provide flexible mapping between inputs and outputs. The arrangement of the simple units into a multilayer framework produces a map between inputs and outputs that is consistent with any underlying functional relationship regardless of its "true" functional form. Having a general map between the input and output vectors eliminates the need for unjustified priori restrictions that are needed in conventional statistical and econometric modeling. Therefore, a neural network is often viewed as a "universal approximator", i.e. a flexible functional form that can approximate any arbitrary function arbitrarily well, given sufficient middle-layer units and properly adjusted weights (Hornik et al., 1989; White, 1990). Both theoretical proof and empirical applications have confirmed that a three-layer BP neural network (BPNN) model with an identity transfer function in the output unit and logistic functions in the middle-layer units is adequate for foreign exchange rates forecasting, which is our research focus in this book. Therefore, a three-layer BP neural network model with identity activation function in the output unit and logistic function in the middle-layer units is used throughout this book except specially specified.

2.2 Basic Structure of the BPNN Model

Consider a three-layer BPNN, which has p nodes of the input layer, q nodes of the hidden layer and k nodes of the output layer. Mathematically, the basic structure of the BPNN model is described by (see derivation later)

$$\hat{Y}(t+1) = F_2[V^T(t)F_1(W(t)X(t))] \tag{2.1}$$

where $X=(x_0,x_1,\cdots,x_p)^T \in R^{(p+1)\times 1}$ are the inputs of BPNN, $\hat{Y}=(\hat{y}_0,\hat{y}_1,\cdots,\hat{y}_k)^T \in R^{k\times 1}$ is the outputs of the BPNN,

$$W = \begin{pmatrix} w_{10} & w_{11} & \cdots & w_{1p} \\ w_{20} & w_{21} & \cdots & w_{2p} \\ \cdots & \cdots & \cdots & \cdots \\ w_{q0} & w_{q1} & \cdots & w_{qp} \end{pmatrix} = (W_0, W_1, \cdots, W_p) \in R^{q\times(p+1)},$$

$$V = \begin{pmatrix} v_{10} & v_{20} & \cdots & v_{k0} \\ v_{11} & v_{21} & \cdots & v_{k1} \\ \cdots & \cdots & \cdots & \cdots \\ v_{1q} & v_{2q} & \cdots & v_{kq} \end{pmatrix} = (V_1, V_2, \cdots, V_k) \equiv R^{(q+1)\times k},$$

$F_1(W(t)X(t)) = \left(F_1(net_0(t))\ F_1(net_1(t))\ \cdots\ F_1(net_q(t))\right)^T \in R^{(q+1)\times 1}$, $net_i(t) = \sum_{j=0}^{p} w_{ij}(t)x_j(t)$, $i = 0, 1, \ldots, q$, is the output of the i-th hidden node. $w_{i0}(t)$, $i = 1, \ldots, q$, are the bias of the hidden nodes; $v_{ij}(t)$, $i = 1, \ldots, q$, $j = 1, \ldots, k$, are the weights form the hidden node i to the output node j; $v_{i0}(t)$ is the bias of the output node; $F_1(\bullet)$ and $F_2(\bullet)$ are the activation function, which can be any nonlinear function as long as they are continuous, bounded and differentiable. Typically, $F_1(\bullet)$ is a sigmoidal or hyperbolic tangent function and t is a time factor. For convenience, a symmetric hyperbolic tangent function (i.e., $f_1(x) = \tanh(u_0^{-1}x)$ where u_0 is the shape factor of the activation function) is used as the activation function of the hidden layer (Yu et al., 2005a, b).

Derivation of Equation (2.1):

$$\hat{Y}(t+1) = \begin{bmatrix} \hat{y}_1(t+1) \\ \hat{y}_2(t+1) \\ \cdots \\ \hat{y}_k(t+1) \end{bmatrix} = \begin{bmatrix} f_2[\sum_{i=1}^{q} f_1(\sum_{j=1}^{p} w_{ij}(t)x_j(t) + w_{i0}(t))v_{1i}(t) + v_{10}(t)] \\ f_2[\sum_{i=1}^{q} f_1(\sum_{j=1}^{p} w_{ij}(t)x_j(t) + w_{i0}(t))v_{2i}(t) + v_{20}(t)] \\ \cdots \\ f_2[\sum_{i=1}^{q} f_1(\sum_{j=1}^{p} w_{ij}(t)x_j(t) + w_{i0}(t))v_{ki}(t) + v_{k0}(t)] \end{bmatrix}$$

$$= \begin{bmatrix} f_2[\sum_{i=0}^{q} f_1(\sum_{j=0}^{p} w_{ij}(t)x_j(t))v_{1i}(t)] \\ f_2[\sum_{i=0}^{q} f_1(\sum_{j=0}^{p} w_{ij}(t)x_j(t))v_{2i}(t)] \\ \cdots \\ f_2[\sum_{i=0}^{q} f_1(\sum_{j=0}^{p} w_{ij}(t)x_j(t))v_{ki}(t)] \end{bmatrix} = \begin{bmatrix} f_2[\sum_{i=0}^{q} f_1(net_i)v_{1i}(t)] \\ f_2[\sum_{i=0}^{q} f_1(net_i)v_{2i}(t)] \\ \cdots \\ f_2[\sum_{i=0}^{q} f_1(net_i)v_{ki}(t)] \end{bmatrix}$$

$$= \begin{bmatrix} f_2[V_1^T F_1(W(t)X(t))] \\ f_2[V_2^T F_1(W(t)X(t))] \\ \cdots \\ f_2[V_k^T F_1(W(t)X(t))] \end{bmatrix} = F_2[V^T(t)F_1(W(t)X(t))]$$

where f_1 is the activation function of the hidden nodes and f_2 is the activation function of output nodes and t is a time factor. ■

Through estimating model parameter vector or network connection weights (W, V) via BPNN training and learning, we can realize the corresponding modeling tasks, such as function approximation, pattern recognition, and time series prediction.

2.3 Learning Process of the BPNN Algorithm

The ability to learn and improve its performance from examples is the neural network's fundamental trait. For BPNN, it is a class of supervised learning algorithm in the form of the neural network associative memory. Usually, the back-propagation learning mechanism consists of two phases: forward-propagation and back-propagation phase, as Tam and Kiang (1992) reported.

Suppose we have n samples. Each is described by $X_i = (x_{i1}, x_{i2}, \cdots, x_{ip})$ and $Y_i = (y_{i1}, y_{i2}, \cdots, y_{ik})$ where X_i is an input vector, Y_i is a target output vector and $1 \leq i \leq n$.

In the first phase (forward-propagation phase), X_i is fed into the input layer, and an output $\hat{Y}_i = (\hat{y}_{i1}, \hat{y}_{i2}, \cdots, \hat{y}_{ik})$ is generated based on the current weight vector W. The objective is to minimize an error function E defined as

$$E = \sum_{i=1}^{n} \sum_{j=1}^{k} \frac{(y_{ij} - \hat{y}_{ij})^2}{2} \tag{2.2}$$

Through changing W so that all input vectors are correctly mapped to their corresponding output vectors.

In the second phase (back-propagation phase), a gradient descent in the weight space, W, is performed to locate the optimal solution. The direction and magnitude change Δw_{ij} can be computed as

$$\Delta w_{ij} = -\frac{\partial E}{\partial w_{ij}} \varepsilon \tag{2.3}$$

where $0 < \varepsilon < 1$ is a learning parameter controlling the algorithm's convergence rate.

The total squared error calculated by Equation (2.1) is propagated back, layer by layer, from the output units to the input units in the second phase. Weight adjustments are determined on the way of propagation at each level. The two phases are executed during each iteration of the back-propagation algorithm until E converges.

2.4 Weight Update Formulae of the BPNN Algorithm

Actually, the model parameter vector or neural network weights (W, V) (as defined by Section 2.2) can be obtained by minimizing iteratively a cost function, $E(X\colon W, V)$. In general, $E(X\colon W, V)$ is a sum of the error squares cost function with k output nodes and n training pairs or patterns, that is,

$$\begin{aligned} E(X:W,V) &= \frac{1}{2}\sum_{j=1}^{n}\sum_{i=1}^{k} e_{ij}^2 = \frac{1}{2}\sum_{j=1}^{n} e_j^T e_j \\ &= \frac{1}{2}\sum_{j=1}^{n}[y_j - \hat{y}_j(X:W,V)]^T[y_j - \hat{y}_j(X:W,V)] \end{aligned} \tag{2.4}$$

where y_j is the jth actual value and $y_j(X\colon W,V)$ is the jth estimated value. Given the time factor t, Equation (2.4) can be rewritten as

$$\begin{aligned} E(t) &= \frac{1}{2}\sum_{j=1}^{n}\sum_{i=1}^{k} e_{ij}^2(t) = \frac{1}{2}\sum_{j=1}^{n} e_j^T(t) e_j(t) \\ &= \frac{1}{2}\sum_{j=1}^{n}[y_j(t) - \hat{y}_j(t)]^T[y_j(t) - \hat{y}_j(t)] \end{aligned} \tag{2.5}$$

where $e_j(t) = [e_{1j}(t) \quad e_{2j}(t) \quad \cdots \quad e_{kj}(t)]^T \in R^{k\times 1}$, $j = 1, 2, \cdots, n$, $y_j(t)$ and $\hat{y}_j(t)$ are the jth actual value predicted value at time t, respectively.

By applying the gradient descent rule to the cost function $E(t)$ (as shown in Equation (2.5)) and considering Equation (2.1), we can obtain the weight increment formulae with respect to W and V, respectively (see proof later).

$$\Delta W(t) = -\eta(t)\nabla_W E(t) = \eta(t)\sum_{j=1}^{n} F'_{1(j)} V F'_{2(j)} e_j x_j^T \tag{2.6}$$

$$\Delta V(t) = -\eta(t)\nabla_V E(t) = \eta(t)\sum_{j=1}^{n} F_{1(j)} e_j^T F'_{2(j)} \tag{2.7}$$

where η is the learning rate, ∇ is the gradient operator, $\Delta W(t)$ and $\Delta V(t)$ are the weight adjustment increments at iteration t, respectively.

Suppose $\Delta W(t) = W(t) - W(t-1)$ and $\Delta V(t) = V(t) - V(t-1)$, then the weights update formulae for standard BP learning algorithm with respect to W and V are given by, respectively

$$W(t) = W(t-1) + \eta(t)\sum_{j=1}^{n} F'_{1(j)} V F'_{2(j)} e_j x_j^T \tag{2.8}$$

$$V(t) = V(t-1) + \eta(t)\sum_{j=1}^{n} F_{1(j)} e_j^T F'_{2(j)} \tag{2.9}$$

where Δ is the incremental operator, $F'_{1(j)} = \mathrm{diag}[f'_{1(1)} \ \cdots \ f'_{1(q)}] \in R^{q\times q}$,

$F'_{2(j)} = \mathrm{diag}[f'_{2(1)} \ f'_{2(2)} \ \cdots \ f'_{2(k)}] \in R^{k\times k}$,

$$f'_{1(i)} = f'_1(net_i) = \frac{\partial f_1(net_i)}{\partial net_i}, i = 1,2,\cdots,q,$$

$$f'_{2(i)} = f'_2[v_i^T F_1(WX)] = \frac{\partial f_2[v_i^T F_1(WX)]}{\partial [v_i^T F_1(WX)]}, i = 1,2,\cdots k,$$

$$V = \begin{bmatrix} v_{11} & v_{21} & \cdots & v_{k1} \\ v_{12} & v_{22} & \cdots & v_{k2} \\ \cdots & \cdots & \cdots & \cdots \\ v_{1q} & v_{2q} & \cdots & v_{kq} \end{bmatrix} = [v_1 \ v_2 \ \cdots \ v_k] \in R^{q\times p},$$

$v_i = [v_{i1} \ \cdots \ v_{iq}]^T \in R^{q\times 1}, i = 1,2,\cdots,k$.

In order to prove Equations (2.8) and (2.9), three lemmas must be firstly introduced (Sha and Bajic, 2002).

Lemma 2.1

The derivative of activation function $F_1(WX)$ in hidden layer with respect to the vector *Net* or *WX* is given by

$$F_1'(WX) = \mathrm{diag}\,[f_1'(net_1) \ \ f_1'(net_2) \ \ \cdots \ \ f_1'(net_q)]$$

$$= \frac{1}{u_0}\begin{bmatrix} 1 - f_1^2(net_1) & 0 & \cdots & 0 \\ 0 & 1 - f_1^2(net_2) & \cdots & 0 \\ \cdots & \cdots & \cdots & \cdots \\ 0 & 0 & \cdots & 1 - f_1^2(net_q) \end{bmatrix}$$

$$= \frac{1}{u_0}\begin{bmatrix} 1 - f_1^2(W_1X) & 0 & \cdots & 0 \\ 0 & 1 - f_1^2(W_2X) & \cdots & 0 \\ \cdots & \cdots & \cdots & \cdots \\ 0 & 0 & \cdots & 1 - f_1^2(W_qX) \end{bmatrix}$$

where $F_1(WX) = \left[f_1(net_0) \ \ f_1(net_1) \ \ \cdots \ \ f_1(net_q)\right]^T$, $F_1'(WX)$ is the Jacobian matrix of $F_1(WX)$, $f_1'(net_i) = \partial f_1(net_i)/\partial net_i$, i = 0, 1, …, q, $f_1(x) = \tanh(u_0^{-1}x)$ and its derivative $f_1'(x) = u_0^{-1}[1 - f_1^2(x)]$.

Proof:

Due to $f_1(x) = \tanh(u_0^{-1}x)$ and its derivative $f_1'(x) = u_0^{-1}[1 - f_1^2(x)]$, $net_i = W_iX$, $F_1(WX) = \left[f_1(net_0)\, f_1(net_1) \cdots f_1(net_q)\right]^T$, the derivative of activation function $F_1(WX)$ in hidden layer with respect to the vector *Net* or *WX* can be calculated as

$$F_1'(WX) = \begin{bmatrix} \frac{\partial f_1(net_1)}{\partial net_1} & \cdots & \frac{\partial f_1(net_1)}{\partial net_q} \\ \cdots & \cdots & \cdots \\ \frac{\partial f_1(net_q)}{\partial net_1} & \cdots & \frac{\partial f_1(net_q)}{\partial net_q} \end{bmatrix} = \begin{bmatrix} \frac{\partial f_1(net_1)}{\partial net_1} & \cdots & 0 \\ \cdots & \cdots & \cdots \\ 0 & \cdots & \frac{\partial f_1(net_q)}{\partial net_q} \end{bmatrix}$$

$$= \begin{bmatrix} f_1'(net_1) & \cdots & 0 \\ \cdots & \cdots & \cdots \\ 0 & \cdots & f_1'(net_q) \end{bmatrix} = diag\left[f_1'(net_1) \quad \cdots \quad f_1'(net_q)\right]$$

$$= \begin{bmatrix} u_0^{-1}\left[1 - f_1^2(net_1)\right] & 0 & \cdots & 0 \\ 0 & u_0^{-1}\left[1 - f_1^2(net_2)\right] & \cdots & 0 \\ \cdots & \cdots & \cdots & \cdots \\ 0 & 0 & \cdots & u_0^{-1}\left[1 - f_1^2(net_q)\right] \end{bmatrix}$$

$$= \frac{1}{u_0}\begin{bmatrix} 1 - f_1^2(W_1X) & 0 & \cdots & 0 \\ 0 & 1 - f_1^2(W_2X) & \cdots & 0 \\ \cdots & \cdots & \cdots & \cdots \\ 0 & 0 & \cdots & 1 - f_1^2(W_qX) \end{bmatrix} \qquad \blacksquare$$

Lemma 2.2

The partial derivative of the single hidden output $V^T F_1(WX)$ with respect to the weight matrix W is given by

$$\frac{\partial[V^T F_1(WX)]}{\partial W} = F_1'(WX)VX^T$$

Proof:

$$\frac{\partial[V^T F_1(WX)]}{\partial W} = \left[\frac{\partial[\sum_{i=0}^{q} v_i f_1(\sum_{j=0}^{p} w_{ij}x_j)]}{\partial w_{ij}}\right]_{q\times(p+1)} = [v_i f_1'(net_i)x_j]_{q\times(p+1)}$$

$$= \begin{bmatrix} v_1 f_1'(net_1)x_0 & v_1 f_1'(net_1)x_1 & \cdots & v_1 f_1'(net_1)x_p \\ v_2 f_1'(net_2)x_0 & v_2 f_1'(net_2)x_1 & \cdots & v_2 f_1'(net_2)x_p \\ \cdots & \cdots & \cdots & \cdots \\ v_q f_1'(net_q)x_0 & v_q f_1'(net_q)x_1 & \cdots & v_q f_1'(net_q)x_p \end{bmatrix}$$

$$= \begin{bmatrix} f_1'(net_1) & \cdots & 0 \\ \cdots & \cdots & \cdots \\ 0 & \cdots & f_1'(net_q) \end{bmatrix} \cdot \begin{bmatrix} v_1 \\ \cdots \\ v_q \end{bmatrix} \cdot \begin{bmatrix} x_0 & x_1 & \cdots & x_p \end{bmatrix}$$

$$= F_1'(WX)VX^T \qquad \blacksquare$$

Lemma 2.3

The partial derivative of the single hidden output $V^T F_1(WX)$ with respect to the weight vector V is given by

$$\frac{\partial [V^T F_1(WX)]}{\partial V} = F_1(WX)$$

Proof:

$$\frac{\partial [V^T F_1(WX)]}{\partial V} = \left[\frac{\partial [\sum_{i=0}^{q} v_i f_1(net_i)]}{\partial v_i} \right]_{(q+1)\times 1}$$

$$= [f_1(net_i)]_{(q+1)\times 1} = F_1(WX) \qquad \blacksquare$$

Using Lemmas 2.1-2.3, we can prove the weight update formula in the following. First of all, we derive the gradient of $E(t)$ with respect to weights W and V.

With reference to Equation (2.5) and Lemmas 2.1 and 2.2, the gradient of $E(t)$ with respect to W can be obtained by applying the steepest descent method to $E(t)$,

$$\nabla_W E(t) = \frac{\partial E(t)}{\partial W(t)} = \sum_{j=1}^{n}\sum_{i=1}^{k} e_{ij}(t)\frac{\partial e_{ij}(t)}{\partial W(t)} = -\sum_{i=1}^{k} e_{ij}(t)\frac{\partial \hat{y}_{ij}(t)}{\partial W(t)}$$

$$= -\sum_{j=1}^{n}\sum_{i=1}^{k} e_{ij}(t) F_{2(j)}' \left[v_i^T F_{1(j)}(Wx_j) \right] \frac{\partial [v_i^T F_1(Wx_j)]}{\partial W(t)}$$

$$= -\sum_{j=1}^{n}\sum_{i=1}^{k} e_{ij} F_{2(j)}' F_{1(j)}' VX^T = -\sum_{j=1}^{n} F_{1(j)}' \left(\sum_{i=1}^{k} e_{ij} F_{2(j)}' v_i \right) x_j^T$$

$$= -\sum_{j=1}^{n} F'_{1(j)} \begin{bmatrix} e_{1j} f'_{2(1j)} v_{11} + e_{2j} f'_{2(2j)} v_{21} + \cdots e_{kj} f'_{2(kj)} v_{k1} \\ e_{1j} f'_{2(1j)} v_{12} + e_{2j} f'_{2(2j)} v_{22} + \cdots e_{kj} f'_{2(kj)} v_{k2} \\ \cdots \\ e_{1j} f'_{2(1j)} v_{1q} + e_{2j} f'_{2(2j)} v_{2q} + \cdots e_{kj} f'_{2(kj)} v_{kq} \end{bmatrix} x_j^T$$

$$= -\sum_{j=1}^{n} F'_{1(j)} \begin{bmatrix} v_{11} & v_{21} & \cdots & v_{k1} \\ v_{12} & v_{22} & \cdots & v_{k2} \\ \cdots & \cdots & \cdots & \cdots \\ v_{1q} & v_{2q} & \cdots & v_{kq} \end{bmatrix} \cdot \begin{bmatrix} e_{1j} f'_{2(1j)} \\ e_{2j} f'_{2(2j)} \\ \cdots \\ e_{kj} f'_{2(kj)} \end{bmatrix} x_j^T$$

$$= -\sum_{j=1}^{n} F'_{1(j)} \begin{bmatrix} v_{11} & v_{21} & \cdots & v_{k1} \\ v_{12} & v_{22} & \cdots & v_{k2} \\ \cdots & \cdots & \cdots & \cdots \\ v_{1q} & v_{2q} & \cdots & v_{kq} \end{bmatrix}$$

$$\times \begin{bmatrix} f'_{2(1j)} & 0 & \cdots & 0 \\ 0 & f'_{2(2j)} & \cdots & 0 \\ \cdots & \cdots & \cdots & \cdots \\ 0 & 0 & \cdots & f'_{2(kj)} \end{bmatrix} \cdot \begin{bmatrix} e_{1j} \\ e_{2j} \\ \cdots \\ e_{kj} \end{bmatrix} \cdot x_j^T$$

$$= -\sum_{j=1}^{n} F'_{1(j)} V F'_{2(j)} e_j x_j^T$$

Similarly, the gradient of $E(t)$ with respect to V can also be obtained

$$\nabla_V E(t) = \frac{\partial E(t)}{\partial V(t)} = \sum_{j=1}^{n} \sum_{i=1}^{k} e_{ij}(t) \frac{\partial e_{ij}(t)}{\partial V(t)} = -\sum_{j=1}^{n} \sum_{i=1}^{k} e_{ij}(t) \frac{\partial \hat{y}_{ij}(t)}{\partial V(t)}$$

$$= -\sum_{j=1}^{n} \sum_{i=1}^{k} e_{ij}(t) \begin{bmatrix} \frac{\partial \hat{y}_{ij}}{\partial v_{10}} & \frac{\partial \hat{y}_{ij}}{\partial v_{20}} & \cdots & \frac{\partial \hat{y}_{ij}}{\partial v_{i0}} & \cdots & \frac{\partial \hat{y}_{ij}}{\partial v_{k0}} \\ \frac{\partial \hat{y}_{ij}}{\partial v_{11}} & \frac{\partial \hat{y}_{ij}}{\partial v_{21}} & \cdots & \frac{\partial \hat{y}_{ij}}{\partial v_{i1}} & \cdots & \frac{\partial \hat{y}_{ij}}{\partial v_{k1}} \\ \cdots & \cdots & \cdots & \cdots & \cdots & \cdots \\ \frac{\partial \hat{y}_{ij}}{\partial v_{1q}} & \frac{\partial \hat{y}_{ij}}{\partial v_{2q}} & \cdots & \frac{\partial \hat{y}_{ij}}{\partial v_{iq}} & \cdots & \frac{\partial \hat{y}_{ij}}{\partial v_{kq}} \end{bmatrix}$$

$$= -\sum_{j=1}^{n}\sum_{i=1}^{k} e_{ij}(t) F_2'[V^T F_1(WX)] \begin{bmatrix} 0 & 0 & \cdots & f_{1(0j)} & \cdots & 0 \\ 0 & 0 & \cdots & f_{1(1j)} & \cdots & 0 \\ \cdots & \cdots & \cdots & \cdots & \cdots & \cdots \\ 0 & 0 & \cdots & f_{1(qj)} & \cdots & 0 \end{bmatrix}$$

$$= -\sum_{j=1}^{n} \begin{bmatrix} e_{1j} f'_{2(1j)} f_{1(0j)} & \cdots & e_{ij} f'_{2(ij)} f_{1(0j)} & \cdots & e_{kj} f'_{2(kj)} f_{1(0j)} \\ e_{1j} f'_{2(1j)} f_{1(1j)} & \cdots & e_{ij} f'_{2(ij)} f_{1(1j)} & \cdots & e_{kj} f'_{2(kj)} f_{1(1j)} \\ \cdots & \cdots & \cdots & \cdots & \cdots \\ e_{1j} f'_{2(1j)} f_{1(qj)} & \cdots & e_{ij} f'_{2(ij)} f_{1(qj)} & \cdots & e_{kj} f'_{2(kj)} f_{1(qj)} \end{bmatrix}$$

$$= -\sum_{j=1}^{n} \begin{bmatrix} f_{1(0j)} \\ f_{1(1j)} \\ \cdots \\ f_{1(qj)} \end{bmatrix} \begin{bmatrix} e_{1j} f'_{2(1j)} & e_{2j} f'_{2(2j)} & \cdots & e_{kj} f'_{2(kj)} \end{bmatrix}$$

$$= -\sum_{j=1}^{n} F_{1(j)} \begin{bmatrix} e_{1j} & e_{2j} & \cdots & e_{kj} \end{bmatrix} \begin{bmatrix} f'_{2(1j)} & 0 & \cdots & 0 \\ 0 & f'_{2(2j)} & \cdots & 0 \\ \cdots & \cdots & \cdots & \cdots \\ 0 & 0 & \cdots & f'_{2(kj)} \end{bmatrix}$$

$$= -\sum\nolimits_{j=1}^{n} F_{1(j)} e_j^T F'_{2(j)}$$

Therefore, the weight increments $\Delta W(t)$ and $\Delta V(t)$ can be obtained from the above derivation processes, i.e.,

$$\Delta W(t) = -\eta(t)\nabla_W E(t) = \eta(t)\sum\nolimits_{j=1}^{N} F'_{1(j)} V F'_{2(j)} e_j x_j^T$$

$$\Delta V(t) = -\eta(t)\nabla_V E(t) = \eta(t)\sum\nolimits_{j=1}^{n} F_{1(j)} e_j^T F'_{2(j)}$$

Assume $\Delta W(t) = W(t) - W(t-1)$ and $\Delta V(t) = V(t) - V(t-1)$, then the weights update rule (i.e., Equations (2.8) and (2.9)) for standard BP learning algorithm with respect to W and V can be obtained, i.e.,

$$W(t) = W(t-1) + \eta(t)\sum\nolimits_{j=1}^{n} F'_{1(j)} V F'_{2(j)} e_j x_j^T$$

$$V(t) = V(t-1) + \eta(t)\sum\nolimits_{j=1}^{n} F_{1(j)} e_j^T F'_{2(j)}$$

Using the weight update formulae (i.e., Equations (2.8) and (2.9)), we can train BPNN to perform the corresponding tasks, such as data mining, function approximation and financial time series forecasting.

2.5 Conclusions

In this chapter, some preliminaries about back-propagation neural networks are presented. First of all, a basic architecture of three-layer BPNN model is described in the form of matrix. Then we briefly introduce a basic learning process including forward propagation phase and back-propagation phase for BPNN. Based upon the basic structure and learning process, the weight update rules are derived in terms of steepest gradient descent algorithm. Using the weight update rules, some data analysis tasks are performed.

However, in the neural network applications, an important process, data preparation process, is often neglected by researchers and business users. Although data preparation in neural network data analysis is important, some existing literature about the neural network data preparation are scattered, and there is no systematic study about data preparation for neural network data analysis (Yu et al., 2006a). Therefore, this book tries to develop an integrated data preparation scheme for neural network data analysis, which will be described in the next chapter.

3 Data Preparation in Neural Network Data Analysis

3.1 Introduction

Preparing data is an important and critical step in neural network data analysis and it has an immense impact on the success of a wide variety of complex data analysis, such as data mining and knowledge discovery (Hu, 2003). The main reason is that the quality of the input data into neural network models may strongly influence the results of the data analysis (Sattler and Schallehn, 2001). As Lou (1993) stated, the effect on the neural network's performance can be significant if important input data are missing or distorted. In general, properly prepared data are easy to handle, which makes the data analysis task simple. On the other hand, improperly prepared data may make data analysis difficult, if not impossible. Furthermore, data from different sources and growing amounts of data produced by modern data acquisition techniques have made data preparation a time-consuming task. It has been claimed that 50–70 percent of the time and effort in data analysis projects is required for data preparation (Sattler and Schallehn, 2001; Pyle, 1999). Therefore, data preparation involves enhancing the data in an attempt to improve the performance of data analysis.

In past decades, artificial neural networks (ANNs), as a class of typical intelligent data analysis tool, have been studied extensively in many fields of knowledge, from science (e.g., Gardner and Dorling, 1998) to engineering (e.g., Rafiq et al., 1999) and from management (e.g., Krycha and Wagner, 1999) to control (e.g., Hunt et al., 1992), and many software products, such as *NeuroShell* (http://www.neuroshell com), *BrainMaker* (http://www.calsci.com) and *Neural Network Toolbox of Matlab* (http://www.mathworks.com), has been applied successfully in many practical projects. However, most studies and commercial systems focus almost exclusively on the design and implementation of neural models. Data preparation in neural network modeling has received scant recognition. In almost all theoretical and practical researches about neural networks (Gardner and

Dorling, 1998; Rafiq et al., 1999; Krycha and Wagner, 1999; Hunt et al., 1992; Rumelhart, 1994; Narendra and Parthasarathy, 1990; Rzimi-Sadjadi and Stricker, 1994; Beltratli et al., 1996; Senol and Gouch, 1992; Gately, 1996; Refenes et al., 1996; Smith and Gupta, 2002; Zhang, 2004; Klei and Rossin, 1999), neural network data preparations concentrate on data normalization for transformation and data division for training. Some studies even utilize neural networks for modeling without any data preparation procedure. In these studies there is an implicit assumption that all the data are prepared in advance and the data can be used directly in modeling. In practice, data are not always prepared beforehand for specific data analysis tasks. Although there are some data for a specific project, the quality and completeness of that data is limited. As a result, the complex data analysis process cannot succeed without a serious effort to prepare the data. Strong evidences (Klei and Rossin, 1999; Redman, 1992; Redman, 1996) reveal that data quality has a significant effect on neural network models. In addition, various interpretations have been given to the role and the need for data preparation. Zhang et al. (2003) revealed that data preparation could generate smaller magnitude and higher quality data, which can significantly improve the efficiency of complex data analysis. In the case of neural network learning, data preparation would enable users to decide how to represent the data, which concepts to learn, and how to present the results of data analysis so that it is easier to explain them in the real world (Famili et al., 1997). Data preparation is therefore crucial in neural network data analysis for guaranteeing data quality and completeness.

Although data preparation is useful for any kind of analysis, neural networks, as a class of important intelligent data analysis tool, have some special requirements for data preparation. First of all, neural networks are a kind of novel learning paradigm, but it is a time-consuming data analysis tool, which is different from any other data analysis tool. To speed up the analysis process, data preparation is very important and necessary for complex data analysis tasks. Second, data preparation can largely reduce model complexity for neural network modeling, which is important for complex data analysis tasks relative to other data analysis tools. Third, an effective data preparation can increase generalization ability of data analysis models, especially for neural networks. Basically, the main objective of complex data analysis is to discovery knowledge that will be used to solve problems or make decisions, but problems about the data may prevent this. In most cases, imperfections with the data are not noticed until the data analysis starts. This is the case especially for the neural networks. Therefore, data preparation in neural networks is more important than that of other data analysis tool.

However, in many neural network data preparation studies (Stein, 1993a, 1993b; McAulay and Li, 1992; Nedeljkovic and Milosavljevic, 1992; Sjoberg, 1992; De Noord, 1994; DeWitt, 1994; Joo et al., 2000; Nguyen and Chan, 2004), data preparation is restricted to data cleaning, normalization and division. In their studies, most authors considered that data preprocessing is equivalent to data preparation. Actually, the two terms are different. First of all, they have different significations. According to the American Heritage Dictionary (Pickett, 2000), 'preparation' is defined as 'a preliminary measure that serves to make ready for something', while 'preprocessing' is 'to perform conversion, formatting, or other functions on (data) before further processing'. Second, basic contents covered are different. Data preparation covers more contents than data preprocessing. Generally, data preprocessing only includes data transformation and data formatting, while data preparation contains more contents, such as data collection, data selection, data integration and data validation in addition to data preprocessing. In our study, data preparation is expanded into an integrated scheme with three phases from a systematic perspective. This integrated scheme comprises a data pre-analysis phase, including data collection, data selection and data integration, and a data post-analysis phase, such as data validation and data re-adjustment, as well as a data preprocessing phase. In this sense, our study goes well beyond previous studies (Stein, 1993a, 1993b; McAulay and Li, 1992; Nedeljkovic and Milosavljevic, 1992; Sjoberg, 1992; De Noord, 1994; DeWitt, 1994; Joo et al., 2000; Nguyen and Chan, 2004).

In the overview of this topic, we found that there are three main problems in neural network data preparation. The first is that there is no universal scheme or methodology for neural network data analysis (Problem I). That is, there is no perfect and systematic data preparation framework or architecture for neural network data analysis. Although some related studies are presented, a systematic work about neural network data preparation has not been formulated so far. Most existing studies focused on the data preprocessing. Furthermore, these researchers often confused the difference between data preparation and data preprocessing. Therefore, it is necessary to construct a universal data preparation scheme for neural network data analysis. Second, in preparing data for neural network data analysis, some important issues and dilemmas are often faced and are hard to handle (Problem II). For example, skilled experts may find a good solution, but may also find it difficult to judge whether the preparation chosen is appropriate. Third, data preparation requires extra effort, raising the question of cost versus benefits (Problem III). If they see data preparation as having little impact on final neural network data analysis results, decision-makers may be unwilling to invest in data preparation.

In light of the three problems outlined above, the main motivations of this study are four-fold: (a) to propose an integrated data preparation framework for neural network data analysis; (b) to provide some intelligent solutions to some important issues and dilemmas in the data preparation scheme; (c) to analyze and confirm the effects of the proposed data preparation scheme on a neural network data analysis and (d) to survey the literature about data preparation of neural network data analysis. Based upon the previous problems and motivations, the study first proposes an integrated data preparation scheme for neural network modeling to contribute to the solution of the first main problem, and then presents in detail alternative solutions to the dilemmas of the data preparation framework. For the third problem, a cost-benefit analysis framework for neural network data preparation is proposed. However, empirical evidence of the impact of data preparation on neural network data analysis is not provided here even though it is critical. Interested readers can be referred to Yu et al. (2006a) for empirical analysis. Because this empirical analysis is not related to our topic of foreign exchange rates forecasting, it is not presented in this book.

The remainder of the chapter is organized as follows. A brief description of neural network models for complex data analysis is presented in Section 3.2. In view of the neural network data analysis framework, an integrated data preparation scheme for neural networks is proposed to fill up the gap in the literature. Meanwhile, the steps in every phase, as well as important issues of the integrated scheme, are described and discussed in detail. Accordingly, some intelligent solutions to some important issues and dilemmas are provided in Section 3.3. A comprehensive cost-benefit analysis framework for analyzing the effect of the proposed integrated data preparation scheme on neural network data analysis is proposed in Section 3.4. Finally, some concluding remarks are given in Section 3.5.

3.2 Neural Network for Data Analysis

The foundation of the artificial neural networks (ANNs) paradigm was laid in the 1950s. Since then ANNs have earned significant attention because of the development of more powerful hardware and neural algorithms (Rumelhart, 1994). ANNs have been studied and explored by many researchers, and been applied and manipulated in almost every field, examples being system identification, modeling (Narendra and Parthasarathy, 1990), prediction, and classification (Rzimi-Sadjadi and Stricker, 1994; Beltratli et al., 1996; Senol and Gouch, 1992; Gately, 1996; Refenes et al.,

1996; Smith and Gupta, 2002; Zhang, 2004). Generally, ANNs can be used as an effective intelligent data analysis tool for their unique learning capability. In this section, an entire process of neural network data analysis is presented, as shown in Fig. 3.1.

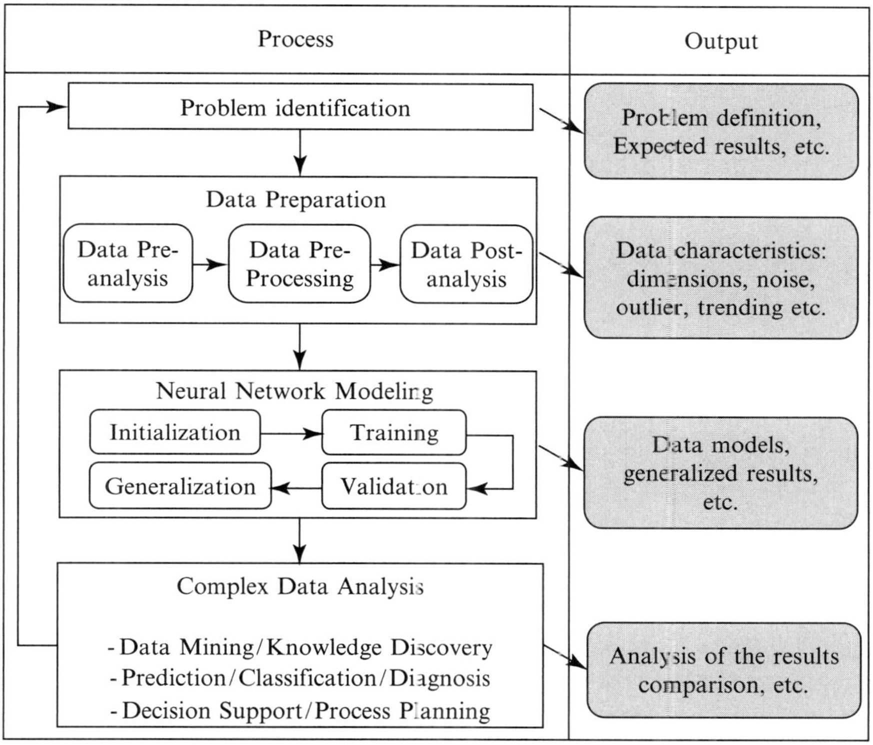

Fig. 3.1. The process of neural network data analysis

As can be seen from Fig. 3.1, neural network modeling for complex data analysis has four main processes: problem identification, data preparation, neural network modeling, and data analysis. In the first process, we can identify a problem by analyzing its expected results and consulting the relevant domain experts. Problem definitions and expected results are formulated to guide the subsequent tasks. The aim of the second process, data preparation, which will be described later, is to prepare high-quality data for data analysis so as to obtain satisfactory results. In the third process, after initialization, neural network models are trained iteratively. If the results of the data validation are rational, the generalized results obtained from the trained networks can be used for data analysis. Finally, depending on the generalized results, the goal of the complex data analysis, such as data mining and decision support, can be realized.

3.3 An Integrated Data Preparation Scheme

In this section, we first propose an integrated data preparation scheme focused on the data preparation process of neural network data analysis framework. We then present details of the scheme and overview some related literature. Subsequently, some intelligent solutions to some important issues and dilemmas in the integrated scheme are presented.

3.3.1 Integrated Data Preparation Scheme for Neural Network Data Analysis

As earlier noted, neural network data preparation for complex data analysis is very important. However, no standard data preparation framework for neural network modeling has so far been suggested (Problem I). In view of the importance of data preparation, we propose an integrated data preparation scheme for neural network data analysis, as illustrated in Fig. 3.2.

As shown in Fig. 3.2, the integrated data preparation scheme consists of three phases: data pre-analysis, in which data of interest are identified and collected; data preprocessing, in which data are examined and analyzed, and in which some data may be restructured or transformed to make them more useful, and data post-analysis, in which some data are validated and re-adjusted. In the integrated data preparation scheme, every phase comprises different processing steps. For example, the data pre-analysis phase includes data requirement analysis, data collection, data selection and data integration. Data preprocessing comprises data inspection and data processing. Data post-analysis contains data division and re-division, data validation and data re-adjustment. This phase is called 'post-analysis' because the data preparation tasks may be adjusted in terms of feedback information in the process of neural network training, learning and validation (i.e., modeling process). Because the post-analysis adjusts the data for modeling purpose, the data post-analysis is still seen as the range of data preparation. In almost all existing studies, data preparation includes only the second phase, data preprocessing. Therefore, our proposed data preparation scheme is broader than others, which distinguishes our study from others. This is an important contribution that can fill up the gap in the literature.

In the following subsections, the three phases of the integrated data preparation scheme are described in detail. First, the detailed steps of every phase and some data problems that are normally encountered are discussed, and some existing methods and techniques to overcome these problems are overviewed step-by-step. For some important issues and dilemmas

(Problem II), they are described and some rational intelligent solutions are presented.

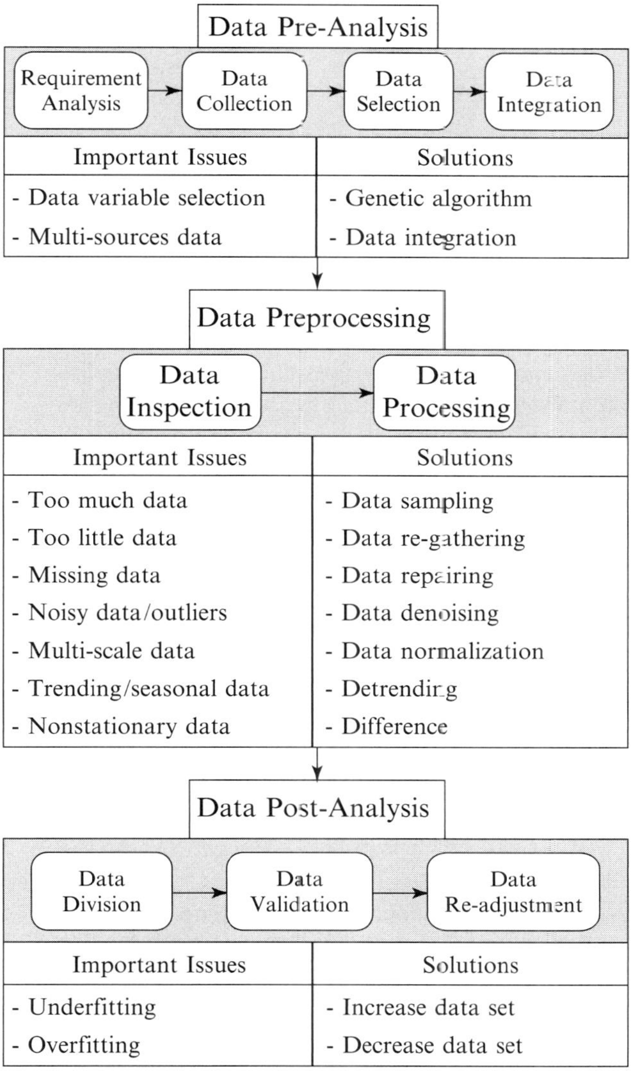

Fig. 3.2. The integrated data preparation scheme for neural network data analysis

3.3.2 Data Pre-Analysis Phase

As seen from Fig. 3.2, this phase consists of four steps: data requirement analysis, data selection, data collection and data integration.

3.3.2.1 Data requirement analysis

For a specific data analysis project, the first step is to understand the data requirements of the project in conjunction with the problem definitions and expected objectives. If the problem is outside one's field of expertise, interviewing specialists or domain experts may provide insight into the underlying process so that some potential problems may be avoided (Stein, 1993a). Questions may include the following: (i) What information would we like to have? (ii) What data are required for a specific task? (iii) Where can the data be found? (iv) What format are the data in? (v) What external sources of data are available?

When understanding the data requirements, data will be collected from various sources.

3.3.2.2 Data collection

This is an important step because the outcome of the step will restrict subsequent phases or steps. Based on the data requirements, all kinds of approaches, such as information retrieval (Ingwersen, 1992) and text mining (Nahm, 2001), will be used to collect various data from various sources. In some situations, some important data may be hard to collect. Thus, surrogate data is useful and necessary.

3.3.2.3 Data variable selection

Once data are collected, determining variables for modeling becomes possible. The goal of any model should be parsimony; i.e., to find the simplest explanation of the facts using the fewest variables. Therefore, it is best to identify the variables that will save modeling time and reduces the problem space (Stein, 1993a). There are many means of variable selection (Lemke and Muller, 2003; Tuv and Runger, 2003; Granger, 1969; Diamantaras and Kung, 1996; Ashley and Allegrucci, 1999). For example, Lemke and Muller (2003) used a modular approach and self-organizing variable selection to realize variable reduction, while Tuv and Runger (2003) presented a non-hierarchical metric clustering method to deal with high-dimensional classification. Some other methods, such as correlation analysis with Granger causality method (Granger, 1969), principal component analysis (PCA)

(Diamantaras and Kung, 1996) and stepwise multiple regression (Ashley and Allegrucci, 1999) are also mentioned in the literature.

3.3.2.4 Data integration

If data are collected from many different sources by several different groups, the data are still in disorder and scattered, and data integration treatment becomes vital (Famili et al., 1997). This is especially true when data contain text and symbolic attributes and have to be combined for further analysis.

In general, data sources can be divided into internal and external sources (Yan et al., 2003). Similarly, data representations can be divided roughly into structural representation and non-structural representation. Therefore, there are different data integration methods for different data representations from multi-sources. Regarding structural data from different sources, we can utilize mature database techniques, such as virtual views and data warehouse techniques (Chaudhuri and Dayal, 1997), to integrate data relations from different sources via join and/or union operations. Semantic and descriptive conflicts can be solved by renaming operations and conversions. Some meta-level conflicts and instance-level conflicts can be solved by advanced schema transformation (e.g., transposition of relations), reconciliation function and user-defined aggregates. More information can be found in (Abiteboul et al., 1999; Chaudhuri and Dayal, 1997). With regard to the integration of non-structural data from different sources, Baumgarten (1999) and Li et al. (2003) presented some solutions.

3.3.2.5 Important issues and dilemmas of this phase

In this phase, two important issues are described and discussed in detail.

A. Important issue I: Data variable selection with genetic algorithms
From the previous analysis, we find that data variable selection is an extremely important issue and many related studies (Lemke and Muller, 2003; Tuv and Runger, 2003; Granger, 1969; Diamantaras and Kung, 1996; Ashley and Allegrucci, 1999) are presented. Here we present an intelligent solution — genetic algorithm (GA) — to this important issue for neural network data analysis.

To date, genetic algorithms (GAs) have become a popular optimization method as they often succeed in finding the best optimum in contrast to most common optimization algorithms. Genetic algorithms imitate the natural selection process in biological evolution with selection, mating reproduction and mutation, and the sequence of the different operations of a genetic algorithm is shown in the left part of Fig. 3.3. The parameters to be optimized are represented by a chromosome whereby each parameter is encoded in a binary string called gene. Thus, a chromosome consists of as many genes as parameters to be optimized. Interested readers can be referred to Holland (1992) and Goldberg (1989) for more details. In the following GA for data variable selection is discussed.

First of all, a population, which consists of a given number of chromosomes, is initially created by randomly assigning "1" and "0" to all genes. In the case of variable selection, a gene contains only a single bit string for the presence and absence of a variable. The top right part of Fig. 3.3 shows a population of four chromosomes for a three-variable selection problem. In this study, the initial population of the GA is randomly generated except of one chromosome, which was set to use all variables. The binary string of the chromosomes has the same size as variables to select from whereby the presence of a variable is coded as "1" and the absence of a variable as "0". Consequently, the binary string of a gene consists of only one single bit. The subsequent work is to evaluate the chromosomes generated by previous operation by a so-called fitness function, while the design of the fitness function is a crucial point in using GA, which determines what a GA should optimize. In the case of a variable selection for neural network data analysis, the goal is to find a small subset of variables, which are most significant for complex data analysis. In this study, the complex data analysis is based on neural networks for modeling the relationship between the input variables and the responses. Thus, the evaluation of the fitness starts with the encoding of the chromosomes into neural networks whereby "1" indicates that a specific variable is used and "0" that a variable is not used by the network. Then the networks are trained with a training data set and after that, a testing data set is predicted. Finally, the fitness is calculated by a so-called fitness function *f*. For a prediction/classification problem, for example, our fitness function for the GA variable selections can use the following form:

$$f = 0.3RMSE_{training} + 0.7RMSE_{testing} - \alpha(1 - n_v / n_{tot}) \tag{3.1}$$

where n_v is the number of variables used by the neural networks, n_{tot} is the total number of variables and *RMSE* is the root mean square error, which is defined in Equation (3.2) with N as total number of samples predicted, y_t as the actual value and $\hat{y}_t$ as the predicted value:

$$RMSE = \sqrt{\frac{1}{N}\sum_{t=1}^{N}(\hat{y}_t - y_t)^2} \tag{3.2}$$

From Equation (3.1), we find that the fitness function can be broken up into three parts. The first two parts correspond to the accuracy of the neural networks. Thereby $RMSE_{training}$ is based on the prediction of the training data used to build the neural networks, whereas $RMSE_{testing}$ is based on the prediction of separate testing data not used for training the neural networks. It was demonstrated by Kupinski and Giger (1999) that using the same data for the variable selection and for the model calibration introduces a bias. Thus, variables are selected based on data poorly representing the true relationship. On the other hand, it was also shown that a variable selection based on a small data set is unlikely to find an optimal subset of variables (Kupinski and Giger, 1999). Therefore, a ratio of 3:7 between the influence of training and testing data was chosen. Although being partly arbitrary this ratio should give as little influence to the training data as to bias the feature selection yet taking the samples of the larger training set partly into account. The third part of the fitness function rewards small networks using only few variables by an amount proportional to the parameter a. The choice of a will influence the number of variables used by the evolved neural networks. A high value of results in only few variables selected for each GA whereas a small value of a results in more variables being selected. In sum, the advantage of this fitness function is that it takes into account not only the testing error of testing data but also partially the training error and primarily the number of variables used to build the corresponding neural networks.

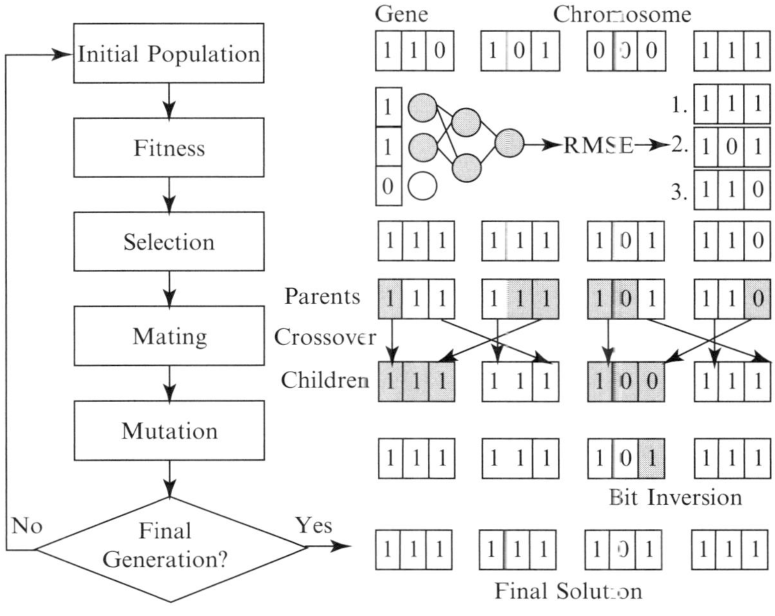

Fig. 3.3. The data variable selection with the genetic algorithm

After evolving the fitness of the population, the best chromosomes with the highest fitness value are selected by means of the roulette wheel. Thereby, the chromosomes are allocated space on a roulette wheel proportional to their fitness and thus the fittest chromosomes are more likely selected.

In the following mating step, offspring chromosomes are created by a crossover technique. Usually, the crossover operator combines a part of one string with a part of another string and is controlled by a crossover probability. Typically, it takes two strings called the parents as inputs, and returns one or two new ones, children or offspring. In this way, we expect to combine the good parts of another string, producing a better string after this operation. Many different kinds of crossover operators have been proposed including single-point, two-point, and uniform crossover (Mahfoud, 1995; Kim and Street, 2004). In this chapter, a so-called one-point crossover technique is employed, which randomly selects a crossover point within the chromosome. Then two parent chromosomes are interchanged at this point to produce two new offspring.

After that, the chromosomes are mutated with a mutation probability. Through introducing new genetic characteristics into the pool of chromosomes, it prevents the gene depletion that might lead a GA to converge into a local optimum. The mutation operator here always randomly changing genes from "0" to "1" and vice versa.

Finally, the final generation will be judged. If yes, then the optimized subsets are selected. If no, then the evaluation and reproduction steps are repeated until a certain number of generations, until a defined fitness or until a convergence criterion of the population are reached. In the ideal case, all chromosomes of the last generation have the same genes representing the optimal solution.

B. Important issue II: The integration of non-structural data

Another important issue is the integration of non-structural data. Non-structural data consist of unstructured data and semi-structural data. As earlier noted, integrating non-structural data from different sources is difficult. Here a three-phase approach for this task is proposed.

The first phase is to extract related features from text data or documents by semantic analysis and formulate an event-specific summary. This extraction makes non-structural data more readable and representative. Some string matching algorithms, such as (Mani and Bloedorn, 1997; Saravanan et al., 2003), can be used. The second phase is to transform the summary into corresponding nominal variables or numerical variables by classification algorithms, such as (Saravanan et al., 2003). This transformation makes modeling easier because the transformed variables can be treated as dummy variables in models. The last phase is to tabulate the transformed variables, making them more easily used by models. Non-structural data can thereby be formulated into a single dataset using previous data integration techniques.

3.3.3 Data Preprocessing Phase

After the data are identified and collected, they must be examined to identify any characteristics that may be unusual or indicative of more complex relationship. This is because the data from the previous phase may be impure, divergent, untrustworthy or even fraudulent. Therefore, data preprocessing is required. In this study, data preprocessing is a transformation, or conditioning, of data designed to make modeling more robust, which include data inspection and processing.

3.3.3.1 Data inspection

The first step of data preprocessing is data inspection. The goal of data inspection is to find problems with data. Data inspection includes data quantity and quality inspection. The former is to check the size of datasets and the latter any unusual dataset patterns. The data quantity inspection can be performed by observation. Generally, there are two main problems with this: a too-large data size or a too-small data size. The data quality inspection can also be performed by statistical methods including the check of data noise, data missing, data scale and data trending and data nonstationarity. There are four approaches to data quality inspection: line graph for checking missing data and data trending, control plot (Shewhart, 1931) for data noise detection, unit root test (Dickey and Fuller, 1979) for data nonstationarity check, and SVM-OD (Wang et al., 2003) for outlier detection.

3.3.3.2 Data processing

The above step (i.e., data inspection) can identify seven main problems: too many data, too few data, noisy data including outliers and errors, missing data, multi-scale data, trending (or seasonal) data, and nonstationary data. Accordingly, several processing techniques – data sampling (Han and Fu, 1994; Fayyad and Irani, 1993), data re-gathering, data denoising (Wang et al., 2003; Srinivasan et al., 1992; John, 1995; Gamberger et al., 2000), data repairing (Pyle, 1999; Batista and Monard, 2003a, 2003b; Little and Murphy, 1987; Ragel and Cremilleux, 1998; Lee et al., 1976; Tseng et al., 2003), data normalization (Lou, 1993; Weigend and Gershenfeld,

1994), and data difference (Weigend and Gershenfeld, 1994; Tseng, et al., 2002; Moody, 1995) – are used to deal with them.

3.3.3.3 Important issues and dilemmas of this phase

In this section, six important issues and two dilemmas about data preprocessing are presented.

A. Important issue I: Too many data and data sampling
In many domains, such as space (e.g., image data) and finance (e.g., stock price data every five minutes), the volume of data and the rate at which data are produced may be a limiting factor in performing on-time data analysis. Furthermore, the amount of data is sometimes beyond the capability of the hardware and software available for data analysis. Therefore, sample space reduction is important. Here, clustering and data discretization are used to treat the problem.

If there are a large number of observations, i.e., a large sample size, a useful approach is to obtain a representative subset of data or a data sampling. An effective way of doing this is to divide the sample by forming clusters of sample observations. Every cluster can then be represented by one observation. This can be (i) one specific observation, (ii) the mean value of all observations in the cluster, (iii) observation which has the lowest distance from all the others, and so on. In addition, data sampling techniques (Han and Fu, 1994; Fayyad and Irani, 1993), which selects a representative subset from a large population of data, can also be used.

If data clustering is difficult, discretization can be used. Discretization is aimed at reducing the number of distinct values for a given attribute, particularly for analysis methods requiring discrete attribute values. Possible methods of discretization are (a) histogram-based discretization, (b) discretization based on concept-hierarchies (Han and Fu, 1994), and (c) entropy-based discretization (Fayyad and Irani, 1993). Here we focus on histogram-based discretization due to its simplicity.

B. Important issue II: Too few data and data re-gathering
Conversely, if too few data are collected, data re-gathering will be necessary for complex data analysis. Nguyen and Chan (2004) found that neural networks perform more badly when few data are available or data are insufficient. As data re-gathering may be difficult, all kinds of information channels and gathering tools should be used for this task.

C. Important issue III: Noisy data and data de-noising
As has been stated (Famili et al., 1997), noise in the data weakens the predictive capability of the features. Therefore, noise reduction or data

denoising is very important for neural network data analysis. In the existing literature, noise elimination has been extensively studied (Wang et al., 2003; Srinivasan et al., 1992; John, 1995; Gamberger et al., 2000). Here we propose a regression-based data denoising approach to eliminate the effect of noise.

In our approach, the first step in noise reduction is noise detection. We use a control plot to detect the noise, as previously mentioned. The second step in noise reduction is noise filtering to eliminate outliers. Here we use regression technique. In linear regression, also known as least square method, the goal is to find a straight line modeling a two-dimensional dataset. This line $y = \alpha x + \beta$ is specified by the parameters α and β, which are calculated from the known values of the attribute x and y. Let

$$\bar{x} = \sum x_i / n \quad \text{and} \quad \bar{y} = \sum y_i / n \tag{3.3}$$

Then

$$\alpha = \sum (x_i - \bar{x})(y_i - \bar{y}) \Big/ \sum (x_i - \bar{x})^2 \quad \text{and} \quad \beta = \bar{y} - \alpha\bar{x} \tag{3.4}$$

The parameters α and β can now be used to remove data items well away from the regression line. For example, this can be decided simply based on the absolute distance or by removing n percent of items with the largest distance as noise or outliers.

D. Important issue IV: Missing data and data repairing

Roughly, missing data can be divided into two types: missing attributes and missing attribute values. Missing or insufficient attributes are examples of data problems that may complicate data analysis tasks, such as learning, and hinder accurate performance of most data analysis systems (Famili et al., 1997). For example, in the case of learning, these data insufficiencies limit the performance of a learning algorithm or statistical tool applied to the collected data, no matter how complex the algorithm is or how many data are used. Furthermore, missing attributes are a source of too few data, as was previously revealed. Therefore, related attribute data should be further re-gathered.

However, in most practical applications, an important problem is the handling of missing attribute values in a data set. Several studies have been done on dealing with missing values with numerous methods (see Pyle, 1999; Batista and Monard, 2003a, 2003b; Little and Murphy, 1987; Ragel and Cremilleux, 1998; Lee et al., 1976; Tseng et al., 2003). The aim of these methods is to recover the missing values that are as close as possible to the original values. The methods of doing this can be categorized into two types: imputation-based and data mining-based methods. The former is primarily for handling missing values of numerical data, while the latter is for category data. The principle of imputation methods is to estimate the missing values by using the existing values as an auxiliary base. The underlying assumption is that there are certain correlations between different data tuples over all attributes. The existing methods include mean imputation (Batista and Monard, 2003a), hot-deck or cold-deck imputation (Batista and Monard, 2003b), regression, and composite imputation (Little and Murphy, 1987). For the data mining-based methods, techniques such as associations (Ragel and Cremilleux, 1998), clustering (Lee et al., 1976) and regression (Tseng et al., 2003) are used to discover similar patterns between data tuples so as to predict the missing values.

E. Important issue V: Multi-scale data and data normalization
In neural network learning, data with different scales often lead to the instability of neural networks (Weigend and Gershenfeld, 1994). At the very least, data must be scaled into the range used by the input neurons in the neural network. This is typically -1 to 1 or zero to 1 (Lou, 1993). Many commercially available generic neural network development programs, such as BrainMaker, automatically scale each input. Moreover, neural networks always require that the range of the data is neither too small nor too large, so that the precision limits of the computer are not exceeded. Otherwise, the data should be scaled. Furthermore, data normalization helps to improve the performance of neural networks (Lou, 1993). Therefore, data normalization is necessary for treating multi-scale data. The main reason is that the neural network models often rely on Euclidean measures, and un-scaled data could bias or interfere with the training process. Line scaling and sigmoidal function normalization are the commonly used methods.

Line scaling method is a simple and effective approach. Let the maximal and minimal value of input range be I_{max} and I_{min}; then the formula for transforming each data D to an input value I is:

$$I = I_{\min} + [(I_{\max} - I_{\min}) \times (D - D_{\min})]/(D_{\max} - D_{\min}) \tag{3.5}$$

where D_{max} and D_{min} are the maximal and minimal value of a given input range. This method of normalization will scale input data into the appropriate range.

In addition, a logistic function can be used as a data normalization method, depending on the characteristics of the data. Here, a sigmoidal function is utilized in the following:

$$I(x) = \frac{r_i}{1 + \exp[-p_i(x_i - q_i)]}, i = 1,2,\cdots,n. \tag{3.6}$$

where r_i is used to constrain the range for the ith transformed element of the n-element data set, q_i can be selected as the smallest in the ith element of data set, and p_i decides the sharpness of the transferred function. Also, Equation (3.6) can compress the abnormal data to a specific range. Note that different continuous and differentiable transformation function can also be selected.

F. Important issue VI: Trending data, seasonal and nonstationary data

For a neural predictor, the presence of a trend may have undesired effects on the prediction performance (Weigend and Gershenfeld, 1994). Similarly, researchers (Tseng et al., 2002; Moody, 1995) have demonstrated that seasonal data have a significant impact on neural network prediction. As to univariate time series analysis with neural networks, nonstationarity is a problem for time series analysis (Moody, 1995). Therefore, data detrending and deseasonalization and data stationarity are also important issues in complex data analysis. For trending and seasonal data and nonstationary data, difference or log-difference (Weigend and Gershenfeld, 1994; Tseng et al., 2002; Moody, 1995) is a simple and effective treatment method that is widely used.

G. Dilemma I: Data sampling and sample representative trade-off

In this phase, the first dilemma is data sampling size and sample representative trade-off problem. Generally, as a larger data sample is taken, the variability of the sample tends to fluctuate even less between the smaller and larger samples. To resolve the trade-off problem, this study presents a novel convergence approach.

The convergence approach has two types: incremental and decremental. In the incremental type, a random sample is first selected and the distribution properties (such as mean, standard deviation, skewness and kurtosis) are calculated. Then, the sample distribution is tested repeatedly by adding additional instances. If the sample distribution is recalculated as each additional instance is added, a low number of instances will appear in the sample; that is, each addition will make a large impact on the shape of the curve. However, when the number of instances in the sample is modest, the overall shape of the curve will settle down and will change little as new instance values are added. This settling down of the overall curve is the key to deciding the 'convergence' between two different data sets. The decremental method is the opposite of the incremental method.

H. Dilemma II: Noise and nonstationarity trade-off

The second dilemma of this phase is the so-called 'noise–nonstationarity trade-off (Moody, 1995)' for neural network univariate time series models. That is, when there is noise and nonstationarity in the time series at the same time, neural network training on older data sets (longer training window) can induce biases in predictions because of nonstationarity, whereas using a shorter training window can increase estimation error (too much model variance) because of the noise in the limited data set. Moody (1995) suggested using the testing error against the training window length in the choice of the optimal training window. We followed this suggestion.

3.3.4 Data Post-Analysis Phase

3.3.4.1 Data division and re-division

Following data preprocessing, data obtained from the previous phase is used for network training and generalization. The first main data preparation task in this phase is to split data into subsets for neural network learning. Usually, a data set is divided into training data and testing data (sometimes there is a third data set – a validation set). So far, there is no universal rule to determine the size of either a training data set or a testing data set. Brainmaker software randomly selects ten percent of the facts from the data set and uses them for testing. Yao and Tan (2000) suggested that historical data be divided into three sets: training, validation and testing. The training set contains seventy percent of the collected data, while the validation and the testing sets contain twenty percent and ten percent respectively. Sometimes, through feedback of modeling results, data re-division is required.

3.3.4.2 Data validation

Generally, the training error always decreases with an increase in the number of cycles or epochs. In contrast, the testing error does not have a continuously decreasing trend where a minimum value is found on the curve. Thus, there are two classes of problems (overfitting and underfitting) to disturb the network. In overfitting a network usually performs worse instead of better after a certain point during training. This is because such long training may make the network memorize the training patterns, including all of their peculiarities. Underfitting results from insufficient training. This makes the network's generalization very poor. The solution to two problems is to validate the data with a section of extra data by cross-validation.

3.3.4.3 Data Re-adjustment

Depending on the feedback of data validation and model training results, data re-adjustment is often necessary. Different feedbacks mean different adjustments. For example, if the training result of neural network is not satisfactory, data re-division may be necessary. If there is overfitting or underfitting in the data validation, data re-division is also required. After data adjustment is completed, new learning begins once more.

3.3.4.4 Important issues and dilemmas of this phase

In this phase, an important issue and two dilemmas are described in the following.

A. Important issue I: Overfitting, underfitting and model complexity
Neural networks are often referred to as universal function approximators since theoretically any continuous function can be approximated to a prescribed degree of accuracy by increasing the number of neurons in the hidden layer of a feedforward network (Hornik et al., 1989). Yet in reality, the objective of a data analysis (e.g., prediction) is not to approximate a data set with an ultimate accuracy, but to find a suitable model with the best possible generalizing ability (Esposito, 2000). The gap between the approximation of a data set and the model generalization ability becomes the more problematic the higher the number of variables and the smaller the data set, which will be explained below.

For a prediction problem, the best measure for the generalizing ability is the prediction error of as many independent separate validation data as possible. According to the left side of Fig. 3.4 the prediction error is composed of two main contributions, the remaining interference error and the estimation error (Martens and Naes, 1989). The interference error is the systematic error (bias) due to un-modeled interference in the data, as the data analysis model (e.g., a prediction model) is not complex enough to capture all the interferences of the relationship. The estimation error is caused by modeling measured random noise of various kinds. The optimal prediction is obtained, when the remaining interference error and the estimation error balance each other, as shown in Fig. 3.4. The effect of the prediction error increasing due to a too simple model is called underfitting whereas the effect of the increased prediction error due to a too complex model is called overfitting.

In the right side of Fig. 3.4 it is shown that the optimal complexity of the model highly depends on the size and quality of the data set. For data sets, which are noisy and limited in size, a simple model is needed to prevent the overfitting. Neural networks, which are too complex (too big), are

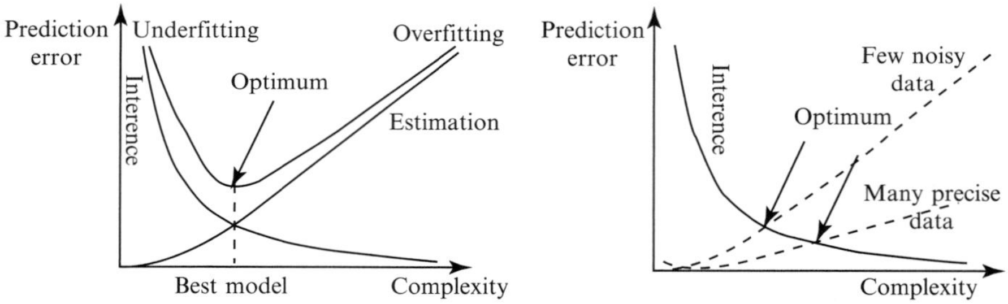

Fig. 3.4. Overfitting, underfitting and model complexity

in danger of learning these data by heart and consequently model noise of the data. For big data sets, which contain only little noise, the best model is more complex resulting in an overall smaller prediction error for the same functional relationship. Consequently, for each data set an optimal model complexity has to be found whereby the complexity of the models is directly related with the number of data variables utilized by the model.

B. Dilemma I: Training set size and model fitting

In this phase, the first dilemma is the training set size and model fitting dilemma; that is, how to determine the rational training data size. Generally, training data size is not too large or too small. A too-large training set may lead to a long training time and slower training speed, particularly when the entire training set is presented to a network between weights updates, and even lead to overfitting. When a training set size is too small, the network cannot learn effectively; this leads to underfitting and weak generalization. To solve this problem, cross-validation, such as *k*-fold and leave-one-out, can be used.

C. Dilemma II: Training epochs and network generalization

Another dilemma is the number of training epochs or cycles and network generalization. Usually, if the number of training cycles increases, the training error rate should decrease. The error rate on test cases should begin to decrease, and then it will eventually turn upward. This corresponds to the dynamic of underfitting and then overfitting. So far there is no universal rule to determine training epochs, except trial and error with incremental algorithms.

A key issue of our study is, however, whether the integrated data preparation scheme is of value in neural network data analysis given that data preparation is time-consuming. In the next section, we will discuss this issue from a general viewpoint.

3.4 Costs–Benefits Analysis of the Integrated Scheme

Data preparation requires extra time and effort; hence, the question of costs versus benefits arises (Problem III). Decision-makers may not be willing to invest in data preparation if it has little impact on the final data analysis results of neural network models. This problem is analyzed from three aspects:

(a) ***Total time saving for neural network modeling***. Although additional time is needed to prepare data, the learning time of neural network data analysis may decrease sharply. As has been stated (Rajas, 1996), every hour invested in data preparation may save many days in training a network. Therefore, our scheme will result in an overall time saving. Empirical evidence can be found in the work of Yu et al. (2006a).

(b) ***Model complexity reduction for neural network modeling***. Usually, the model complexity of neural networks can be referred to as the number of parameters, namely the number of weights and the number of biases. The complexity of neural network models can be mainly reduced to the number of adjustable parameters. Generally, the complexity of neural network models can be calculated as:

$$C(n) = n_h(n_i + 1) + n_o(n_h + 1) \tag{3.7}$$

where $C(n)$ is the model complexity, n_i is the number of input nodes, n_h is the number of hidden neurons and n_o is the number of output nodes. From Equation (3.7), we can see that the model complexity can be reduced by appropriate data preparation work, e.g., variable selection.

(c) ***Performance improvement for complex data analysis***. In practice, many applications (Beltratli et al., 1996; Senol and Gouch, 1992; Gately, 1996; Refenes et al., 1996; Smith and Gupta, 2002; Zhang, 2004; Nedeljkovic and Milosavljevic, 1992; Sjoberg, 1992; De Noord, 1994; DeWitt, 1994; Joo, et al., 2000; Nguyen and Chan, 2004) also revealed that data preparation minimizes error. To explain this further, a bias-variance-noise decomposition achieved by extending the bias-variance decomposition originally proposed by (Geman et al., 1992) is used to analyze performance improvement.

Considering a classification or prediction problem, $y = f(x)$ of an unknown target t, provided by a learner f on input x, with $x \in R^d$ and $y \in R$, the classical decomposition of the error in bias and variance for the specific loss function is defined as $L(t, y)$. (Here loss function is not limited in squared error $L(t, y) = (t - y)^2$, other functions, e.g., absolute error

$L(t, y) = |t - y|$ and zero-one loss $L(t, y) = 0$ when $t = y$; and 1 otherwise, are also used.).

Assuming that optimal prediction is $y^* = \arg\min_{y'}(E_t[L(t, y')])$ and the main prediction is $y_m = \arg\min_{y'}(E_D[L(y, y')])$ with D a set of training set. Then expected loss $E_{y,t}[L(y,t)]$ with D a set of training set can be decomposed into three parts: bias, variance and noise.

Bias: $B(x) = L(y^*, y_m)$. The loss of this part is from the difference between main prediction and the optimal prediction and mainly caused by the learning algorithm.

Variance: $V(x) = E_y[L(y_m, y)]$. The loss originates from the difference between any prediction and the main prediction, or overfitting. A high variance tells us that classifications differ a lot if we use different training sets.

Noise: $N(x) = E_t[L(y^*, t)]$. The loss is very small and from the difference between the optimal prediction and target. Sometimes the loss is unavoidable.

To summarize, the general error can be represented as

$$\begin{aligned} E_{y,t}[L(y,t)] &= c_1 \cdot E_t[L(y^*, t)] + L(y^*, y_m) + c_2 \cdot E_y[L(y, y_m)] \\ &= c_1 \cdot N(x) + B(x) + c_2 \cdot V(x) \end{aligned} \tag{3.8}$$

It should be note that c_1 and c_2 take on different values for different loss functions. For further illustration, mean squared error is defined as a loss function of a neural network model. Assume that there is a true function, $y = f(x) + \varepsilon$, where ε is normally distributed with zero mean and standard deviation σ. Given a set of training sets D: $\{(x_i, y_i)\}$, we fit the unknown function $h(x) = w \bullet x + \varepsilon$ to the data by minimizing the squared error $\sum_i [y_i - h(x_i)]^2$. Given a new data point x^* with the observed value $y^* = f(x^*) + \varepsilon$, the expected error $E[(y^* - h(x^*))^2]$ can be decomposed into bias, variance and noise below:

$$\begin{aligned} E[(h(x^*) - y^*)^2] &= E[(h(x^*))^2 - 2h(x^*)y^* + (y^*)^2] \\ &= E[(h(x^*))^2] - 2E[h(x^*)]E(y^*) + E[(y^*)^2] \\ &\qquad (\because E(Z - \bar{Z})^2 = E(Z^2) - \bar{Z}^2) \\ &= E[(h(x^*) - \bar{h}(x^*))^2] + (\bar{h}(x^*))^2 - 2h(x^*)f(x^*) \\ &\quad + E[(y^* - f(x^*))^2] + (f(x^*))^2 \\ &= E[(h(x^*) - \bar{h}(x^*))^2] + E[(y^* - f(x^*))^2] + (\bar{h}(x^*) - f(x^*))^2 \end{aligned}$$

$$= Var\ (h(x^*)) + E(\varepsilon^2) + Bias^2(h(x^*))$$
$$= Var\ (h(x^*)) + \sigma^2 + Bias^2(h(x^*)) \tag{3.9}$$

As revealed in Equation (3.9), we can improve the data analysis performance by three-fold data preparation work. First, noise reduction and filtering can alleviate the effects of noise because the noise does not always follow the normal distribution. Next, data division, data validation and data re-grouping can effectively eliminate the effects of bias. Finally, every data preprocessing technique can reduce the effects of variance. However, this is only a theoretical discussion. The impact of the data preparation on neural network data analysis can refer to the work of Yu et al. (2006a) for more details.

3.5 Conclusions

In this chapter, a comprehensive data preparation scheme with three-phase data processing is proposed to obtain better performance for specific neural network data analysis task. The novel integrated data preparation scheme proposed in this paper will enhance the neural network learning process and reduce the complexity of neural network models and will be of immense help for complex data analysis. Through theoretical analysis and investigations, several important conclusions were obtained: (i) the integrated data preparation scheme can definitely speed up data analysis process, reduce model complexity and improve the performance of data analysis tasks; (ii) the scheme is necessary and beneficial in data preparation for neural network data analysis. As the proposed integrated data preparation scheme has been proved to be very effective in the performance improvement of neural network data analysis, as revealed by Yu et al. (2006a). This point also leads to the final conclusion: (iii) the proposed integrated data preparation scheme can be used as a promising solution to improve the performance of neural network data analysis, which is worth being generalized in the future.

To summarize, the main contributions of this study reflects the following four-fold. One contribution is to propose an integrated data preparation scheme with three-phase data processing for neural network data analysis. Although some related studies about data preparation are presented in the literature, a systematic study of data preparation has not been formulated so far. Another contribution is to present an overview of data preparation. In neural network data analysis, we have discussed a number of data preparation techniques in three phases and accordingly some practical solutions to some dilemmas are provided. Although many techniques are shown in the previous literature, such as work of Fayyad et al. (1996), there are some distinct difference between our work and the previous work. First of all, the previous work, such as (Zhang et al., 2003; Fayyad et al., 1996), only focused on the data preprocessing, while our work are broader, and mainly focus on all data processing including data pre-analysis and data post-analysis in addition to data preprocessing. Second, their work does not present a systematic study for data preparation. Their works about data preparation are scattered and non-systematical. Comparatively speaking, our proposed integrated scheme presents a more comprehensive study for data preparation. The third contribution of our integrated data preparation scheme is to present a comprehensive survey about neural network data preparation, which is different from others. In addition, some new data preparation techniques, such as GA for variable selection, are suggested for neural network data preparation. The final contribution is to provide a full cost-benefit analysis framework for integrated data preparation scheme. These contributions fill up the gap in previous studies.

However, there are still some important issues to be considered for future research on the data preparation for complex data analysis:

(i) The scope of this study is limited to neural network data analysis. Future research should extend more data analysis models such as data mining and knowledge discovery models.

(ii) The study only presents some limited data preparation techniques in the integrated data preparation scheme. More new data preparation techniques in all steps of the integrated data preparation scheme are worth exploring.

(iii) To perform meaningful data preparation, either the domain expert should be a member of the data analysis team or the domain should be extensively studied before the data are preprocessed. Involvement of the domain expert would lead to useful feedback for verifying and validating the use of particular data preparation techniques. Thus, the integration of expert opinions into a neural network data preparation framework is an important issue.

(iv) A module for analyzing the effects of data preparation should be added to neural network software packages so that users working in other domains can more easily understand the impact of data preparation techniques on their work.

Part III: Individual Neural Network Models with Optimal Learning Rates and Adaptive Momentum Factors for Foreign Exchange Rates Prediction

4 Forecasting Foreign Exchange Rates Using an Adaptive Back-Propagation Algorithm with Optimal Learning Rates and Momentum Factors

4.1 Introduction

Foreign exchange rates prediction is regarded as a rather difficult and challenging task due to its high volatility and noisy market environment (Yu et al., 2005c). Usually, the difficulty in forecasting exchange rates is attributed to the limitation of many conventional forecasting models as many statistical and econometric models are unable to produce significantly better predictions than the random walk (RW) model (Ince and Trafalis, 2005; Chen and Leung, 2004), which has also encouraged academic researchers and business practitioners to develop more predictable forecasting models. Recent studies provide some evidence that nonlinear models are able to produce better predictive results. Of the various nonlinear models, the artificial neural network (ANN) model has emerged as a strong alternative for predicting exchange rates (Yu et al., 2005c, d). As claimed by Grudnitski and Osburn (1993), neural networks are particularly well suited for finding accurate solutions in an environment characterized by complex, noisy, irrelevant or partial information. Furthermore, neural networks have been proved to be universal function approximators that can map any nonlinear function without any a priori assumption about the data (Haykin, 1999). Unlike traditional statistical models, neural networks are a class of data-driven, non-parametric weak models, and they let the data "speak for themselves". So neural networks are less susceptible to the model misspecification problem than most of the parametric models, and they are more powerful in describing the dynamics of financial time series than traditional statistical models (Kaastra and Boyd, 1995; Zhang and Hu, 1998). Literature documenting this research effort is quite diverse and involves different design strategies. In earlier studies, Refenes et al. (1993) applied multilayer feed-forward neural network (MLFNN) to predict the price change of currency exchange rate. Similarly, Zhang and Hu (1998) also use a MLFNN model for GBP/USD exchange rate prediction. Tenti (1996) utilized recurrent neural network (RNN) to forecast foreign exchange rates, while Hsu et al. (1995) developed a clustering neural network (CNN) model to predict the direction of movements in the USD/DEM exchange rate. Their experimental results suggested that their proposed model achieved better forecasting performance relative to other indicators. Chen and Leung (2004) used an error correction neural network (ECNN) model to predict exchange rates and good forecasting results can be obtained with their model. To understand the good or bad of these single neural network models mentioned above, De Matos (1994) compared the strength of a MLFNN with that of a RNN based on the forecasting of Japanese yen futures. In the same way, Kuan and Liu (1995) also provided a comparative evaluation of the performance of the MLFNN and the RNN on the prediction of an array of commonly traded exchange rates. In a recent study by Leung et al. (2000), the forecasting accuracy of MLFNN was compared with the general regression neural network (GRNN). The results obtained showed that the GRNN possessed a greater forecasting strength relative to MLFNN with respect to a variety of currency rates.

Recently, some hybrid forecasting models for foreign exchange rate prediction have been developed that integrate neural network techniques with many conventional econometrical models or time series models to improve prediction accuracy. For example, Lai et al. (2006b) hybridized neural network and exponential smoothing to predict the exchange rate of EUR/USD and JPY/USD and obtained good performance. Similarly, Ince and Trafalis (2005) proposed a two-stage hybrid model integrating parametric models, e.g., auto-regressive integrated moving average (ARIMA), vector auto-regression (VAR), co-integration model and nonparametric model, e.g., ANN and support vector regression (SVR) to predict foreign exchange rates. Likewise, Zhang (2003) also proposed a hybrid methodology that combined both ARIMA and ANN models taking advantage of the unique strengths of these models in linear and nonlinear modeling for time series forecasting. Empirical results with real data sets indicated that the hybrid model could provide an effective way to improve the forecasting accuracy achieved by either of the models used separately. However, in the study of Yu et al. (2005c), a hybrid model integrating ANN and generalized linear auto-regression (GLAR) for foreign exchange rates prediction reported a worse performance than individual ANN model.

Instead of using single network architecture and hybrid forecasting models, Zhang and Berardi (2001) and Yu et al. (2005c) investigated the use of ensemble methods in exchange rate forecasting. Essentially, they proposed using ensemble models to predict foreign exchange rates. Experiment results indicated that the ensemble model could consistently outperform single network models and hybrid forecasting models. A good recent survey about neural network in foreign exchange rates forecasting can be referred to Yu et al. (2005e) for more details.

Although different neural network models have been used to predict foreign exchange rates, almost all neural network models do not use the optimal learning rates and momentum factors. In their studies, the learning rates and momentum factors are usually set to a fixed value when training. It is, however, critical to determine a proper fixed learning rate and momentum factor for the neural network applications. For learning rate, if it is large, learning may occur quickly, but it may also become un-stable and may even not learn at all. To ensure stable learning, the learning rate must be sufficiently small, but a small learning rate may lead to a long learning time and a slow convergence speed. Also, it is unclear just how small the learning rate should be. In addition, the best fixed learning rate is problem-independent, and it varies with different neural network structure for different problem applications (Sha and Bajic, 2002; Yu et al., 2005a, 2006c). Therefore, a suitable learning rate is important for any learning algorithm. Similarly, the original motivation of introducing the momentum term is to help dampen the oscillation of the weight adjustment. But existing studies do not give any clue as to how the momentum factor should be set. In the literature, it is usually determined empirically. Based on the same reason of learning rate, finding a suitable momentum factor is hard in neural network training. Furthermore, an improper choice of the momentum factor can actually slow the convergence rate of neural network (Yu et al., 1995). In a sense, a proper momentum factor is also equally important.

A common problem for learning rate and momentum factor mentioned above is a non-optimal choice problem of the learning rate and momentum factor. A feasible solution is to derive optimal learning rate formulae for neural networks and then allow an adaptive change at each iteration step during the training process. The resulting algorithm will eliminate the need for a search for the proper fixed learning rate and momentum factor and thus providing a fast convergence rate.

Due to the highly nonlinearity of neural networks, it is difficult to obtain the optimum learning rate and momentum factor. In this paper, a new method based on gradient descent rule and optimization techniques is proposed to derive optimal learning rate and momentum factor of multi-layer BP algorithm and construct an adaptive BP learning algorithm to train neural networks. To test the effective-ness and efficiency of the proposed algorithm, three major international currencies (British pounds, euros and Japanese yen against US dollar, which is abbreviated as GBP/USD,

EUR/USD and JPY/USD for short, respectively) are chosen as the forecasting targets. In order to provide a fair and robust evaluation, the forecasting performance of the proposed BP learning algorithm is compared with those of the standard BP algorithm and several popular weight update algorithms, such as the Levenberg-Marquardt algorithm (Hagan and Menhaj, 1994) and extended Kalman filter algorithm (Iiguni et al., 1992; Ruck et al., 1992), which are used as the benchmark models.

The rest of this chapter is organized as follows. In Section 4.2, an adaptive BP learning algorithm with optimal learning rate and momentum factor is first proposed in terms of the gradient descent rule and optimization techniques. For verification and illustration, Section 4.3 gives the experiments about three foreign exchange rates prediction and reports their results. Finally, conclusions and future directions are given in Section 4.4.

4.2 BP Algorithm with Optimal Learning Rates and Momentum Factors

In this section, an adaptive BP learning algorithm with optimal learning rate and momentum factor is proposed. First of all, optimized adaptive learning rates are derived in term of gradient descent rule and optimization technique. Then optimal momentum factors are derived from gradient descent direction of line search during BP learning. Finally, the proposed adaptive BP learning algorithm with optimal learning rate and momentum factor is constructed.

4.2.1 Optimal Learning Rates Determination

In this subsection, we use a matrix-vector notation of the neural network description in order to be able to express later by simple formula. Consider a three-layer BPNN, which has p nodes in the input layer, q nodes in the hidden layer and k nodes in the output layer. As shown in Chapter 2, the basic structure of the BPNN model can be represented as

$$\hat{Y}(t+1) = F_2[V^T(t)F_1(W(t)X(t))] \tag{4.1}$$

where all notations are the same to Chapter 2.

Usually, by estimating model parameter vectors or network connection weights (W, V) (as shown in Equation (4.1)) via BPNN training and learning, we can realize the corresponding tasks such as pattern recognition, function approximation and system prediction. In fact, the model parameter vector (W, V) can be obtained by iteratively minimizing a cost function $E(X: W, V)$. In general, $E(X: W, V)$ is a sum of the error squares cost function with k output nodes and n training pairs or patterns, i.e.,

$$
\begin{aligned}
E(X:W,V) &= \frac{1}{2}\sum_{j=1}^{n}\sum_{i=1}^{k} e_{ij}^{2} = \frac{1}{2}\sum_{j=1}^{n} e_j^T e_j \\
&= \frac{1}{2}\sum_{j=1}^{n}[y_j - \hat{y}_j(X:W,V)]^T[y_j - \hat{y}_j(X:W,V)]
\end{aligned}
\tag{4.2}
$$

where y_j is the jth actual value and $y_j(X: W,V)$ is the jth estimated value. Given the time factor t, Equation (4.2) can be rewritten as

$$
\begin{aligned}
E(t) &= \frac{1}{2}\sum_{j=1}^{n}\sum_{i=1}^{k} e_{ij}^{2}(t) = \frac{1}{2}\sum_{j=1}^{n} e_j^T(t) e_j(t) \\
&= \frac{1}{2}\sum_{j=1}^{n}[y_j(t) - \hat{y}_j(t)]^T[y_j(t) - \hat{y}_j(t)]
\end{aligned}
\tag{4.3}
$$

where $e_j(t) = [e_{1j}(t) \quad e_{2j}(t) \quad \cdots \quad e_{kj}(t)]^T \in R^{k\times 1}$, $j = 1,2,\cdots,n$, $y_j(t)$ and $\hat{y}_j(t)$ are the jth actual value predicted value at time t, respectively.

By applying the gradient descent rule to the cost function $E(t)$ (as shown in Equation (4.3)) and considering Equation (4.1), we can obtain the weight increment formulae with respect to W and V, respectively (Proofs can be refer to Chapter 2)

$$
\Delta W(t) = -\eta(t)\nabla_W E(t) = \eta(t)\sum_{j=1}^{n} F'_{1(j)} V F'_{2(j)} e_j x_j^T \tag{4.4}
$$

$$
\Delta V(t) = -\eta(t)\nabla_V E(t) = \eta(t)\sum_{j=1}^{n} F_{(j)} e_j^T F'_{2(j)} \tag{4.5}
$$

where η is the learning rate, ∇ and Δ are the gradient operator and incremental operator, $\Delta W(t)$ and $\Delta V(t)$ are the weight adjustment increments at iteration t, respectively.

To derive the optimal learning rate, consider the following error increment equation:

$$
\Delta e(t+1) = e(t+1) - e(t) = y(t+1) - \hat{y}(t+1) - y(t) + \hat{y}(t) \tag{4.6}
$$

Let $\Delta y(t+1) = y(t+1) - y(t)$ be the change of the actual value and let $\Delta\hat{y}(t+1) = \hat{y}(t+1) - \hat{y}(t)$ be the change of the neural network output. Here we assume that the absolute value of the change of the actual value is much smaller than the absolute value of the change of the neural network output, i.e., $|\Delta y(t+1)| << |\Delta\hat{y}(t+1)|$. This implies that the value $y(t)$ can approximate $y(t+1)$ locally during the training process, that is to say, the change of the actual value can be ignored comparing with the change of neural network output during the learning process. This assumption is realistic for many processes due to energy constraints in practical systems, while no constraints are imposed to the output of the neural networks (Sha and Bajic, 2002, Yu et al., 2005a, 2006c). Also, if this condition is not satisfied, then the neural network prediction system will not be able to adapt sufficiently fast to change in the practical systems and the prediction of the practical systems will be impossible. With the above assumption, the increment in Equation (4.6) can be approximated as

$$\Delta e(t+1) = e(t+1) - e(t) = \Delta y(t+1) - \Delta \hat{y}(t+1) \approx -\Delta \hat{y}(t+1) \tag{4.7}$$

Usually, in BPNN learning algorithm, the change of output of networks can be represented as

$$\Delta \hat{y}(t+1) = \eta \xi(t) e(t) \tag{4.8}$$

where $\xi(t) = \mathbf{F}_2'[(\mathbf{F}_1^T \mathbf{F}_1)\mathbf{I}_{k^2} + (\mathbf{X}^T \mathbf{X})(V^T \mathbf{F}_1' \mathbf{F}_1' V)]\mathbf{F}_2'$ with

$$\mathbf{F}_2' = \begin{bmatrix} F'_{2(1)} & 0 & \cdots & 0 \\ 0 & F'_{2(2)} & \cdots & 0 \\ \cdots & \cdots & \cdots & \cdots \\ 0 & 0 & \cdots & F'_{2(N)} \end{bmatrix}, \quad \mathbf{F}_1^T \mathbf{F}_1 = \begin{bmatrix} F_{1(1)}^T F_{1(1)} & F_{1(1)}^T F_{1(2)} & \cdots & F_{1(1)}^T F_{1(N)} \\ F_{1(2)}^T F_{1(1)} & F_{1(2)}^T F_{1(2)} & \cdots & F_{1(2)}^T F_{1(N)} \\ \cdots & \cdots & \cdots & \cdots \\ F_{1(N)}^T F_{1(1)} & F_{1(N)}^T F_{1(2)} & \cdots & F_{1(N)}^T F_{1(N)} \end{bmatrix},$$

$$\mathbf{X}^T \mathbf{X} = \begin{bmatrix} x_1^T x_1 & x_1^T x_2 & \cdots & x_1^T x_N \\ x_2^T x_1 & x_2^T x_2 & \cdots & x_2^T x_N \\ \cdots & \cdots & \cdots & \cdots \\ x_N^T x_1 & x_N^T x_2 & \cdots & x_N^T x_N \end{bmatrix}, \quad \mathbf{F}_1' \mathbf{F}_1' = \begin{bmatrix} F'_{1(1)} F'_{1(1)} & F'_{1(1)} F'_{1(2)} & \cdots & F'_{1(1)} F'_{1(N)} \\ F'_{1(2)} F'_{1(1)} & F'_{1(2)} F'_{1(2)} & \cdots & F'_{1(2)} F'_{1(N)} \\ \cdots & \cdots & \cdots & \cdots \\ F'_{1(N)} F'_{1(1)} & F'_{1(N)} F'_{1(2)} & \cdots & F'_{1(N)} F'_{1(N)} \end{bmatrix}.$$

In order to prove Equation (4.8), a lemma must be firstly introduced.

Lemma 4.1

The total time derivative of the BPNN single hidden output $V^T F_1(WX)$ can be represented as

$$\frac{d[V^T F_1(WX)]}{dt} = F_1(WX)\frac{dV}{dt} + V^T F_1'(WX)\frac{dW}{dt}X = \frac{dV^T}{dt}F_1(WX) + V^T F_1'(WX)\frac{dW}{dt}X$$

where $V^T F_1(WX)$ is the single output of BPNN, $\frac{dW}{dt}$ and $\frac{dV}{dt}$ are the derivatives with respect to time t, W, V are the weight vectors, X is the input vector and $X = [x_0, x_1, \cdots, x_p]^T$, $F_1(WX) = [f_1(net_0), f_1(net_1), \cdots, f_1(net_q)]^T$,

$$F_1'(WX) = \begin{bmatrix} f'(net_1) & \cdots & 0 \\ \cdots & \cdots & \cdots \\ 0 & \cdots & f'(net_q) \end{bmatrix}, \quad \frac{dW}{dt} = \begin{bmatrix} \frac{dw_{00}}{dt} & \frac{dw_{01}}{dt} & \cdots & \frac{dw_{0p}}{dt} \\ \frac{dw_{10}}{dt} & \frac{dw_{11}}{dt} & \cdots & \frac{dw_{1p}}{dt} \\ \cdots & \cdots & \cdots & \cdots \\ \frac{dw_{q0}}{dt} & \frac{dw_{q1}}{dt} & \cdots & \frac{dw_{qp}}{dt} \end{bmatrix},$$

$$V = [v_1, v_2, \cdots, v_q]^T, \quad \frac{dV}{dt} = \begin{bmatrix} \frac{dv_0}{dt} & \frac{dv_1}{dt} & \cdots & \frac{dv_q}{dt} \end{bmatrix}^T.$$

Proof: Derivation of $\frac{d[V^T F_1(WX)]}{dt}$ is as follows:

$$\frac{d[V^T F_1(WX)]}{dt} = \frac{d\left[\sum_{i=0}^{q} v_i f_1\left(\sum_{j=0}^{p} w_{ij} x_j\right)\right]}{dt}$$

$$= \sum_{i=0}^{q} \frac{\partial\left[\sum_{i=0}^{q} v_i f_1\left(\sum_{j=0}^{p} w_{ij} x_j\right)\right]}{\partial v_i} \frac{dv_i}{dt} + \sum_{i=0}^{q} \sum_{j=0}^{p} \frac{\partial\left[\sum_{i=0}^{q} v_i f_1\left(\sum_{j=0}^{p} w_{ij} x_j\right)\right]}{\partial w_{ij}} \frac{dw_{ij}}{dt}$$

$$= \sum_{i=0}^{q} f_1\left(\sum_{j=0}^{p} w_{ij} x_j\right) \frac{dv_i}{dt} + \sum_{i=0}^{q} \sum_{j=0}^{p} v_i f_1'\left(\sum_{j=0}^{p} w_{ij} x_j\right) x_j \frac{dw_{ij}}{dt}$$

$$= \left[f_1(net_0) \frac{dv_0}{dt} + f_1(net_1) \frac{dv_1}{dt} + \cdots + f_1(net_q) \frac{dv_q}{dt} \right]$$

$$+ v_0 f_1'(net_0) \left[x_0 \frac{dw_{00}}{dt} + x_1 \frac{dw_{01}}{dt} + \cdots + x_p \frac{dw_{0p}}{dt} \right]$$

$$+ v_1 f_1'(net_1) \left[x_0 \frac{dw_{10}}{dt} + x_1 \frac{dw_{11}}{dt} + \cdots + x_p \frac{dw_{1p}}{dt} \right]$$

$$+ \cdots + v_q f_1'(net_q) \left[x_0 \frac{dw_{q0}}{dt} + x_1 \frac{dw_{q1}}{dt} + \cdots + x_p \frac{dw_{qp}}{dt} \right]$$

$$= \left[f_1(net_0) \quad f_1(net_1) \quad \cdots \quad f_1(net_q) \right] \begin{bmatrix} \frac{dv_0}{dt} & \frac{dv_1}{dt} & \cdots & \frac{dv_q}{dt} \end{bmatrix}^T$$

$$+\begin{bmatrix} v_1 & v_2 & \cdots & v_q \end{bmatrix} \begin{bmatrix} f'(net_1) & \cdots & 0 \\ \cdots & \cdots & \cdots \\ 0 & \cdots & f'(net_q) \end{bmatrix} \begin{pmatrix} \text{due to } f(net_0) \equiv 1, \\ f'(net_0) = 0 \end{pmatrix}$$

$$\times \begin{bmatrix} \frac{dw_{00}}{dt} & \frac{dw_{01}}{dt} & \cdots & \frac{dw_{0p}}{dt} \\ \frac{dw_{10}}{dt} & \frac{dw_{11}}{dt} & \cdots & \frac{dw_{1p}}{dt} \\ \cdots & \cdots & \cdots & \cdots \\ \frac{dw_{q0}}{dt} & \frac{dw_{q1}}{dt} & \cdots & \frac{dw_{qp}}{dt} \end{bmatrix} \begin{bmatrix} x_0 \\ x_1 \\ \cdots \\ x_p \end{bmatrix}$$

$$= F_1(WX)\frac{dV}{dt} + V^T F_1'(WX)\frac{dW}{dt}X = \frac{dV^T}{dt}F_1(WX) + V^T F_1'(WX)\frac{dW}{dt}X \qquad \blacksquare$$

The subsequent task is to prove Equation (4.8). First of all, we consider the change in output of the BPNN for the *m*th pattern. The above Lemma 4.1 together with Equations (4.4) and (4.5) can give the following equations:

$$\Delta \hat{y}^m(t+1) \approx \left(\frac{d\hat{y}^m(t+1)}{dt} \right) \Delta t = \begin{pmatrix} \frac{d\hat{y}_1^m(t+1)}{dt} \\ \frac{d\hat{y}_2^m(t+1)}{dt} \\ \cdots \\ \frac{d\hat{y}_k^m(t+1)}{dt} \end{pmatrix} \Delta t$$

$$= \begin{pmatrix} f'_{2(1),m} \cdot \left(F_{1,m}^T \frac{dV_1}{dt} + V_1^T F'_{1,m} \frac{dW}{dt} x_m \right) \\ f'_{2(2),m} \cdot \left(F_{1,m}^T \frac{dV_2}{dt} + V_2^T F'_{1,m} \frac{dW}{dt} x_m \right) \\ \cdots \\ f'_{2(k),m} \cdot \left(F_{1,m}^T \frac{dV_k}{dt} + V_k^T F'_{1,m} \frac{dW}{dt} x_m \right) \end{pmatrix} \Delta t$$

$$\approx \begin{pmatrix} f'_{2(1),m} \cdot \left(F_{1,m}^T \frac{\Delta V_1}{\Delta t} + V_1^T F'_{1,m} \frac{\Delta W}{\Delta t} x_m \right) \\ f'_{2(2),m} \cdot \left(F_{1,m}^T \frac{\Delta V_2}{\Delta t} + V_2^T F'_{1,m} \frac{\Delta W}{\Delta t} x_m \right) \\ \cdots \\ f'_{2(k),m} \cdot \left(F_{1,m}^T \frac{\Delta V_k}{\Delta t} + V_k^T F'_{1,m} \frac{\Delta W}{\Delta t} x_m \right) \end{pmatrix} \Delta t$$

$$= \begin{bmatrix} f'_{2(1),m} \cdot (F_{1,m}^T \Delta V_1 + V_1^T F'_{1,m} \Delta W x_m) \\ f'_{2(2),m} \cdot (F_{1,m}^T \Delta V_2 + V_2^T F'_{1,m} \Delta W x_m) \\ \cdots \\ f'_{2(k),m} \cdot (F_{1,m}^T \Delta V_k + V_k^T F'_{1,m} \Delta W x_m) \end{bmatrix}$$

$$= \begin{bmatrix} f'_{2(1),m} \cdot (F_{1,m}^T \sum_{j=1}^N \Delta V_{1j} + V_1^T F'_{1,m} \Delta W x_m) \\ f'_{2(2),m} \cdot (F_{1,m}^T \sum_{j=1}^N \Delta V_{2j} + V_2^T F'_{1,m} \Delta W x_m) \\ \cdots \\ f'_{2(k),m} \cdot (F_{1,m}^T \sum_{j=1}^N \Delta V_{kj} + V_k^T F'_{1,m} \Delta W x_m) \end{bmatrix}$$

$$= \begin{bmatrix} f'_{2(1),m} \cdot [F_{1,m}^T \left(\eta \sum_{j=1}^N F_{1j} e_{1j} f'_{2(1),j}\right) + V_1^T F'_{1,m} \left(\eta \sum_{j=1}^N F'_{1j} V F'_{2j} e_j x_j^T\right) x_m] \\ f'_{2(2),m} \cdot [F_{1,m}^T \left(\eta \sum_{j=1}^N F_{1j} e_{2j} f'_{2(2),j}\right) + V_2^T F'_{1,m} \left(\eta \sum_{j=1}^N F'_{1j} V F'_{2j} e_j x_j^T\right) x_m] \\ \cdots \\ f'_{2(k),m} \cdot [F_{1,m}^T \left(\eta \sum_{j=1}^N F_{1j} e_{kj} f'_{2(k),j}\right) + V_k^T F'_{1,m} \left(\eta \sum_{j=1}^N F'_{1j} V F'_{2j} e_j x_j^T\right) x_m] \end{bmatrix}$$

$$= \eta \sum_{j=1}^N \begin{bmatrix} f'_{2(1),m} \cdot (F_{1,m}^T F_{1j} e_{1j} f'_{2(1),j} + V_1^T F'_{1,m} F'_{1j} V F'_{2j} e_j x_j^T x_m) \\ f'_{2(2),m} \cdot (F_{1,m}^T F_{1j} e_{2j} f'_{2(2),j} + V_2^T F'_{1,m} F'_{1j} V F'_{2j} e_j x_j^T x_m) \\ \cdots \\ f'_{2(k),m} \cdot (F_{1,m}^T F_{1j} e_{kj} f'_{2(k),j} + V_k^T F'_{1,m} F'_{1j} V F'_{2j} e_j x_j^T x_m) \end{bmatrix}$$

$$= \eta \sum_{j=1}^N \begin{bmatrix} f'_{2(1),m} & 0 & \cdots & 0 \\ 0 & f'_{2(2),m} & \cdots & 0 \\ \cdots & \cdots & \cdots & \cdots \\ 0 & 0 & \cdots & f'_{2(k),m} \end{bmatrix}$$

$$\cdot \begin{bmatrix} F_{1,m}^T F_{1j} e_{1j} f'_{2(1),j} + V_1^T F'_{1,m} F'_{1j} V F'_{2j} e_j x_j^T x_m \\ F_{1,m}^T F_{1j} e_{2j} f'_{2(2),j} + V_2^T F'_{1,m} F'_{1j} V F'_{2j} e_j x_j^T x_m \\ \cdots \\ F_{1,m}^T F_{1j} e_{kj} f'_{2(k),j} + V_k^T F'_{1,m} F'_{1j} V F'_{2j} e_j x_j^T x_m \end{bmatrix}$$

$$= \eta \sum_{j=1}^N F'_{2,m} \cdot (F'_{2j} e_j F_{1,m}^T F_{1j} + V^T F'_{1,m} F'_{1j} V F'_{2j} e_j x_j^T x_m)$$

$$= \eta \sum_{j=1}^N F'_{2,m} \cdot (F_{1,m}^T F_{1j} I_{k^2} + V^T F'_{1,m} F'_{1j} V x_j^T x_m) \cdot F'_{2j} e_j$$

Thus, the total change caused by all patterns is

$$
\Delta\hat{y}(t+1)=\begin{bmatrix}\Delta\hat{y}_1(t+1)\\ \Delta\hat{y}_2(t+1)\\ \cdots\\ \Delta\hat{y}_m(t+1)\\ \cdots\\ \Delta\hat{y}_N(t+1)\end{bmatrix}\approx\begin{bmatrix}\eta\sum_{j=1}^{N}F'_{2,1}\cdot(F_{1,1}^{T}F_{1j}I_{k^2}+V^{T}F'_{1,1}F'_{1j}Vx_j^{T}x_1)\cdot F'_{2j}e_j\\ \eta\sum_{j=1}^{N}F'_{2,2}\cdot(F_{1,2}^{T}F_{1j}I_{k^2}+V^{T}F'_{1,2}F'_{1j}Vx_j^{T}x_2)\cdot F'_{2j}e_j\\ \cdots\\ \eta\sum_{j=1}^{N}F'_{2,m}\cdot(F_{1,m}^{T}F_{1j}I_{k^2}+V^{T}F'_{1,m}F'_{1j}Vx_j^{T}x_m)\cdot F'_{2j}e_j\\ \cdots\\ \eta\sum_{j=1}^{N}F'_{2,N}\cdot(F_{1,N}^{T}F_{1j}I_{k^2}+V^{T}F'_{1,N}F'_{1j}Vx_j^{T}x_N)\cdot F'_{2j}e_j\end{bmatrix}
$$

$$
=\eta\begin{bmatrix}F'_{2,1}&0&\cdots&0\\ 0&F'_{2,2}&\cdots&0\\ \cdots&\cdots&\cdots&\cdots\\ 0&0&\cdots&F'_{2,N}\end{bmatrix}\cdot\begin{bmatrix}(F_{1,1}^{T}F_{11}I_{k^2}+V^{T}F'_{1,1}F'_{11}Vx_1^{T}x_1)&(F_{1,1}^{T}F_{12}I_{k^2}+V^{T}F'_{1,1}F'_{12}Vx_1^{T}x_2)&\cdots&(F_{1,1}^{T}F_{1N}I_{k^2}+V^{T}F'_{1,1}F'_{1N}Vx_1^{T}x_N)\\ (F_{1,2}^{T}F_{11}I_{k^2}+V^{T}F'_{1,2}F'_{11}Vx_2^{T}x_1)&(F_{1,2}^{T}F_{12}I_{k^2}+V^{T}F'_{1,2}F'_{12}Vx_2^{T}x_2)&\cdots&(F_{1,2}^{T}F_{1N}I_{k^2}+V^{T}F'_{1,2}F'_{1N}Vx_2^{T}x_N)\\ \cdots&\cdots&\cdots&\cdots\\ (F_{1,N}^{T}F_{11}I_{k^2}+V^{T}F'_{1,N}F'_{11}Vx_N^{T}x_1)&(F_{1,N}^{T}F_{12}I_{k^2}+V^{T}F'_{1,N}F'_{12}Vx_N^{T}x_2)&\cdots&(F_{1,N}^{T}F_{1N}I_{k^2}+V^{T}F'_{1,N}F'_{1N}Vx_N^{T}x_N)\end{bmatrix}
$$

$$
\times\begin{bmatrix}F'_{21}e_1\\ F'_{22}e_2\\ \cdots\\ F'_{2N}e_N\end{bmatrix}=\eta\begin{bmatrix}F'_{21}&0&\cdots&0\\ 0&F'_{22}&\cdots&0\\ \cdots&\cdots&\cdots&\cdots\\ 0&0&\cdots&F'_{2N}\end{bmatrix}\left\{\begin{bmatrix}F_{11}^{T}F_{11}I_{k^2}&F_{11}^{T}F_{12}I_{k^2}&\cdots&F_{11}^{T}F_{1N}I_{k^2}\\ F_{12}^{T}F_{11}I_{k^2}&F_{12}^{T}F_{12}I_{k^2}&\cdots&F_{12}^{T}F_{1N}I_{k^2}\\ \cdots&\cdots&\cdots&\cdots\\ F_{1N}^{T}F_{11}I_{k^2}&F_{1N}^{T}F_{12}I_{k^2}&\cdots&F_{1N}^{T}F_{1N}I_{k^2}\end{bmatrix}\right.
$$

$$
\left.+\begin{bmatrix}x_1^{T}x_1&x_1^{T}x_2&\cdots&x_1^{T}x_N\\ x_2^{T}x_1&x_2^{T}x_2&\cdots&x_2^{T}x_N\\ \cdots&\cdots&\cdots&\cdots\\ x_N^{T}x_1&x_N^{T}x_2&\cdots&x_N^{T}x_N\end{bmatrix}\times\left(V^{T}\begin{bmatrix}F'_{11}F'_{11}&F'_{11}F'_{12}&\cdots&F'_{11}F'_{1N}\\ F'_{12}F'_{11}&F'_{12}F'_{12}&\cdots&F'_{12}F'_{1N}\\ \cdots&\cdots&\cdots&\cdots\\ F'_{1N}F'_{11}&F'_{1N}F'_{12}&\cdots&F'_{1N}F'_{1N}\end{bmatrix}\overline{V}\right)\right\}
$$

$$
\times\begin{bmatrix}F'_{21}&0&\cdots&0\\ 0&F'_{22}&\cdots&0\\ \cdots&\cdots&\cdots&\cdots\\ 0&0&\cdots&F'_{2N}\end{bmatrix}\begin{bmatrix}e_1\\ e_2\\ \cdots\\ e_N\end{bmatrix}
$$

$$
=\eta\mathbf{F}'_2[(\mathbf{F}_1^{T}\mathbf{F}_1)\mathbf{I}_{k^2}+(\mathbf{X}^{T}\mathbf{X})(V\mathbf{F}'_1\mathbf{F}'_1V)]\mathbf{F}'_2e(t)=\eta\xi(t)e(t) \qquad \blacksquare
$$

Substituting Equation (4.8) into Equation (4.7), we obtain

$$e(t+1) \approx e(t) - \eta\xi(t)e(t) \tag{4.9}$$

Because our goal is to derive an optimal learning rate η, thus, at iteration t, an optimal value of the learning rate $\eta^*(t)$ is obtained by minimizing error function $E(t+1)$. Define the error function:

$$E(t+1) = \frac{1}{2}e^T(t-1)e(t+1) \tag{4.10}$$

Using Equation (4.9), Equation (4.10) may be rewritten as

$$E(t+1) = \frac{1}{2}e^T(t+1)e(t+1) = \frac{1}{2}\left[e(t) - \eta\xi(t)e(t)\right]^T\left[e(t) - \eta\xi(t)e(t)\right] \tag{4.11}$$

This gives the error e, at iteration $t+1$, as a function of the learning rate η. By minimizing $E(t+1)$, we can obtain an optimal learning rate. From Equation (4.11), the first and second order conditions can be calculated as

$$\left.\frac{dE(t+1)}{d\eta}\right|_{\eta=\eta^*(t)} = -\frac{1}{2}[\xi(t)e(t)]^T[e(t) - \eta^*(t)\xi(t)e(t)] - \frac{1}{2}[e(t) - \eta^*(t)\xi(t)e(t)]^T\xi(t)e(t) = 0$$

$$\left.\frac{d^2E(t+1)}{d\eta^2}\right|_{\eta=\eta^*(t)} = e^T(t)\xi^T(t)\xi(t)e(t) > 0$$

Because $\xi(t)$ is positively definite, the second condition can be met and the optimum value of the learning rate, $\eta^*(t)$, can be obtained from the first order condition, as illustrated in Equation (4.12). Interestedly, the optimal learning rate that we obtained is distinctly different from the results produced by the previous several studies (Yu et al., 1995; Sha and Bajic, 2002).

$$\eta^*(t) = \frac{e^T(t)\xi^T(t)e(t)}{e^T(t)\xi^T(t)\xi(t)e(t)} \tag{4.12}$$

Relying on the optimization technique, we can get the optimized learning rate. This result about optimal learning rate eliminates the need for a search for the proper fixed learning rate and provides a solution to fast convergence for training BPNN. In addition, the optimal learning rate obtained from Equation (4.12) is derived from typical three-layer BPNN model, but it can be easily extended to multilayer BP neural networks.

4.2.2 Determination of Optimal Momentum Factors

The BP algorithm with momentum factors has been extensively reported in the literature (Allred and Kelly, 1990; Ooyen, 1992). The main aim of introduction of the momentum term is to lower the oscillation of the weight adjustment and to accelerate convergence rate of BP algorithm. However, literature does not give any implication as to how the momentum factor should be set, as is similar to the learning rate. In most applications, it is set to a fixed value determined by trial and error. Usually, an improper choice of the fixed momentum factor can actually slow the convergence rate of BP algorithm (Yu et al., 1995). In such a situation, we try to derive the optimal momentum factor to training BP neural network. In the sequel we will derive the optimal momentum factor using gradient descent direction of line search of BPNN learning algorithm.

Usually, the momentum term of BP algorithm is proportion to the previous weight increment, the weight update formulae can be represented by

$$\Delta W(t) = -\eta(t)\nabla_W E(t) + \mu(t)\Delta W(t-1) \tag{4.13}$$

$$\Delta V(t) = -\eta(t)\nabla_V E(t) + \mu(t)\Delta V(t-1) \tag{4.14}$$

where $\Delta W(t)$ and $\Delta V(t)$ are the weight increments, $\nabla_W E(t)$ and $\nabla_V E(t)$ are the error gradients with respect to W and V, $\mu(t)\Delta W(t-1)$ and $\mu(t)\Delta V(t-1)$ are the momentum terms and $\mu(t)$ is the momentum factor. For simplification, we only derive the optimal momentum factor with respect to W. The main reason reflects the following two aspects. First of all, the derivation process of optimal momentum factor with respect to V is the same to that of optimal momentum factor with respect to W. Second, the optimal momentum factor obtained from two derivation processes is identical.

Let $d(t)$ be the gradient descent direction, we have $d(t) = -\nabla_W E(t)$ and thus $\Delta W(t) = \eta^*(t)d(t)$ with optimal learning rate $\eta^*(t) > 0$. This leads to the following equation:

$$\Delta W(t) = \eta^*(t)d(t) = -\eta(t)\nabla_W E(t) + \mu(t)\eta^*(t-1)d(t-1) \tag{4.15}$$

From Equation (4.15), we can get the following representation:

$$d(t) = -\frac{\eta(t)}{\eta^*(t)}\nabla_W E(t) + \frac{\mu(t)\eta^*(t-1)}{\eta^*(t)}d(t-1) \tag{4.16}$$

In the BPNN, a solution can be obtained by minimizing the error function. Particularly, when the error function is not specified analytically, its minimization along descent direction $d(t)$ can be accomplished through a numerical line search for learning rate $\eta(t)$ or through numerical differentiation as noted in Equation (15) proposed in Abraham (2004), i.e., $d(t)$ can be given by

$$d(t) = -\nabla_W E(t) - \eta^*(t) H_W(t) d(t-1) \tag{4.17}$$

where $H_W(t)$ is a Hessian matrix, defined by $H_W(t) = \nabla_W^2 E(t) = \frac{\partial^2 E(t)}{\partial W^2(t)}$ (Abraham, 2004). Comparing Equations (4.16) and (4.17), we can get the optimal momentum factor:

$$\eta(t) = \eta^*(t) \text{ and } \mu(t) = -\frac{(\eta^*(t))^T H_W(t)\eta^*(t)}{\eta^*(t-1)} \tag{4.18}$$

Applying the optimal learning rate $\eta(t)$ and optimal momentum factor $\mu(t)$, the adaptive BP learning algorithm may achieve the fast convergence rate. Accordingly, the final weights update formulae can be represented as

$$\begin{aligned}\Delta W(t) &= -\eta(t)\nabla_W E(t) + \mu(t)\Delta W(t-1) \\ &= \left(\frac{e^T(t)\xi^T(t)e(t)}{e^T(t)\xi^T(t)\xi(t)e(t)}\right)\left(\sum_{j=1}^N F'_{1(j)} V F'_{2(j)} e_j x_j^T\right) \\ &\quad - \left(\frac{(\eta^*(t))^T H_W(t)\eta^*(t)}{\eta^*(t-1)}\right)(\Delta W(t-1))\end{aligned} \tag{4.19}$$

$$\begin{aligned}\Delta V(t) &= -\eta(t)\nabla_V E(t) + \mu(t)\Delta V(t-1) \\ &= \left(\frac{e^T(t)\xi^T(t)e(t)}{e^T(t)\xi^T(t)\xi(t)e(t)}\right)\left(\sum_{j=1}^N F_{1(j)} e_j^T F'_{2(j)}\right) \\ &\quad - \left(\frac{(\eta^*(t))^T H_W(t)\eta^*(t)}{\eta^*(t-1)}\right)(\Delta V(t-1))\end{aligned} \tag{4.20}$$

Until now, we complete the derivation process of achieving optimal learning rates and momentum factors. Using the updated weight formulae, an adaptive BP learning algorithm is generated. For convenience, the proposed adaptive BP learning algorithm is summarized as follows, as illustrated in Fig. 4.1. Note that in neural networks, a single pass over the input data set is usually called as an epoch, as shown in Fig. 4.1.

```
Set input nodes to the number of input vectors
Set target value for a specific pattern
Initialize weight vectors
While i < Epoch_num
    For j = 1 to N_records
        Compute net_i (i = 1,..., q), y_j (j = 1,..., k) using Equation (1);
        Compute error gradients ∇_w E(t) and ∇_v E(t) using Equations (4.4)-
        (4.5), and optimal learning rate η(t) and momentum factor μ(t) us-
        ing Equations (4.12) and (4.18), respectively;
    Endfor
    Update the weight vectors with Equations (4.19)-(4.20).
Wend
```

Fig. 4.1. Outline of the proposed adaptive BP learning algorithm

To verify the effectiveness of the proposed BP learning algorithm, three major international currencies (GBP/USD, EUR/USD and JPY/USD) are used as testing targets. The detailed experiments are presented in the following section.

4.3 Experiment Study

In this section, we first describe the research data and experiment design and then report the experimental results.

4.3.1 Data Description and Experiment Design

We use three different datasets in our forecast performance analysis. The data used in this study are daily and are obtained from Pacific Exchange Rate Service (http://fx.sauder.ubc.ca/), provided by Professor Werner Antweiler, University of British Columbia, Vancouver, Canada. The data consist of the US dollar exchange rate against each of the three main currencies (EUR, GBP and JPY) with which it has been studied in this study. The entire data set covers the period from January 1, 2000 to December 31, 2005 with a total of 1505 observations. The data sets are divided into two periods: the first period covers January 1, 2000 to December 31, 2004 with 1255 observations, while the second period is from January 1, 2005 to

December 31, 2005 with 250 observations. The first period, which is assigned to in-sample estimation, is used for network learning, i.e., training set. The second period, which is reserved for out-of-sample evaluation, is used for validation, i.e., testing set. For space limitation, the original data are not listed in this paper, and detailed data can be obtained from the website of Pacific Exchange Rate Service (http://fx.sauder.ubc.ca/).

For comparison, four related algorithms, standard BP algorithm (Haykin, 1999), BP algorithm only with optimal learning rate (Yu et al., 2005a, 2006c), Levenberg-Marquart (LM) based learning algorithm (Hagan and Menhaj, 1994), and extended Kalman filter (EKF) based learning algorithm (Iiguni et al., 1992; Ruck et al., 1992), are employed in this study. For standard BP learning algorithm, the weight update formulae are shown in Equations (4.4)-(4.5), and the learning rate η is fixed at 0.30, more details about standard BP learning algorithm can be referred to Haykin (1999). In the BP learning algorithm only with optimal learning rate, there are no momentum terms in weights update formulae and the Equation (4.12) are used to obtain optimal learning rates.

The Levenberg-Marquart (LM) based algorithm (Hagan and Menhaj, 1994) is a kind of effective learning algorithm which has the little computation time for per iteration. Basically, the link weights of the neural network are updated based on the Jacobian matrix, J, collecting the partial derivatives of the neural network error e with respect to the weights. In other words, the update increment ΔW collecting the corrections of the weights in matrix W is computed by

$$\Delta W = -[J^T J + \mu I]^{-1} J^T e \tag{4.21}$$

$$J = \begin{bmatrix} \frac{\partial e}{\partial w_{11}} & \frac{\partial e}{\partial w_{12}} & \cdots & \frac{\partial e}{\partial w_{1n}} \\ \frac{\partial e}{\partial w_{21}} & \frac{\partial e}{\partial w_{22}} & \cdots & \frac{\partial e}{\partial w_{2n}} \\ \cdots & \cdots & \cdots & \cdots \\ \frac{\partial e}{\partial w_{m1}} & \frac{\partial e}{\partial w_{m2}} & \cdots & \frac{\partial e}{\partial w_{mn}} \end{bmatrix} \tag{4.22}$$

It is worth noting that the LM-based algorithm is rather flexible. If μ is sufficiently large, the above weight update algorithm is similar to the gradient descent algorithm. If μ is equal to zero, the above algorithm will be a Gaussian-Newton algorithm. In this sense, the LM-based algorithm has the characteristics of both the gradient descent algorithm and the Gaussian-Newton algorithm.

The extended Kalman filter (EKF) based algorithm (Iiguni et al., 1992; Ruck et al., 1992) is a novel weight adjustment algorithm for BPNN. In this algorithm, the Kalman filter is used to update the weight vector of BPNN. The generic principle of EKF-based algorithm is that the EKF can modify the weight parameters to maximize the posterior probability of current instance with respect to its predicted probability distribution of weight parameters. Recent work proposed by Ruck et al. (1992) has revealed that the BPNN algorithm is actually a degenerated form of the EKF. Due to its excellent convergence properties, a lot of successful applications have been reported. Basically, the EKF-based weight adjustment formulae are illustrated as follows.

$$W(t) = W(t-1) + K(t)[y(t) - \hat{y}(t)] \tag{4.23}$$

$$K(t) = P(t-1)[\nabla E^T(t)][\nabla E(t)P(t-1)\nabla E^T(t) + R(t)]^{-1} \tag{4.24}$$

$$P(t) = P(t-1) - K(t)[\nabla E(t)]P(t-1) \tag{4.25}$$

where $W(t)$ is the connect weight of BPNN, $K(t)$ is called the Kalman gain, $y(t)$ is the actual value, $\hat{y}(t)$ is the predicted value produced by neural networks, $P(t)$ is the error variance-covariance matrix, defined by $P(t) = E\{[y(t) - \hat{y}(t)]^T [y(t) - \hat{y}(t)]\}$ and $\nabla E(t)$ is the gradient, defined by Equations (4.4)-(4.5). Usually, the system actual output $y(t) = \hat{y}(t) + \varepsilon(t)$, $\varepsilon(t)$ is assumed to be white noise vector with covariance $R(t)$ regarded as a modeling error. For more details, please refer to Iiguni et al. (1992) and Ruck et al. (1992).

In all the neural network predictors, nine input nodes are determined by auto-regression testing. The appropriate number of hidden nodes is set to 21 determined by trial and error. The training epochs are set to 2000 due to the problem complexity. Finally, to examine the forecasting performance, the root mean square error (*NMSE*) and directional change statistics (D_{stat}) (Yu et al., 2005c) of time series are employed as the performance measurement of the testing set.

4.3.2 Experimental Results

When the data are prepared, we begin to perform experiments according to the previous experiment design. First of all, the prediction results with five different algorithms are reported. Figs. 4.2-4.4 give graphical representations of the forecasting results for three typical foreign exchange rates using different learning algorithms. Tables 4.1-4.2 show a detailed prediction performance of the different algorithms in terms of both the level measurement (*NMSE*) and direction measurement (D_{stat}). From the figures and tables, we can generally find that the prediction results of the proposed BP learning algorithm with optimal learning rate and momentum factor are very promising for three foreign exchange rates under study either where the measurement of forecasting performance is the goodness-of-fit such as *NMSE* or where the forecasting performance criterion is the D_{stat}.

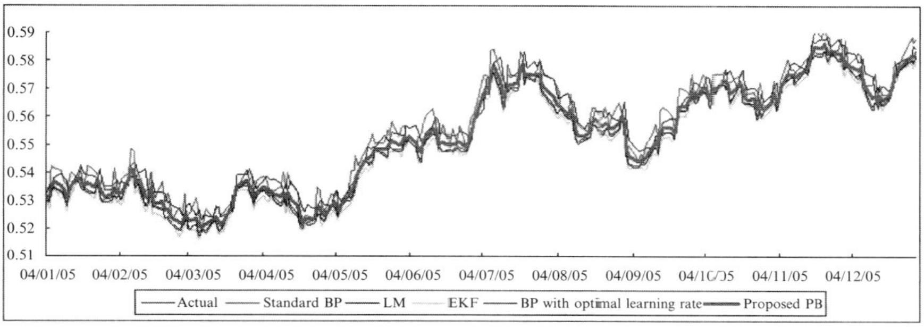

Fig. 4.2. Forecasting results with different learning algorithms for GBP/USD

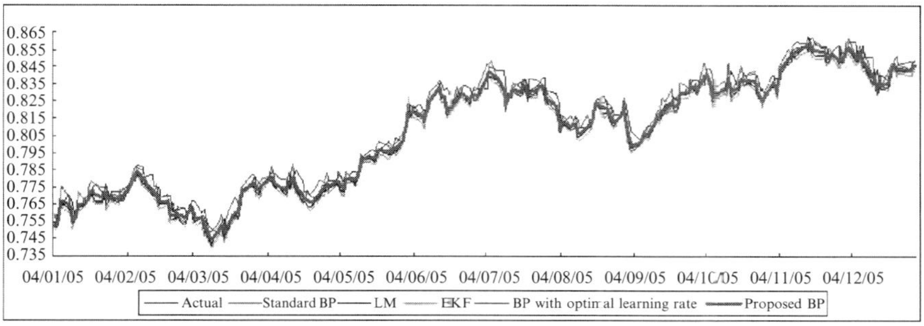

Fig. 4.3. Forecasting results with different BP learning algorithms for EUR/USD

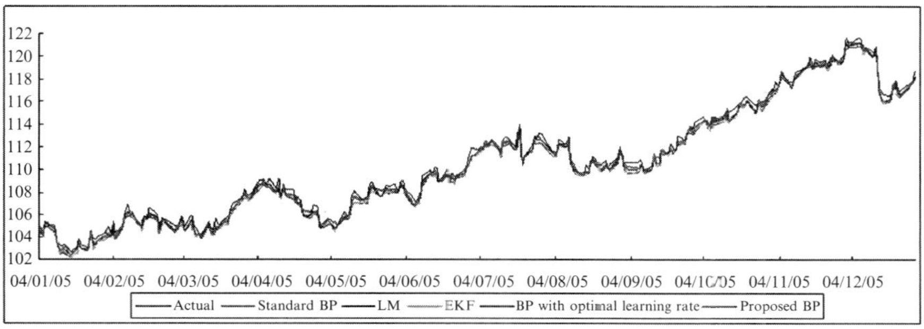

Fig. 4.4. Forecasting results with different BP learning algorithms for JPY/USD

In detail, Fig. 4.2 shows the prediction comparison of the GBP/USD between the proposed BP learning algorithm and the other four learning algorithm. Similarly, it can be seen from Fig. 4.3 that forecasting accuracy of the EUR/USD has significantly improved using the proposed BP learning algorithm with optimal learning rate and momentum factor. Fig. 4.4 provides a graphical comparison about the prediction results of the JPY/USD using different algorithms. The graphical results roughly indicate that the proposed BP learning algorithm performs better than the other algorithms presented here. This indication can also be further verified by some concrete prediction evaluation results in terms of *NMSE* and D_{stat}, as given by Tables 4.1-4.2.

Table 4.1. A *NMSE* comparison with different algorithms for three currencies

Algorithms	GBP/USD		EUR/USD		JPY/USD	
	NMSE	Rank	*NMSE*	Rank	*NMSE*	Rank
Standard BP learning	0.0837	5	0.0816	4	0.2124	4
LM-based learning	0.0731	4	0.0837	5	0.2182	5
EKF-based learning	0.0678	3	0.0745	3	0.1637	2
BP with optimal learning rates	0.0442	2	0.0639	2	0.1728	3
BP with optimal learning rates and momentum factors	0.0385	1	0.0554	1	0.1356	1

Table 4.2. A D_{stat} comparison with different algorithms for three currencies

Algorithms	GBP/USD		EUR/USD		JPY/USD	
	D_{stat}(%)	Rank	D_{stat}(%)	Rank	D_{stat}(%)	Rank
Standard BP learning	58.64	5	62.34	5	66.45	4
LM-based learning	63.71	4	64.18	4	65.32	5
EKF-based learning	70.22	3	71.46	3	72.44	3
BP with optimal learning rates	74.57	2	75.63	2	77.63	2
BP with optimal learning rates and momentum factors	79.87	1	80.67	1	83.48	1

Focusing on the *NMSE* indicator, the proposed BP learning algorithm performs the best in all the testing cases, followed by the BP learning algorithm with optimal learning rate, EKF-based learning algorithm, standard BP learning algorithm and LM-based learning algorithm. Comparing with standard BP learning algorithm, the *NMSE* of the proposed BP learning algorithm is much smaller. Interestingly, the *NMSE*s of the LM-based learning algorithm are not better than those of the standard BP learning algorithm except the testing case of GBP/USD. The possible reason is that the

EUR/USD and JPY/USD are more volatile than GBP/USD in this testing period. In addition, in the testing case of JPY/USD, the EKF-based learning algorithm performs better than the BP learning algorithm with optimal learning rate, as compared by the testing case of GBP/USD and EUR/USD. This suggests that there may be some additional factors that need to be explored in relation to Japanese yen. Two possible reasons might explain this phenomenon. One reason for this may be that the Japanese yen exchange rate is more volatile than that of the British pound and euros; another might be that the market for yen is bigger and more efficient than the market for British pounds. Although the BP learning algorithm with optimal learning rate is slightly worse than the EKF-based learning algorithm, the proposed BP learning algorithm with optimal learning rate and momentum factor consistently outperform the other different algorithms presented here in terms of *NMSE*.

However, the low *NMSE* does not necessarily mean that there is a high hit rate of forecasting direction for foreign exchange movement direction prediction. A typical example is EUR/USD. In Table 4.1, the *NMSE* of the standard BP learning algorithm is lower than that of the LM-based learning algorithm. But in Table 4.2, the D_{stat} of the LM-based learning algorithm is higher than that of the standard BP learning algorithm. Thus, the D_{stat} comparison is necessary. Furthermore, from the business practitioners' point of view, D_{stat} is more important than *NMSE* because the former is an important decision criterion. Focusing on D_{stat} indicator of Table 4.2, we find the proposed BP learning algorithm with optimal learning rate and optimal momentum factor still performs much better than the other models according to the rank. Furthermore, the differences between the different models are very significant. For example, for the GBP/USD testing case, the D_{stat} for the standard BP learning algorithm is 58.64%, for the LM-based learning algorithm it is 63.71%, and for the EKF-based learning algorithm D_{stat} is only 70.22%; while for the proposed BP learning algorithm with optimal learning rate and optimal momentum factor, D_{stat} reaches 79.87%. Relative to the standard BP learning algorithm, the performance improvement of the proposed BP learning algorithm arrives at 21.23% (79.87%-58.64%). This implies that the proposed adaptive BP learning algorithm with optimal learning rate and optimal momentum factor is an effective and feasible forecasting solution to foreign exchange rates prediction.

In summary, from the experiments presented in this study we can draw the following conclusions. (1) The experimental results show that the performance of the proposed BP learning algorithm with optimal learning rate and optimal momentum factor is superior to the standard BP learning algorithm, LM-based learning algorithm, EKF-based learning algorithm as well as the BP learning algorithm only with optimal learning rate for the testing cases of three main currencies in terms of level measurement (*NMSE*) and directional change statistics (D_{stat}), as can be seen from Tables 4.1 and 4.2. (2) The empirical results reveal that the proposed BP learning algorithm is able to improve forecasting accuracy significantly. In other words, the prediction performance and computational efficiency of the proposed BP learning algorithm are better than those of other learning algorithm, especially for standard BP learning algorithm in terms of *NMSE* and D_{stat}. This point leads to the final conclusion. (3) The proposed BP learning algorithm with optimal learning rate and optimal momentum factor can be used as an effective tool and solution to foreign exchange rate forecasting to obtain greater forecasting accuracy and further improve the computational efficiency.

4.4 Concluding Remarks

In this study, the optimal learning rate and optimal momentum factor are first determined by means of optimal technique and gradient descent rule. Then an adaptive BP learning algorithm with optimized learning rates and optimal momentum factors is proposed for time series forecasting. Finally, this exploratory research examines the potential of using the proposed BP learning algorithm with optimized learning rates and optimal momentum factors to predict three main foreign exchange rates – GBP/USD, EUR/USD and JPY/USD. Our empirical results obtained reveal that the proposed BP learning algorithm with optimized learning rates and optimal momentum factors indeed can provide much better forecasts than the other four learning algorithms. This implies that the proposed BP learning algorithm with optimized learning rates and optimal momentum factors is very suitable for foreign exchange rate prediction.

However, our work also highlights some problems that still need to be addressed. For example, the accuracy of forecasting is still worth improving for certain currencies, such as Japanese yen. In addition, some possible reasons that affect foreign exchange rate predictions are required to be further explored. Of course, the above problems show possible directions for future work in formulating a fast adaptive BP learning algorithm for exchange rate prediction as follows:

(i) As foreign exchange markets are a class of very complex dynamic markets, more factors that influence the exchange rate movement should be considered in future research.

(ii) A new adaptive BP learning algorithm with optimized learning rates and optimal momentum factors aimed at improving the traditional BP learning algorithm should be added to neural network software packages so that users working in other domains can more easily utilize this new neural network learning algorithm in their work.

5 An Online BP Learning Algorithm with Adaptive Forgetting Factors for Foreign Exchange Rates Forecasting

5.1 Introduction

The foreign exchange market is a complex, evolutionary, and nonlinear dynamical system. Foreign exchange rate series as a kind of financial time series are inherently noisy, non-stationary, and deterministically chaotic (Yaser and Atiya, 1996). This means that the distribution of foreign exchange rate series changes over time. Not only is a single data series non-stationary in the sense of the mean and variance of the series, but the relationship of the data series to other related data series may also be continuously changing (Yu et al., 2006b). Modeling such dynamical and non-stationary time series is a challenging task. Over the past few years, neural networks have been successfully applied to foreign exchange rates prediction and achieve promising results, as Chapters 1 and 4 indicated.

Among these neural network models, the three-layer back-propagation neural network (BPNN) is widely used for foreign exchange rate prediction (as indicated in Chapter 1) due to its approximations to nonlinear functions and its self-learning capability (Haykin, 1999). However, the BPNN has several limitations. For example, the convergence speed of the BPNN algorithm is often slow because the learning rate is fixed (Iiguni, et al., 1992), thus increasing the network learning time. Therefore, some fast training algorithms, such as adaptive learning algorithms (Tollenaere, 1990; Park, 1991), real-time learning algorithms (Iiguni, et al., 1992; Sha and Bajic, 2002) and other fast learning algorithms (Jacobs, 1988; Brent, 1991; Hagan and Menhaj, 1994), have been developed in an attempt to reduce these shortcomings. But two main limitations still remain so far.

Firstly, most BPNN models do not use the optimized instantaneous learning rates except the work of Yu et al. (1995) and Sha and Bajic (2002). In most studies in which these are introduced, the learning rate is set to a fixed value. It is, however, critical to determine a proper fixed learning rate for the BPNN applications, as described in Chapter 4.

Secondly, in the existing literature, almost all fast algorithms are batch learning algorithms. Although neural network batch learning is highly effective, the computation involved in each learning step is very big, especially when large sample data sets are presented. Furthermore, the neural networks must re-learn from the beginning as new data become available. Therefore, this may overly affect the overall computational efficiency of batch learning. In this sense, batch learning is unsuitable for real-time or online prediction when neural networks are used as a predictor.

For the first problem, an optimized instantaneous learning rate is derived from the gradient descent rule based on optimization techniques. For the second problem, an online learning algorithm should be created to overcome the drawbacks of batch learning algorithm. Actually, there is a difference between online learning algorithm and batch learning algorithm in the neural networks models. In the online learning algorithm, the weight vectors are updated recursively after the presentation of each input vector. While in the batch learning algorithm, the weight vectors of neural networks are updated only at the end of each epoch, which will be further illustrated later. Usually, in the neural networks, a single pass over the input data set is called as an epoch. Furthermore, the weight sequence should be chosen to given a higher weight to more recent data in the time series prediction. So an adaptive forgetting factor is also introduced into the proposed online learning algorithm. In order to verify the effectiveness and efficiency of the proposed online learning algorithm, two typical foreign exchange rate series, Canadian dollars (CAD) and Swiss francs (CHF) against US dollars (USD), are chosen for testing.

The rest of this chapter is organized as follows. In Section 5.2, the proposed online learning algorithm with adaptive forgetting factor is first presented in terms of the gradient descent algorithm and optimization techniques. For further illustration, an empirical analysis is then given in Section 5.3. Finally, some concluding remarks are drawn in Section 5.4.

5.2 An Online BP Learning Algorithm with Adaptive Forgetting Factors

Similar to the previous chapters, a typical three-layer BP neural network is used. Accordingly, the basic architecture of the BP neural network is also described by

$$\hat{Y}(t+1) = F_2[V^T(t)F_1(W(t)X(t))] \tag{5.1}$$

where all notations are the same to Chapter 2.

In the same way, the data analysis tasks can be performed through BP neural network learning. The BP training is actually an iteration process by minimizing an error function $E(t)$, i.e.,

$$\begin{aligned} E(t) &= \frac{1}{2}\sum_{j=1}^{n}\sum_{i=1}^{k} e_{ij}^2(t) = \frac{1}{2}\sum_{j=1}^{n} e_j^T(t)e_j(t) \\ &= \frac{1}{2}\sum_{j=1}^{n}[y_j(t) - \hat{y}_j(t)]^T[y_j(t) - \hat{y}_j(t)] \end{aligned} \tag{5.2}$$

where $e_j(t) = [e_{1j}(t) \quad e_{2j}(t) \quad \cdots \quad e_{kj}(t)]^T \in R^{k\times 1}, j = 1,2,\cdots,n$, $y_j(t)$ and $\hat{y}_j(t)$ are the jth actual value predicted value at time t, respectively; n is the number of training patterns and k is the number of output nodes.

However, the learning algorithm based on Equation (5.2) is batch learning of neural networks. As earlier mentioned, the computation of the batch learning algorithm is very large if big sample data sets are given. Also, the neural networks must re-learn when new data are available. To overcome the shortcomings, the neural network learning should be iterative or recursive, allowing the network parameters to be updated at each sample interval as new data become available. This idea will be activated to create a new online learning algorithm. In addition, a weighting sequence should be chosen to give a higher weight for more recent data in order to perform online prediction. To arrive at this goal, an adaptive forgetting factor is introduced to this problem. In this study, an exponential forgetting mechanism in the error function, like a recursive algorithm with the forgetting factor, is used, and then Equation (5.2) can be rewritten as

$$\begin{aligned} E(t) &= \frac{1}{2}\sum_{j=1}^{t}\lambda^{t-j}\sum_{i=1}^{k} e_i^2(j) = \frac{1}{2}\sum_{j=1}^{t}\lambda^{t-j} e^T(j)e(j) \\ &= \frac{1}{2}\sum_{j=1}^{t}\lambda^{t-j}[y(j) - \hat{y}(j)]^T[y(j) - \hat{y}(j)] \end{aligned} \tag{5.3}$$

where λ is the forgetting factor, $0 < \lambda \le 1$, $e(j) = [e_1(j), e_2(j), \cdots, e_k(j)]^T \in R^{k\times 1}, j =$ 1, …, t; t is a time factor, representing the number of training pairs here.

By applying the steepest descent method to the error cost function $E(t)$ (i.e., Equation (5.3)), we can obtain the gradient of $E(t)$ with respect to parameters W and V, respectively.

$$\nabla_W E(t) = \frac{\partial E(t)}{\partial W(t)} = \sum_{j=1}^{t} \lambda^{t-j} \sum_{i=1}^{k} e_i(j) \frac{\partial e_i(j)}{\partial W(j)}$$
$$= -\sum_{j=1}^{t} \lambda^{t-j} \sum_{i=1}^{k} e_i(j) \frac{\partial \hat{y}_i(j)}{\partial W(j)} = -\sum_{j=1}^{t} \lambda^{t-j} F_1'(j)V(j)F_2'(j)e(j)x^T(j)$$
$$= \lambda \nabla_W E(t-1) - F_1'(t)V(t)F_2'(t)e(t)x^T(t) \tag{5.4}$$

$$\nabla_V E(t) = \frac{\partial E(t)}{\partial V(t)} = \sum_{j=1}^{t} \lambda^{t-j} \sum_{i=1}^{k} e_i(j) \frac{\partial e_i(j)}{\partial V(j)}$$
$$= -\sum_{j=1}^{t} \lambda^{t-j} \sum_{i=1}^{k} e_i(j) \frac{\partial \hat{y}_i(j)}{\partial V(j)} = -\sum_{j=1}^{t} \lambda^{t-j} F_1(j)e^T(j)F_2'(j)$$
$$= \lambda \nabla_V E(t-1) - F_1(t)e^T(t)F_2'(t) \tag{5.5}$$

So, the updated formulae of weights are given by, respectively

$$\Delta W(t) = -\eta \nabla_W E(t) = -\eta\left(\lambda \nabla_W E(t-1) - F_1'(t)V(t)F_2'(t)e(t)x^T(t)\right) \tag{5.6}$$

$$\Delta V(t) = -\eta \nabla_V E(t) = -\eta\left(\lambda \nabla_V E(t-1) - F_1(t)e^T(t)F_2'(t)\right) \tag{5.7}$$

where η is the learning rate; λ is the forgetting factor; Δ is the incremental operator; ∇ is the gradient operator; ΔW and ΔV are the weight adjustment increments; $F_{1(j)}' = diag[f_{1(1)}' \; f_{1(2)}' \; \cdots \; f_{1(q)}'] \in R^{q \times q}$; $F_2' = diag[f_{2(1)}' \; f_{2(2)}' \; \cdots \; f_{2(k)}'] \in R^{k \times k}$; $f_{1(i)}' = f_1'(net_i) = \dfrac{\partial f_1(net_i)}{\partial net_i}, i = 1,2,\cdots,q$; $f_{2(i)}' = f_2'[v_i^T F_1(WX)] = \dfrac{\partial f_2[v_i^T F_1(WX)]}{\partial [v_i^T F_1(WX)]}, i = 1,2,\cdots k$;

$$\overline{V} = \begin{bmatrix} v_{11} & v_{21} & \cdots & v_{k1} \\ v_{12} & v_{22} & \cdots & v_{k2} \\ \cdots & \cdots & \cdots & \cdots \\ v_{1q} & v_{2q} & \cdots & v_{kq} \end{bmatrix} = [v_1 \; v_2 \; \cdots \; v_k] \in R^{q \times p}, v_i = [v_{i1} \; \cdots \; v_{iq}]^T \in R^{q \times 1}, i = 1,2,\cdots,q.$$

To derive the optimal learning rate for online learning, consider the following error increment equation:

$$\Delta e(t+1) = e(t+1) - e(t) = y(t+1) - \hat{y}(t+1) - y(t) + \hat{y}(t) \tag{5.8}$$

Let $\Delta y(t+1) = y(t+1) - y(t)$ be the change of the actual value and let $\Delta \hat{y}(t+1) = \hat{y}(t+1) - \hat{y}(t)$ be the change of the neural network output. Here we assume that the absolute value of the change of the actual value is much smaller than the absolute value of the change of the neural network output, i.e., $|\Delta y(t+1)| << |\Delta \hat{y}(t+1)|$. This implies that the value $y(t)$ can approximate $y(t+1)$ locally during the training process, that is to say, the change of the actual value can be ignored comparing with the change of neural network output during the learning process. This assumption is realistic for many processes due to energy constraints in practical systems, while no constraints are imposed to the output of the neural networks (Sha and Bajic, 2002). Also, if this condition is not satisfied, then the neural network prediction system will not be able to adapt sufficiently fast to change in the practical systems and the prediction of the practical systems will be impossible. With the above assumption, the increment in Equation (5.8) can be approximated as

$$\Delta e(t+1) = e(t+1) - e(t) = \Delta y(t-1) - \Delta \hat{y}(t+1) \approx -\Delta \hat{y}(t+1) \tag{5.9}$$

According to Lemma 4.1 and Equations (5.6) and (5.7), we can get

$$\Delta\hat{y}(t+1) \approx \left(\frac{d\hat{y}(t+1)}{dt}\right)\Delta t = \begin{pmatrix} \frac{d\hat{y}_1(t+1)}{dt} \\ \frac{d\hat{y}_2(t+1)}{dt} \\ \cdots \\ \frac{d\hat{y}_k(t+1)}{dt} \end{pmatrix}\Delta t = \begin{pmatrix} f'_{2(1)}\left(F_1^T\frac{dV_1}{dt} + V_1^T F_1'\frac{dW}{dt}X\right) \\ f'_{2(2)}\left(F_1^T\frac{dV_2}{dt} + V_2^T F_1'\frac{dW}{dt}X\right) \\ \cdots \\ f'_{2(k)}\left(F_1^T\frac{dV_k}{dt} + V_k^T F_1'\frac{dW}{dt}X\right) \end{pmatrix}\Delta t$$

$$\approx \begin{pmatrix} f'_{2(1)}\cdot\left(F_1^T\frac{\Delta V_1}{\Delta t} + V_1^T F_1'\frac{\Delta W}{\Delta t}X\right) \\ f'_{2(2)}\cdot\left(F_1^T\frac{\Delta V_2}{\Delta t} + V_2^T F_1'\frac{\Delta W}{\Delta t}X\right) \\ \cdots \\ f'_{2(k)}\cdot\left(F_1^T\frac{\Delta V_k}{\Delta t} + V_k^T F_1'\frac{\Delta W}{\Delta t}X\right) \end{pmatrix}\Delta t = \begin{bmatrix} f'_{2(1)}\cdot(F_1^T\Delta V + V_1^T F_1'\Delta WX) \\ f'_{2(2)}\cdot(F_1^T\Delta V_2 + V_2^T F_1'\Delta WX) \\ \cdots \\ f'_{2(k)}\cdot(F_1^T\Delta V_k + V_k^T F_1'\Delta WX) \end{bmatrix}$$

$$= \begin{bmatrix} f'_{2(1)}\cdot(F_1^T[-\eta\nabla_{V_1}E(t)] + V_1^T F_1'[-\eta\nabla_W E(t)]X) \\ f'_{2(2)}\cdot(F_1^T[-\eta\nabla_{V_2}E(t)] + V_2^T F_1'[-\eta\nabla_W E(t)]X) \\ \cdots \\ f'_{2(k)}\cdot(F_1^T[-\eta\nabla_{V_k}E(t)] + V_k^T F_1'[-\eta\nabla_W E(t)]X) \end{bmatrix}$$

$$= -\eta\begin{bmatrix} f'_{2(1)}\cdot(F_1^T\nabla_{V_1}E(t) + V_1^T F_1'\nabla_W E(t)X) \\ f'_{2(2)}\cdot(F_1^T\nabla_{V_2}E(t) + V_2^T F_1'\nabla_W E(t)X) \\ \cdots \\ f'_{2(k)}\cdot(F_1^T\nabla_{V_k}E(t) + V_k^T F_1'\nabla_W E(t)X) \end{bmatrix}$$

$$= -\eta\begin{bmatrix} f'_{2(1)} & 0 & \cdots & 0 \\ 0 & f'_{2(2)} & \cdots & 0 \\ \cdots & \cdots & \cdots & \cdots \\ 0 & 0 & \cdots & f'_{2(k)} \end{bmatrix}\begin{bmatrix} F_1^T[\lambda\nabla E_{V_1}(t-1) - e_1 f'_{2(1)}F_1] + V_1^T F_1'\nabla_W E(t)X \\ F_1^T[\lambda\nabla E_{V_2}(t-1) - e_2 f'_{2(2)}F_1] + V_2^T F_1'\nabla_W E(t)X \\ \cdots \\ F_1^T[\lambda\nabla E_{V_k}(t-1) - e_k f'_{2(k)}F_1] + V_k^T F_1'\nabla_W E(t)X \end{bmatrix}$$

$$= -\eta F_2'\left[\lambda\nabla_V^T E(t-1)F_1 - F_2'eF_1'F_1 + V^T F_1'(\lambda\nabla E_W(t-1) - F_1'VF_2'eX^T)X\right]$$

$$= -\eta\lambda F_2'\left[\nabla_V^T E(t-1)F_1 + V^T F_1'\nabla E_W(t-1)X\right] + \eta F_2'(F_2'eF_1'F_1 - V^T F_1'F_1'VF_2'eX^T X)$$

$$= -\eta\lambda F_2'\left[\nabla_V^T E(t-1)F_1 + V^T F_1'\nabla E_W(t-1)X\right] + \eta F_2'(F_1'F_1 I_{k^2} - V^T F_1'F_1'VX^T X)F_2'e$$

$$= -\eta\lambda\zeta(t-1) + \eta\xi(t)e(t)$$

Therefore, in this recursive algorithm, the change of output of neural networks with adaptive forgetting factors can be represented as

$$\Delta\hat{y}(t+1) = -\eta\lambda\zeta(t-1) + \eta\xi(t)e(t) \tag{5.10}$$

where $\zeta(t-1) = F_2'[\nabla_V^T E(t-1)F_1 + V^T F_1'\nabla_W E(t-1)X]$ and $\xi(t) = F_2'[(F_l^T F_l)I_{k^2} + V^T F_l' F_l' V X^T X]F_2'$ with

$$\mathbf{F}_2' = \begin{bmatrix} F_{2(1)}' & 0 & \cdots & 0 \\ 0 & F_{2(2)}' & \cdots & 0 \\ \cdots & \cdots & \cdots & \cdots \\ 0 & 0 & \cdots & F_{2(N)}' \end{bmatrix}, \quad \mathbf{F}_1^T\mathbf{F}_1 = \begin{bmatrix} F_{1(1)}^T F_{1(1)} & F_{1(1)}^T F_{1(2)} & \cdots & F_{1(1)}^T F_{1(N)} \\ F_{1(2)}^T F_{1(1)} & F_{1(2)}^T F_{1(2)} & \cdots & F_{1(2)}^T F_{1(N)} \\ \cdots & \cdots & \cdots & \cdots \\ F_{1(N)}^T F_{1(1)} & F_{1(N)}^T F_{1(2)} & \cdots & F_{1(N)}^T F_{1(N)} \end{bmatrix},$$

$$\mathbf{X}^T\mathbf{X} = \begin{bmatrix} x_1^T x_1 & x_1^T x_2 & \cdots & x_1^T x_N \\ x_2^T x_1 & x_2^T x_2 & \cdots & x_2^T x_N \\ \cdots & \cdots & \cdots & \cdots \\ x_N^T x_1 & x_N^T x_2 & \cdots & x_N^T x_N \end{bmatrix}, \quad \mathbf{F}_1'\mathbf{F}_1' = \begin{bmatrix} F_{1(1)}' F_{1(1)}' & F_{1(1)}' F_{1(2)}' & \cdots & F_{1(1)}' F_{1(N)}' \\ F_{1(2)}' F_{1(1)}' & F_{1(2)}' F_{1(2)}' & \cdots & F_{1(2)}' F_{1(N)}' \\ \cdots & \cdots & \cdots & \cdots \\ F_{1(N)}' F_{1(1)}' & F_{1(N)}' F_{1(2)}' & \cdots & F_{1(N)}' F_{1(N)}' \end{bmatrix}.$$

Substituting (5.10) into (5.9), we obtain

$$e(t+1) \approx e(t) + \eta\lambda\zeta(t-1) - \eta\xi(t)e(t) \tag{5.11}$$

The objective here is to derive an optimal learning rate η. That is, at iteration t, an optimal value of the learning rate, $\eta^*(t)$, which minimizes $E(t+1)$, is obtained. Define the error function:

$$E(t+1) = 0.5e^T(t+1)e(t+1) \tag{5.12}$$

Using Equation (5.11), Equation (5.12) may be written as

$$E(t+1) = 0.5\left[e(t) + \eta\lambda\zeta(t-1) - \eta\xi(t)e(t)\right]^T \left[e(t) + \eta\lambda\zeta(t-1) - \eta\xi(t)e(t)\right] \tag{5.13}$$

In Equation (5.13), the first and second order conditions are as

$$\left.\frac{dE(t+1)}{d\eta}\right|_{\eta=\eta^*(t)} = -0.5[\xi(t)e(t) - \lambda\zeta(t-1)]^T[e(t) - \eta^*(t)\xi(t)e(t) + \eta^*(t)\lambda\zeta(t-1)]$$
$$-0.5[e(t) - \eta^*(t)\xi(t)e(t) + \eta^*(t)\lambda\zeta(t-1)]^T[\xi(t)e(t) - \lambda\zeta(t-1)] = 0$$

$$\left.\frac{d^2E(t+1)}{d\eta^2}\right|_{\eta=\eta^*(t)} = [\xi(t)e(t) - \lambda\zeta(t-1)]^T[\xi(t)e(t) - \lambda\zeta(t-1)] > 0$$

Since $\xi(t)$ and $\zeta(t-1)$ is positively defined, the second condition is met, the optimum learning rate can be obtained from the first order condition, as illustrated in Equation (5.14). Interestedly, the optimized learning rate that we obtained is distinctly different from the result produced by the work (Sha and Bajic, 2002; Yu et al., 1995; Yu et al., 2005a, 2006c).

$$\eta^*(t) = \frac{[\xi(t)e(t) - \lambda\zeta(t-1)]^T e(t)}{[\xi(t)e(t) - \lambda\zeta(t-1)]^T[\xi(t)e(t) - \lambda\zeta(t-1)]} \tag{5.14}$$

Finally, the increments of the neural network parameters, found using the optimal learning rate, are obtained by replacing the η^* given by Equation (5.14) to Equations (5.6) and (5.7), which yield

$$\begin{aligned}\Delta W(t) = &-\left(\frac{[\xi(t)e(t) - \lambda\zeta(t-1)]^T e(t)}{[\xi(t)e(t) - \lambda\zeta(t-1)]^T[\xi(t)e(t) - \lambda\zeta(t-1)]}\right)\\ &\times\left(\lambda\nabla_W E(t-1) - \bar{F}_1'(t)\bar{V}(t)F_2'(t)e(t)x^T(t)\right)\end{aligned} \tag{5.15}$$

$$\begin{aligned}\Delta V(t) = &-\left(\frac{[\xi(t)e(t) - \lambda\zeta(t-1)]^T e(t)}{[\xi(t)e(t) - \lambda\zeta(t-1)]^T[\xi(t)e(t) - \lambda\zeta(t-1)]}\right)\\ &\times\left(\lambda\nabla_V E(t-1) - F_1(t)e^T(t)F_2'(t)\right)\end{aligned} \tag{5.16}$$

It is worth noting that the forgetting factor λ is adaptive. During the neural network learning process, if the prediction error $e(t)$ grows, this may mean that the neural network parameters have changed. This implies that the network model is incorrect and needs adjustment. So we should reduce the forgetting factor and allow the neural network model to adapt. An adaptive forgetting factor that allows this is

$$\lambda(t) = s(t-1)/s(t) \tag{5.17}$$

where $s(t)$ is a weighted average of the past values of e^Te and is calculated by

$$s(t) = [(\tau-1)/\tau]s(t-1) + (e^Te/\tau) \tag{5.18}$$

τ is the time constant of the forgetting factor determining how fast $\lambda(t)$ changes.

Using the updated weight formula with optimal learning rates and adaptive forgetting factors, a new online recursive learning algorithm is generated. For convenience, the proposed online learning algorithm is summarized as follows, as shown in Fig. 5.1.

To verify the effectiveness of the proposed online learning algorithm, two typical foreign exchange rate time series, Canadian dollars (CAD) and Swiss francs (CHF) against US dollars (USD), are used as testing targets. The simulation experiments are presented in the following section.

```
Set input nodes to the number of input vectors
Set target value for a specific pattern
Intialize weight vectors
While i < Epoch_num
   For j = 1 to N_records
     Compute output ŷ(t) and error E(t) using Equation (5.1) and (5.3);
      Compute error gradients ∇_w E(t) and ∇_v E(t) using Equations (5.4)-
      (5.5), and optimal learning rate η using Equation (5.14);
      Update the weight vectos with Equations (5.15)-(5.16).
   Endfor
Wend
```

Fig. 5.1. Outline of the proposed online learning algorithm

5.3 Experimental Analysis

In this section, there are two main motivations: (1) to evaluate the performance of the proposed online learning algorithm, and (2) to compare the efficiency of the proposed online learning algorithm with other similar algorithms. To perform the two motivations, two real-world data experiments are carried out. In this section, we first describe the research data and experiment design and then report the experimental results.

5.3.1 Data Description and Experiment Design

In the experiments, two typical foreign exchange rates, Canadian dollars against US dollars (CAD/USD) and Swiss francs against US dollars (CHF/USD), are used for testing purpose. The historical data are daily and are obtained from Federal Reserve Bank Reports of Wharton Research Data Service (WRDS), provided by Wharton School of the University of Pennsylvania. The entire data set covers the period from January 1, 2000 to December 31, 2004 with a total of 1257 observations. The data sets are divided into two periods: the first period covers January 1, 2000 to December 31, 2003 with 1004 observations, while the second period is from January 1, 2004 to December 31, 2004 with 253 observations. The first period, which is assigned to in-sample estimation, is used for network learning, i.e., training set. The second period, which is reserved for out-of-sample evaluation, is used for validation, i.e., testing set. For space limitation, the original data are not listed in this paper, and detailed data can be obtained from the WRDS.

For comparison, four related algorithms, standard BPNN algorithm (Haykin, 1999), batch learning algorithm (Yu et al., 2005a), Levenberg-Marquart (LM) based learning algorithm (Hagan and Menhaj, 1994), and extended Kalman filter (EKF) based learning algorithm (Iiguni et al., 1992; Ruck et al., 1992), are employed in this study. For standard BPNN learning algorithm, the learning rate is fixed at 0.3, more details about standard BPNN learning algorithm can be referred to Haykin (1999). In the batch learning algorithm, the weights are updated only at the end of each epoch (Yu et al., 2005a). Similar to the online learning algorithm, the batch learning algorithm can also be summarized as follows, as illustrated in Fig. 5.2.

```
Set input nodes to the number of input vectors
Set target value for a specific pattern
Intialize weight vectors
While i < Epoch_num
   For j = 1 to N_records
     Compute output ŷ(t) and error E(t) using Equation (5.1) and (5.3);
     Compute error gradients ∇_w E(t) and ∇_v E(t) using Equations (5.4)-
     (5.5), and optimal learning rate η using Equation (5.14);
   Endfor
   Update the weight vectos with Equations (5.15)-(5.16).
Wend
```

Fig. 5.2. Outline of the batch learning algorithm

For Levenberg-Marquart (LM) based algorithm (Hagan and Menhaj, 1994) and the extended Kalman filter (EKF) based algorithm (Iiguni et al., 1992; Ruck et al., 1992), Chapter 4 has described some basic ideas about the two methods and more details can be referred to the work of Hagan and Menhaj (1994), Iiguni et al. (1992) and Ruck et al. (1992).

In all the neural network predictors, five input nodes are determined by auto-regression testing. The appropriate number of hidden nodes is set to 12 in terms of trial and error. The training epochs are set to 3000 due to trial and error and the problem complexity.

To examine the forecasting performance, the root mean square error (*RMSE*) and directional change statistics (D_{stat}) (Yu et al., 2005c) of financial time series are employed as the performance measurement of the testing set. In addition, training time and training mean square error (*TMSE*) are used as the efficiency measurement of different algorithms.

5.3.2 Experimental Results

When the data are prepared, we begin to perform experiments according to the previous experiment design. First of all, the prediction results with five algorithms are reported. Figs. 5.3-5.4 give graphical representations of the forecasting results for two typical foreign exchange rate time series using different learning algorithms. Tables 5.1-5.2 show the detailed prediction performance of the different learning algorithms in terms of both the level measurement (*RMSE*) and direction measurement (D_{stat}). From the figures and tables, we can generally find that the prediction results of the proposed online learning algorithm are very promising for two typical foreign exchange rate series under study either where the measurement of forecasting performance is the goodness-of-fit such as *RMSE* or where the forecasting performance criterion is the D_{stat}.

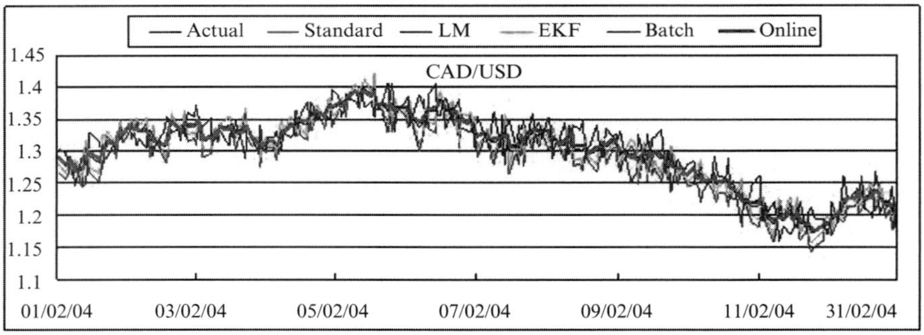

Fig. 5.3. The forecasting results with different learning algorithm for CAD/USD

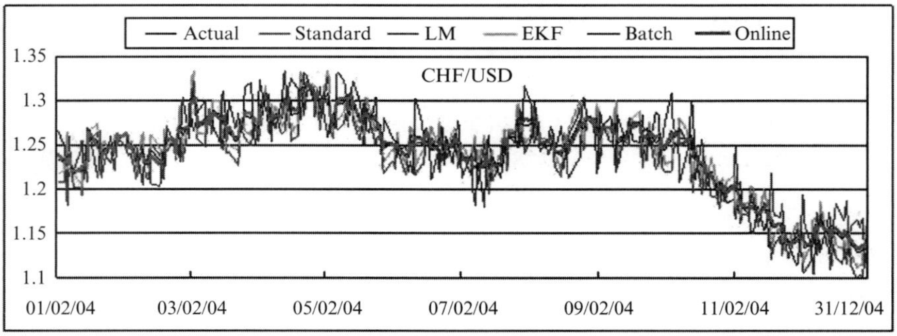

Fig. 5.4. The forecasting results with different learning algorithm for CHF/USD

In detail, Fig. 5.3 reveals that the comparison for the CAD/USD of the proposed online learning algorithm versus the other four learning algorithm. Similarly, it can be seen from Fig. 5.4 that the forecasting performance for

CHF/USD has significantly improved using the proposed online learning algorithm. The graphical results indicate that the proposed online learning algorithm performs than the other algorithms presented here.

Subsequently, the concrete prediction performance comparison of various algorithms for two different financial time series via *RMSE* and D_{stat} are given in Tables 5.1-5.2.

For the exchange rate of CAD/USD, the proposed online learning algorithm outperforms the other four learning algorithms in terms of both *RMSE* and D_{stat}. Focusing on the *RMSE* indicator, the proposed online learning algorithm performs the best, followed by batch learning, EKF-based learning, LM-based learning and Standard BPNN learning algorithm. Comparing with standard BPNN learning algorithm, the RMSE of the proposed online learning algorithm is much smaller. From the viewpoint of D_{stat}, the performance of the proposed online learning algorithm is the best of the all. Relative to the standard BPNN learning algorithm, the performance improvement arrives at 21.35% (84.59%-63.24%). While the performance of the proposed online learning algorithm is slightly improved relative to batch learning algorithm, EKF-based learning algorithm and LM-based algorithm.

Table 5.1. *RMSE* comparisons of five neural network learning algorithms

Algorithm	CAD/USD		CHF/USD	
	RMSE	Rank	*RMSE*	Rank
Online learning	0.0621	1	0.0736	1
Batch learning	0.1042	2	0.1118	2
EKF-based learning	0.1607	3	0.1541	3
LM-based learning	0.1988	4	0.2021	5
Standard BPNN	0.2004	5	0.1986	4

Table 5.2. D_{stat} comparisons of five neural network learning algorithms

Algorithm	CAD/USD		CHF/USD	
	D_{stat} (%)	Rank	D_{stat} (%)	Rank
Online learning	84.59	1	86.96	1
Batch learning	80.24	3	80.63	3
EKF-based learning	81.42	2	83.79	2
LM-based learning	70.75	4	71.94	4
Standard BPNN	63.24	5	66.40	5

For the exchange rate of CHF/USD, the performance of the proposed online algorithm is the best, similar to the results of the CAD/USD. Likewise, the proposed online algorithm has gained much improvement relative to the standard BPNN learning algorithm. Interestedly, the *RMSE* of the batch learning algorithm is slightly better than that of the EKF-based learning algorithm, but the directional performance (i.e., D_{stat}) of the batch learning is somewhat worse than that of the EKF-based learning algorithm. The possible reasons are required to be further addressed later.

In summary, we can conclude that (1) the proposed online learning algorithm with adaptive forgetting factors performs consistently better than other comparable learning algorithm for two testing foreign exchange rates; (2) the evaluation value of the two criteria of the proposed online learning is much better than that of the standard BPNN learning algorithm, indicating that the proposed online learning algorithm can effectively reflect error changes and significantly improve network learning performance. One possible reason for this is that the optimal learning rate and adaptive forgetting factors are used in the online learning algorithm.

In addition, the computation speed of the proposed online algorithm is very fast during the experiments when using a personal computer (PC) and the training performance is also well in predicting time series, indicating that the proposed learning algorithm is an efficient online algorithm. For comparison purpose, Tables 5.3-5.4 report the comparison of the computation time and training performance between the proposed online learning algorithm and the other four learning algorithm presented in this study.

Table 5.3. Comparisons of the computational efficiency

Algorithm	CAD/USD		CHF/USD	
	Time (*Seconds*)	Rank	*Time* (*Seconds*)	Rank
Online learning	221	3	216	3
Batch learning	189	2	186	2
EKF-based learning	175	1	178	1
LM-based learning	527	5	532	5
Standard BPNN	253	4	261	4

Table 5.4. Comparisons of the training performance

Algorithm	CAD/USD		CHF/USD	
	TMSE	Rank	*TMSE*	Rank
Online learning	1.26×10^{-3}	1	3.35×10^{-3}	1
Batch learning	6.15×10^{-3}	2	7.56×10^{-3}	2
EKF-based learning	1.02×10^{-2}	3	9.19×10^{-3}	3
LM-based learning	1.77×10^{-2}	4	1.85×10^{-2}	4
Standard BPNN	1.86×10^{-2}	5	2.18×10^{-2}	5

From Tables 5.3-5.4, we can find the following conclusions. First of all, for both CAD/USD and EUR/USD series, the computational time of the EKF-based learning algorithm is the smallest and the LM-based learning algorithm is the largest. The results reported here are basically consistent with the work of Iiguni et al. (1992). The main reason is that the number of iterations of LM-based learning algorithm is much larger than that of EKF-based learning algorithm, although the computation time per iteration of the EKF-based learning algorithm is larger than that of the LM-based learning algorithm (Iiguni et al., 1992).

Secondly, the computation time of the batch learning algorithm is smaller than that of the online learning algorithm due to the batch processing of the data. However, relative to the standard BPNN learning and LM-based learning algorithms, the computation time of the online learning algorithm is much smaller. The main reason may be that the proposed online learning algorithm adopts optimal learning rate, resulting in the increase of convergence speed.

Thirdly, although the computation time of the proposed online learning algorithm is not the best, the training performance (refer to *TMSE* presented by Table 5.4) and the generalized performance (refer to *RMSE* and D_{stat} reported by Tables 5.1 and 5.2) is the best among the entire learning algorithms presented in this study. The possible reason is that the adaptive forgetting factor helps improve the performance of this proposed algorithm.

Finally, since the difference of the computation time between the proposed online learning algorithm and EKF-based and batch learning algorithm is marginal, and the difference of the performance between the proposed online learning algorithm and EKF-based and batch learning is significant, in this sense, the computational efficiency of the proposed online learning algorithm is satisfactory when forecasting foreign exchange rate time series. In general, the experimental results reveal that the proposed online learning algorithm provide a feasible solutions to foreign exchange rates online prediction.

5.4 Conclusions

In this study, an online learning algorithm with optimized learning rates and adaptive forgetting factors is first proposed. This exploratory research examines the potential of using the proposed online learning algorithm to predict two typical foreign exchange rates series – Canadian dollars against US dollars (CAD/USD) and Swiss francs against US dollars (CHF/USD). Our empirical results suggest that the online learning algorithm may provide much better forecasts than the other four comparable learning algorithms. Furthermore, the learning efficiency is also satisfactory relative to the learning performance. This implies that the proposed online learning algorithm with adaptive forgetting factors is very suitable for online foreign exchange rate prediction.

6 An Improved BP Algorithm with Adaptive Smoothing Momentum Terms for Foreign Exchange Rates Prediction

6.1 Introduction

Back-propagation neural network (BPNN) is one of the most popular neural networks which have been widely applied to many fields such as prediction and pattern recognition due to their strong capability to approximate any arbitrary function arbitrarily well and to provide flexible nonlinear mapping between inputs and outputs (Hornik et al., 1989; White, 1990). The basic learning rule of BPNN is based on the gradient descent optimization method and the chain rule (Widrow and Lehr, 1990). However, some typical drawbacks of the BPNN learning rule based on the gradient descent method are its slowness and its frequent confinement to local minima and over-fitting (Yu, 1992; Lawrence et al., 1997; Yu et al., 2006a). For these reasons, some global optimization algorithms, such as genetic algorithm (GA) (Jain et al., 1996) and simulated annealing (SA) (Karaboga and Pham, 2000), are proposed for escaping local minima. Similarly, bias-variance trade-off (Yu et al., 2006a, 2006e) and cross validation (i.e., *k*-fold) technique (Ng, 1997) are used to avoid over-fitting. For the slowness of BP algorithm, there are several other ways to accelerate the BPNN learning.

The first way is the second-order gradient methods, such as fast adaptive learning (Tollenaere, 1990), conjugate gradient algorithm (Brent, 1991) and Levenberg-Marquardt algorithm (Hagan and Menhaj, 1994). The crucial drawbacks of these methods, however, are that in many applications computational demands are so large that their effective use in on-line predictions is not feasible (Yu et al., 2005a, 2006c).

Another class of fast learning algorithms is the extended Kalman filter (EKF) based techniques, such as Chen and Ogmen (1993), Iiguni et al. (1992), Ruck et al. (1992) and Shah et al. (1992). They have an improved convergence performance, but their numerical stability is not guaranteed and may degrade learning convergence and increase the learning time (Sha and Bajic, 2002; Zhang and Li, 1999).

Recently a class of layer-by-layer optimizing algorithms (Parisi et al., 1996; Wang and Chen, 1996; Yam and Chow, 1997; Ergezinger and Thomsen, 1995) was proposed to accelerate BP neural network learning. The basic idea of their methods is that each layer of the BPNN is decomposed into both a linear part and a non-linear part and the linear part of each is solved by way of the least squares method. These algorithms show fast convergence with reduced computational complexity relative to conjugate gradient algorithm. However, if the targets for the hidden layer cannot be linearly separated, then the mean squared error (MSE) cannot be sufficiently reduced at both the hidden layer and the output layer.

In such situations, a common method to overcome the slowness is to add a momentum term. In the existing literature, the BP algorithm with a momentum term has been extensively reported, such as Allred and Kelly (1990) and Ooyen (1992). Although introduction of the momentum may help dampen the oscillation of the weight adjustment, an improper choice of the momentum can actually slow the convergence rate. Furthermore, no clue has been given to how the momentum should be set. In most cases, it is determined empirically. In the existing literature, the momentum term is proportion to the previous weight increment. This way may increase network convergence speed if error change direction is consistent, but it may accelerate the network to trap into local minima because the weight adjustment (i.e., momentum) does not take the error change direction into consideration at all. To overcome this limitation, an adaptive smoothing momentum is introduced into the weight update formulae of a BPNN learning algorithm based on adaptive smoothing technique and "3σ limits theory" (Shewhart, 1931; Chase et al., 1998; Yu et al., 2005d).

In this chapter, we propose an improved BPNN learning algorithm with adaptive smoothing momentum. In this new algorithm, adaptive smoothing technique is used to adjust the momentum of weight updating formula automatically by tracking error signals in terms of "3σ limits theory". For illustration and verification purposes, the proposed BPNN learning algorithm is applied to foreign exchange rates prediction.

The rest of this chapter is organized as follows. In Section 6.2, an improved BPNN learning algorithm with adaptive smoothing momentum terms is proposed in detail. In order to verify the effectiveness of the proposed algorithms, an exchange rate index prediction experiment of trading-weighted US dollar against currencies of major US trading partners is conducted and the corresponding results are reported in Section 6.3. In addition, Section 6.4 compares the proposed model with the two similar single neural network models proposed by Chapters 4 and 5. Finally, some concluding remarks are drawn in Section 6.5.

6.2 Formulation of the Improved BP Algorithm

In this section, an adaptive smoothing momentum is first introduced into weight updating formulae in terms of "3σ limits theory". And then an improved BPNN learning algorithm with adaptive smoothing momentum terms can be formulated relying on the previous work.

6.2.1 Determination of Adaptive Smoothing Momentum

To determine the adaptive smoothing momentum term, an adaptive smoothing technique is used to adjust the momentum parameter automatically by tracking error signals in terms of "3σ limits theory (Shewhart, 1931; Chase et al., 1998; Yu et al., 2005d)". This technique incorporates quality control (QC) concept into BPNN and produces an adaptive smoothing momentum. Although some previous studies (e.g., Allred and Kelly, 1990; Ooyen, 1992) introduced momentum into BPNN algorithm for accelerating convergence, but an improper choice of the momentum can actually slow the convergence rate (Yu et al., 1995). Furthermore, no clue has been given to how the momentum should be set. In most cases, it is determined empirically. As shown in Chapter 4, the momentum is proportion to the previous weight increment, which reflects the weight update formulae in the following:

$$\Delta W(t) = -\eta(t)\nabla_W E(t) + \mu(t)\Delta W(t-1) \tag{6.1}$$

$$\Delta V(t) = -\eta(t)\nabla_V E(t) + \mu(t)\Delta V(t-1) \tag{6.2}$$

where $\Delta W(t)$ and $\Delta V(t)$ are the weight increments, $\nabla_W E(t)$ and $\nabla_V E(t)$ are the error gradients with respect to W and V, $\mu(t)\Delta W(t-1)$ and $\mu(t)\Delta V(t-1)$ are the momentums and $\mu(t)$ is the momentum factor. In the above weight update formulae with momentum, the momentum term is only a part of previous weight increment. Although the introduction of such a momentum term may increase network convergence speed if the network output change direction is consistent with actual data change path, it may also accelerate the network to trap into local minima because the momentum does not consider the error change (i.e., the difference between actual value and predicted value) direction at all. In such situations, this study proposes a new adaptive smoothing momentum to accelerate the BPNN learning and convergence.

In standard BPNN algorithm, model errors are usually either squared error or mean squared error (*MSE*). But it is difficult for such error metrics to capture any deviation between the actual value and network output value and it is thus hard to adjust the direction of network learning and searching (Here BPNN learning is actually an optimization process by searching optimal point in error surface). To overcome this problem, In the process of BPNN learning, an adaptive smoothing technique that utilize ordinary error (*OE*) and mean absolute deviation (*MAD*) as a supplementary error measure is to construct a momentum so as to adjust the network's parameters (i.e., learning weights). With the cumulative ordinary error (*COE*), *MAD*, and error tracking signal (*ETS*), an adaptive smoothing momentum can be generated. Since the momentum is obtained by adaptive smoothing technique, which is based on error tracking signal and "3σ limits theory", we call it as "adaptive smoothing momentum".

Consistent with notations in Chapters 4 and 5, the error function and *COE* are defined as

$$E(t) = \frac{1}{2}\sum_{j=1}^{N}(y_j(t) - \hat{y}_j(t))^2 \tag{6.3}$$

$$COE(t) = \sum_{j=1}^{N}(y_j(t) - \hat{y}_j(t)) \tag{6.4}$$

where $y_j(t)$ and $\hat{y}_j(t)$ are the actual and the predicted value, N is the number of training samples.

Let E_j and COE_j be a squared error and an ordinary error connected with j sample respectively, then $E_j = (y_j - \hat{y}_j)^2$ and $COE_j = (y_j - \hat{y}_j)$. Clearly, $COE(t) = COE(t-1) + COE_t$. Thus, the *MAD* and *ETS* can be defined as

$$MAD(t) = \frac{\sum_{j=1}^{t}\left|y_j - \hat{y}_j\right|}{t} \tag{6.5}$$

$$ETS = \frac{COE(t)}{MAD(t)} \tag{6.6}$$

If *ETS* is "large", this means that *COE*(*N*) is large relative to the mean absolute deviation *MAD*(*N*). This in turn says that the network output is producing errors that are either consistently positive or consistently negative. That is, a large value of *ETS* implies that the network output is producing output values that are either consistently smaller or consistently larger than the actual values that are being forecast. Since an "accurate" learning system should produce roughly one half positive errors and one half negative errors, a large *ETS* value indicates that the network output is not reliable. In practice, if *ETS* exceeds a control limit, denoted by θ, for two or more consecutive periods, this is taken as a strong indication that the network learning errors have been larger than an accurate learning system that can reasonably be expected to produce. In our study, the control limit θ is generally taken to be 3σ for a neural network model in terms of the "3σ limits theory" proposed by Shewhart (1931).

If the error signal indicates that adjustment is needed, there are several possibilities. One possibility is that the BPNN model needs to be changed. To do this, input variables may be added or deleted to obtain a better representation. Another possibility is that the BPNN model being used does not need to be changed, but the estimates of the model's parameters need to be changed. When using a neural network model, this is accomplished by changing parameters, i.e., the model weights of every layer.

Since the momentum, proportion to the previous weight increment, does not take into account the error change direction and thus it may accelerate BPNN to trap into local minima. To overcome this problem, we construct a new momentum to replace the standard momentum, which is related to the previous weight increment. In this study, the gradient of ordinary error is seen as one of the new momentum in terms of error tracking signal and Equation (4.1), i.e.,

$$\begin{aligned}\nabla_W COE(t) &= \frac{\partial COE(t)}{\partial W(t)} = -\sum_{j=1}^{N}\sum_{i=1}^{k}\frac{\partial \hat{y}_{ij}(t)}{\partial W(t)} = -\sum_{j=1}^{N} F'_{2(j)}[V^T F_{1(j)}(WX)]' \\ &= -\sum_{j=1}^{N} F'_{1(j)}(WX)VF'_{2(j)}X(t) = -\sum_{j=1}^{N} F'_{1(j)}VF'_{2(j)}X(t)\end{aligned} \tag{6.7}$$

$$\begin{aligned}\nabla_V COE(t) &= \frac{\partial COE(t)}{\partial V(t)} = -\sum_{j=1}^{N}\sum_{i=1}^{k}\frac{\partial \hat{y}_{ij}(t)}{\partial W(t)} = -\sum_{j=1}^{N} F'_{2(j)}[V^T F_{1(j)}(WX)]' \\ &= -\sum_{j=1}^{N} F_{1(j)}(WX)F'_{2(j)} = -\sum_{j=1}^{N} F_{1(j)}F'_{2(j)}\end{aligned} \tag{6.8}$$

Now a new weight adjustment related to momentum $\Delta W_m(t)$ and $\Delta V_m(t)$ can be determined in terms of error tracking signal and "3σ limits theory", i.e.,

$$\Delta W_m(t) = \begin{cases} \nabla_W COE(t), & |ETS| > \theta; \\ 0, & |ETS| \le \theta. \end{cases} \tag{6.9}$$

$$\Delta V_m(t) = \begin{cases} \nabla_V COE(t), & |ETS| > \theta; \\ 0, & |ETS| \le \theta. \end{cases} \tag{6.10}$$

where $\theta = 3\sigma$ (Shewhart, 1931) or $\theta = 4 \cdot MAD$ (Chase et al., 1998). The objective of using the above proposed step function based on error tracking signal and "3σ limits theory" is to guarantee that the BPNN can escape from local minima as possible and speed up the convergence (Yu et al., 2005d). The possible reason is that the proposed momentum based upon error tracking signal and the gradient of ordinary error considers the effect of error change direction. In a sense, our proposed momentum does not only accelerate neural network learning speed, but also adjusts the network search path in the limited space (i.e., 3σ area, here we call it as the "trust region"), as shown in Fig. 6.1 and thus the BPNN may escape from possible local minima.

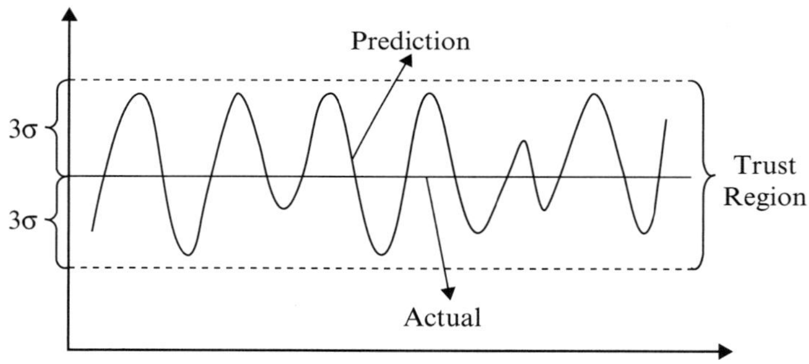

Fig. 6.1. A graphical representation of "3σ limits theory"

To measure the effect of the proposed momentum, we will conduct a series of experiments and verify the effectiveness in the following section. As for the theory proof about the advantage of the proposed momentum is not present here, which is required to be further addressed later.

6.2.2 Formulation of the Improved BPNN Algorithm

Based upon the optimal learning rates, optimal momentum factors and adaptive smoothing momentum term, an improved BPNN algorithm with superior convergence can be formulated. Combining with Equations (4.4), (4.5), (4.12), (6.9) and (6.10), the weight update formulae can be represented by

$$\Delta W(t) = -\eta(t)\nabla_W E(t) + \Delta W_m(t) = \frac{e^T(t)\xi^T(t)e(t)}{e^T(t)\xi^T(t)\xi(t)e(t)} \sum_{j=1}^{N} F'_{1(j)} V F'_{2(j)} e_j x_j^T + \Delta W_m(t) \tag{6.11}$$

$$\Delta V(t) = -\eta(t)\nabla_V E(t) + \Delta V_m(t) = \frac{e^T(t)\xi^T(t)e(t)}{e^T(t)\xi^T(t)\xi(t)e(t)} \sum_{j=1}^{N} F_{1(j)} e_j^T F'_{2(j)} + \Delta V_m(t) \tag{6.12}$$

where $\Delta W_m(t) = \begin{cases} \nabla_W COE(t), & |ETS| > \theta; \\ 0, & |ETS| \le \theta. \end{cases}$ and $\Delta V_m(t) = \begin{cases} \nabla_V COE(t), & |ETS| > \theta; \\ 0, & |ETS| \le \theta. \end{cases}$.

It is worth noting that the BPNN learning algorithm based on the above weight update formulae (i.e., Equations (6.11)-(6.12)) is rather flexible. If *ETS* is always within the trust region during learning, then above weight update formulae is similar to those of the standard BP algorithm, i.e., there is no momentum term. The weight update formulae become the following forms:

$$\Delta W(t) = -\eta(t)\nabla_W E(t) = \frac{e^T(t)\xi^T(t)e(t)}{e^T(t)\xi^T(t)\xi(t)e(t)} \sum_{j=1}^{N} F'_{1(j)} V F'_{2(j)} e_j x_j^T \tag{6.13}$$

$$\Delta V(t) = -\eta(t)\nabla_V E(t) = \frac{e^T(t)\xi^T(t)e(t)}{e^T(t)\xi^T(t)\xi(t)e(t)} \sum_{j=1}^{N} F_{1(j)} e_j^T F'_{2(j)} \tag{6.14}$$

If some specified applications are required to decrease the oscillation of the weight adjustment, an adaptive smoothing factor, or called as momentum factor, can also be introduced, similar to momentum version of BPNN algorithm proposed by Ooyen (1992). Then the weight update formulae can be represented as

$$\Delta W(t) = -\eta(t)\nabla_W E(t) + \mu(t)\Delta W_m(t) = \frac{e^T(t)\xi^T(t)e(t)}{e^T(t)\xi^T(t)\xi(t)e(t)} \sum_{j=1}^{N} F'_{1(j)} V F'_{2(j)} e_j x_j^T + \mu(t)\Delta W_m(t) \tag{6.15}$$

$$\Delta V(t) = -\eta(t)\nabla_V E(t) + \mu(t)\Delta V_m(t) = \frac{e^T(t)\xi^T(t)e(t)}{e^T(t)\xi^T(t)\xi(t)e(t)} \sum_{j=1}^{N} F_{1(j)} e_j^T F'_{2(j)} + \mu(t)\Delta V_m(t) \tag{6.16}$$

where $\Delta W_m(t) = \begin{cases} \nabla_W COE(t), & |ETS| > \theta; \\ 0, & |ETS| \le \theta. \end{cases}$ and $\Delta V_m(t) = \begin{cases} \nabla_V COE(t), & |ETS| > \theta; \\ 0, & |ETS| \le \theta. \end{cases}$,

$\mu(t)$ is a momentum rate, $0 \le \mu(t) \le 1$. Similarly, the optimal momentum factors can be obtained from Equation (4.18). Of course, we can also use constant momentum rate to perform experiments for simplification.

Relying on Equations (6.15) and (6.16), the proposed BPNN learning algorithm with optimized learning rate and adaptive smoothing momentum can be summarized, as illustrated in Fig. 6.2. Note that in neural networks, a single pass over the input data set is usually called as an epoch.

```
Set the architecture of BPNN
Set target value for a specific pattern
Intialize weight vectors and adaptive smoothing factor
While i < Epoch_Num
    For j =1 to N_records
     Compute error gradients ∇_w E(t) and ∇_v E(t) using Equations (4.4)-(4.5)
      Compute the optimal learning rate η(t) and optimal momentum factors μ (t)
      using Equations (4.12) and (4.18)
      Compute the MAD and ETS with Equations (6.5)-(6.6)
      Compute the control limit θ, adaptive smoothing momentums ΔW_w(t)
      and ΔV_W(t) using Equations (6.7)-(6.10)
    EndFor
    Compare the ETS and control limit θ
    Update the weight vectors with Equations (6.15)-(6.16)
Wend
```

Fig. 6.2. Outline of the proposed improved BPNN learning algorithm

6.3 Empirical Study

In this section, the data description and experiment design are firstly given. Then the foreign exchange index prediction results and the comparisons with the similar algorithms are reported. Finally, the effects of different learning rates, momentum rates, error propagation methods, hidden neuron number and hidden neuron activation function on the prediction performance are explored.

6.3.1 Data Description and Experiment Design

In this chapter, an exchange rate index of trading-weighted US dollar against currencies of major US trading partners is used as a testing target. The historical data used in this study are daily and are obtained from Federal Reserve Bank Reports of Wharton Research Data Service (WRDS), provided by Wharton School of the University of Pennsylvania. The entire data set covers the period from January 4, 1995 to December 27, 2005 with a total of 2765 observations. The data sets are divided into two periods: the first period covers January 4, 1995 to December 31, 2003 with 2263 observations, while the second period is from January 2, 2004 to December 31, 2004 with 502 observations. The first period, which is assigned to in-sample estimation, is used for network training, i.e., training set. The second period, which is reserved for out-of-sample evaluation, is used for validation, i.e., testing set. For space limitation, the original data are not listed in this paper, and detailed data can be obtained from the WRDS.

For simplification, the proposed improved BPNN algorithm with optimal learning rate and optimal momentum rate is abbreviated as "BPNN I" algorithm, as illustrated in Fig. 6.2 for more details. To show the importance of the introduced momentum term, the proposed BPNN algorithm with optimal learning rate and adaptive smoothing momentum (BPNN II for short) are also used and the weight update formulae are shown in Equations (6.11)-(6.12). For further comparison, the standard BPNN algorithm (BPNN III for short), the standard BPNN with optimal learning rate (BPNN IV for short), momentum version of standard BPNN algorithm (BPNN V for short) are also used to foreign exchange rate index prediction. In BPNN I, the learning rates and momentum factors are optimally adaptive, which can be obtained from Equations (4.12) and (4.18). Different from BPNN I, the momentum factor of BPNN II is set to 1.0 directly. In BPNN III, the weight update formulae are Equations (6.13)-(6.14) with a constant learning rate η. In BPNN IV, Equations (6.13)-(6.14) are weight update formulae with optimally adaptive learning rate. In BPNN V, weight update formulate are shown in Equations (6.1)-(6.2) and both learning rate and momentum are optimal constants obtained by trial and error. In addition, the mean squared error (*MSE*) during the learning process is used as the evaluation criterion.

6.3.2 Forecasting Results and Comparisons

In this subsection, we applied the improved three-layer BPNN learning algorithm to predict the exchange rate index of trade-weighted US dollar against currencies of major US trading partners. In all the neural network predictors, eight input nodes are determined by auto-regression testing. The appropriate number of hidden nodes is set to 15 determined by trial and error. The training epochs are set to 5000 by trial and error and the activation functions of both hidden neurons are the symmetric hyperbolic tangent functions.

The prediction of the exchange rate index is carried out using the proposed improved BPNN algorithm with optimized learning rate and adaptive smoothing momentum. The prediction results of the exchange rate index are shown in Fig. 6.3. Fig. 6.3 is the plot for actual and predicted value. Roughly speaking, the proposed BPNN algorithm performs well in this exchange rate index prediction problem and the prediction performance is rather high. Note that the blue line in the figure represents the actual value and the red line represents the predicted value.

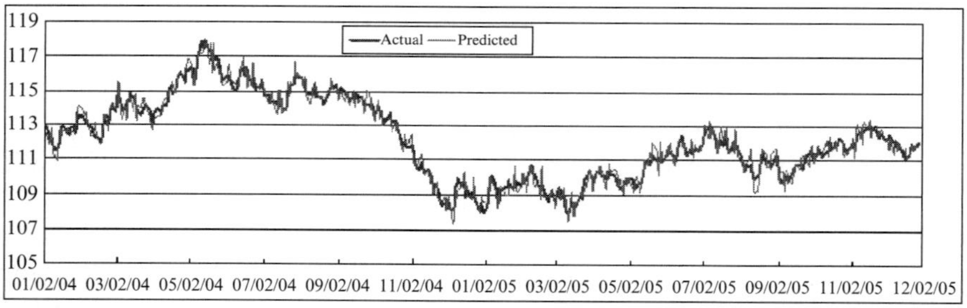

Fig. 6.3. The graphical representation of exchange rate index prediction

Besides the good prediction performance, the convergence of the proposed BPNN algorithm is very fast when conducting an experiment in a PC platform. For verification, Table 6.1 reports the detailed comparison of the results between the proposed BPNN learning algorithm and the other four similar BPNN algorithms, assuming that other conditions (e.g., hidden neuron number, activation function and training epochs, etc.) are same.

Table 6.1. Comparisons of different algorithm for exchange rate index prediction

Algorithm	Learning rate	Momentum factor	Time (s)	MSE
BPNN I	Optimal adaptive	Optimal adaptive	170	0.1598
BPNN II	Optimal adaptive	1.0	130	0.1832
BPNN III	0.25 (Empirical)	N.A.	420	0.2176
BPNN IV	Optimal adaptive	N.A.	293	0.1845
BPNN V	0.25 (Empirical)	0.35 (Trial and error)	340	0.1885

In BPNN I, the learning rates and momentum factors are optimal. In BPNN II, the weight update formulae are Equations (6.11)-(6.12) with optimal adaptive learning rates and adaptive smoothing momentum terms. In BPNN III, Equations (6.13)-(6.14) are weight update formulae with a constant learning rate η. In BPNN IV, Equations (6.13)-(6.14) are weight update formulae with optimal adaptive learning rate. In BPNN V, weight update formulate are shown in Equations (6.1)-(6.2) and both learning rate and momentum are constants.

As can be seen from Table 6.1, we can have the following findings:

(1) The proposed improved BPNN algorithms with optimal learning rate and adaptive smoothing momentum (BPNN I and II) have both fast convergence and good performance relative to other three BPNN algorithm, while the standard BPNN algorithm with a constant learning rate perform worst in both training time and identification performance. Furthermore, the BPNN algorithms with optimal adaptive learning rate (e.g., BPNN I, II and IV) perform consistently better than those without optimal adaptive learning rate (e.g., BPNN III and V) from the performance evaluation. This implies that the optimal adaptive learning rate has a significant impact on the convergence speed and performance, which will be further addressed later.

(2) Comparing BPNN I with BPNN II, the BPNN algorithm with a large momentum rate perform faster than the BPNN with a small momentum rate, but the identification performance of the BPNN II is slightly worse than that of the BPNN I, demonstrating that an improper momentum selection may dampen the oscillation of weight adjustment (Yu et al., 1995) and thus affecting the performance of the BPNN algorithm. But the appropriate small momentum rate used here is obtained by trial and error and it is set to a fixed value during the learning process, this may further influence the performance of the BPNN algorithm. Furthermore, the determination of the appropriate momentum rate is trivial, as the same to the determination of the learning rate in standard BPNN algorithm. This implies that it is necessary and meaningful to derive an optimal adaptive momentum rate although the process is rather difficult.

(3) The BPNN algorithms with momentum (e.g., BPNN I, II and V) perform consistently faster than those without momentum (e.g., BPNN III and IV) from the viewpoint of training time. This reveals the momentum has also an important effect on the convergence speed of BPNN algorithm, which will be addressed later. Interestedly, the speed of the BPNN V (with momentum) is faster than that of the BPNN IV (without momentum), but the performance of the BPNN V is slightly worse than that the BPNN IV, implying the adaptive learning rate is very important.

6.3.3 Comparisons of Different Learning Rates

For comparative purpose, different learning rates – including the optimized adaptive learning rate – are used in the network training process. The main goal of using different learning rate is to give a good idea of what learning rates are the most suitable for training a BPNN on the foreign exchange rate prediction. For convenience, a three-layer BPNN with the structure of "8-15-1" are adopted here. The activation functions of both hidden neurons are the symmetric hyperbolic tangent functions. Some representative learning rates (i.e., η = 0.01, 0.05, 0.10, 0.15, 0.20, 0.25, 0.30, 0.40, 0.50, 0.60, 0.70, 0.80, 0.90) and the optimal adaptive learning rate are tested. Note that the momentum term is assumed to be omitted for simplification. To measure the performance, every experiment was repeated 10 times. The training cycles are set to 1000. Accordingly, the Average MSE and its standard deviation are reported in Table 6.2.

Table 6.2. Performance comparison of BPNN with different learning rates

Learning rate	0.01	0.05	0.10	0.15	0.20
Average MSE	0.2436	0.2281	0.2003	0.1855	0.1663
S.D. MSE	0.0152	0.0147	0.0163	0.0142	0.0112
Learning rate	0.25	0.30	0.35	0.40	0.50
Average MSE	0.1591	0.1644	0.1632	0.1766	0.1979
S.D. MSE	0.0086	0.0098	0.0124	0.0114	0.0136
Learning rate	0.60	0.70	0.80	0.90	Adaptive
Average MSE	0.2468	0.3135	No convergence	No convergence	0.1732
S.D. MSE	0.0144	0.0314	N.A.	N.A.	0.0091

From Table 6.2, it is easy to find that (1) the performance of BPNN is quite poor at very low learning rates (e.g., 0.01). (2) The error tends to converge more quickly as the learning rate is increased, although this increases the deviation of MSE obtained at around 0.1 learning rate. Particularly, in this case, when learning rate is 0.25, the performance is the best. (3) After that, higher learning rates (e.g., 0.3 and 0.4) mean that although the MSE increases in some cases early on, it is relatively stable. (4) Towards 0.5, and particularly 0.7, the learning become unstable when converging and the MSE fluctuates greatly in these cases. When a much higher learning rate (e.g., 0.8 and 0.9) is used, the BPNN cannot learn at all. (5) Comparing the best fixed learning rate and optimal adaptive learning rate, we find that there is no significant difference between the two. However, the convergence speed of the latter is much faster than that of the former, as mentioned in the previous subsection. The main reasons are that the best fixed learning rate is obtained by trial and error and furthermore the best fixed learning rate must be reset when applying to another problem while the proposed optimal learning rate can adjust in an adaptive way. In a sense, the optimal adaptive learning rate is more advantageous than the best fixed learning rate.

6.3.4 Comparisons with Different Momentum Factors

Similar to the previous subsection, we use different momentum factors to test its effect on the performance of the proposed BPNN algorithm. The basic conditions of the BPNN model (e.g., network structure, activation function and training epochs) are the same to the setting of the previous subsection and the optimal adaptive learning rate is used. Fourteen different momentum rates, i.e., 0.0, 0.01, 0.05, 0.1, 0.15 0.25, 0.35, 0.45, 0.5, 0.6, 0.7, 0.8, 0.9, 1.0, are used here. The corresponding results are illustrated in Fig. 6.4. Note that the blue square in the figure represents the average MSE and the line represents the range of standard deviation.

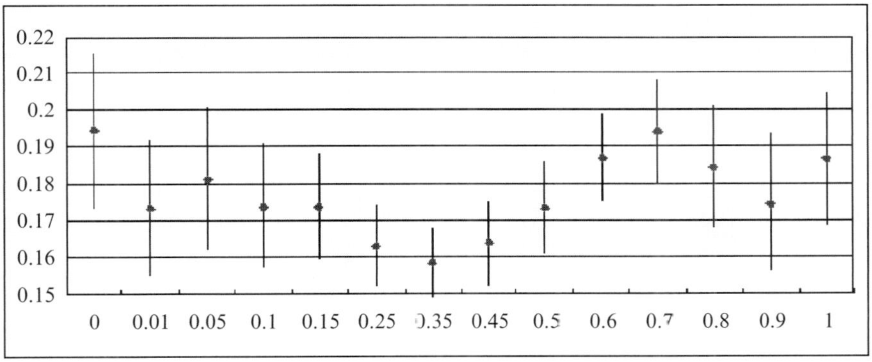

Fig. 6.4. The performance comparison of BPNN with different momentum rates

From Fig. 6.4, we clearly see that (1) The momentum speeds up convergence of training a BPNN model when predicting a foreign exchange rate series because the MSE of all BPNN with momentum is lower than that of the BPNN without momentum (i.e., momentum rate is zero). It prevents large oscillations from occurring that can lead to slow convergence when the error gradient is steep. (2) The MSE tends to decrease more quickly as the momentum rate is increased. Particularly, in this case, when learning rate is 0.35, the performance is the best. (3) From beginning of 0.45, the MSE become larger and larger and the volatility of the MSE is also greater, although this decreases the deviation of MSE obtained at around 0.5 momentum rate. When the momentum rate is 1.0, the performance is basically equivalent to the BPNN without momentum, as revealed in Fig. 6.4.

6.3.5 Comparisons with Different Error Propagation Methods

In this study, we propose a *COE*-based momentum to adjust error propagation and use adaptive smoothing technique to training BPNN, as illustrated in Equations (6.15)-(6.16) and Fig. 6.2. To show the potential of the proposal error adjustment method, we compare it with two other error propagation methods. The first one is resilient back-propagation, or "Rprop" proposed by Riedmiller and Braun (1993), which uses information about the temporal direction of the error gradient of each weight. Each update rule for each weight is therefore allowed to evolve over the course of training. The second one is a more intuitive strategy based on that presented by Fernando and Almeida (1990) which looks at previous changes in the errors. Basically, if the error is seen to decrease from the last two measurements, the learning rate is increase by 0.05. Conversely, if the direction of the gradient of the MSE changes (i.e., fluctuating behavior), then the learning rate is decreased by 0.05. For convenience, this method is abbreviated as "custom BPNN" In addition, the standard BPNN is also used as a benchmark. Similarly, some basic conditions about the BPNN model (e.g., network structure, activation function, learning rate, momentum rate, training epoch, and experiment times etc.) are the same to the setting of the previous subsections. Accordingly, the results of three different methods are reported in Table 6.3.

Table 6.3. Comparisons of four different error propagation methods

Method	Proposed BPNN	Rprop BPNN	Custom BPNN	Standard BPNN
Average MSE	0.1591	0.1686	0.1834	0.2176
S.D. MSE	0.0087	0.0103	0.0147	0.0184

From Table 6.3, we can find that the proposed error propagation method performs the best in the four methods, followed by resilient error propagation method and the custom error propagation method. The worst is the standard error propagation method. As Fahlman (1988) revealed, the standard error propagation method is not an ideal method of estimating the local shape of the error function, nor is it a fast way of facilitating convergence in the shortest amount of time, even though it has been shown the capability to reach good minima. These empirical results further demonstrate that the proposed BPNN algorithm is a very promising method to foreign exchange rate prediction problem.

6.3.6 Comparisons with Different Numbers of Hidden Neurons

From this subsection, we begin to test the effect of the different hidden layer setting (e.g., hidden neuron number and hidden activation function) on the BPNN performance. Since this study adopts three-layer BPNN, we do not discuss the effect of the number of hidden layer, which will be done in the future studies. The effect of different hidden neuron number on the performance of the proposed BPNN algorithm is first explored. Some basic conditions about the BPNN model (e.g., activation function and training epoch, etc.) are the same to the setting of the previous subsections. In addition, the optimal adaptive learning rate is used and the momentum rate is set to 0.35. Here some representative hidden neuron numbers (e.g., 1, 2, 5, 10, 15, 20, 25, 30, 50, 70 and 100) are tested. Accordingly, the results of the different hidden neurons are illustrated in Fig. 6.5.

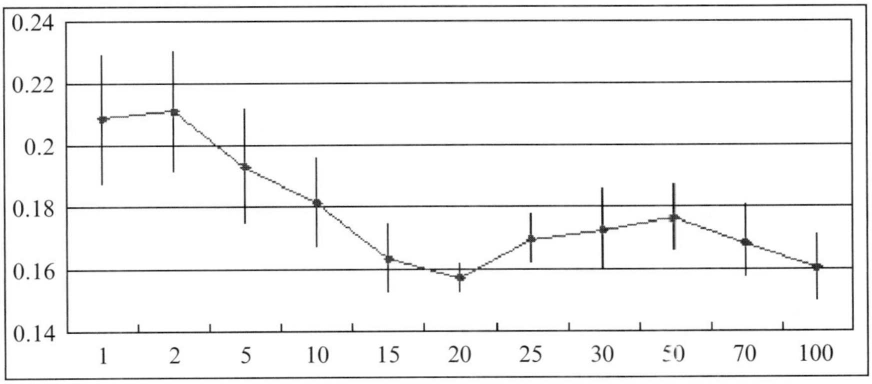

Fig. 6.5. Performance comparisons with different hidden neurons

As revealed by Fig. 6.5, we can find that using only one or two hidden neurons, the network is incapable of learning and the high MSE indicates this. With the increase of the hidden nodes, the performance becomes better. A significant difference can be noticed when fifteen or more neurons are used. Clearly this is the minimum number of hidden neurons that is required for successful learning in foreign exchange rates forecasting. Statistically there seems to be no significant difference between the results where the number of hidden neurons is greater than fifteen. In which case, based on the principle of Occam's Razor (Forster, 2000), we should favor networks of 20 hidden neurons when applying to foreign exchange rate prediction. However, it is widely known that a large number of hidden neurons can lead to overfitting of the training data.

6.3.7 Comparisons with Different Hidden Activation Functions

The activation function of neurons allows non-linearity to be introduced into neural network training and determines the elasticity of weight changes. It therefore can improve convergence in training. In standard BPNN, the logistic sigmoid function is used. To predict some nonlinear time series, such as foreign exchange rate series, the hyperbolic tangent function and Elliot function (Kalman and Kwasny, 1992) are often used. The mathematical expression of the three activation function are given by

$$\text{Logistic sigmoid: } f(x) = \frac{1}{1+e^{-x}} \tag{6.17}$$

$$\text{Hyperbolic tangent: } f(x) = \frac{e^{x} - e^{-x}}{e^{x} + e^{-x}} \tag{6.18}$$

$$\text{Elliot function: } f(x) = \frac{1}{2}\left(\frac{1+x}{1+|x|}\right) \tag{6.19}$$

Now we can compare their performances of the three activation functions experimentally and the corresponding results are reported in Table 6.4.

Table 6.4. Performance comparisons with different hidden activation functions

Activation Function	Logistic sigmoid	Hyperbolic tangent	Elliot function
Average MSE	0.1849	0.1601	0.2105
S.D. MSE	0.0098	0.0089	0.136

As can be seen from Table 6.4, we find that the results show that on average, the hyperbolic tangent function performs better than the other two methods (logistic sigmoid and Elliot). In some cases, logistic sigmoid function may be able to perform better than the other functions in terms of convergence. On average, logistic sigmoid can reach lower MSE than the other methods. Some researchers and our empirical results however have found that hyperbolic tangent function can provide faster convergence as it produces values in a different range (-1 to 1) to that of the logistic sigmoid (0 to 1). Our empirical results also confirm this fact.

6.4 Comparisons of Three Single Neural Network Models

From previous theoretical derivations and empirical analyses, this part (i.e., Part III) proposes three individual neural network models with optimally adaptive learning rate and momentum terms, which are proposed for foreign exchange rates forecasting. For convenience, Table 6.5 lists their similarities and differences.

Table 6.5. Comparisons of three single neural network models

Chapter	Chapter 4	Chapter 5	Chapter 6
Algorithm used	BP algorithm	BP algorithm	BP algorithm
Prediction mode	Level prediction	Level prediction	Level prediction
Prediction range	Short-term	Short-term	Short-term
Characteristics	Optimal learning rates, optimal momentum rates	Optimal learning rates, adaptive forgetting factors	Optimal learning rates, adaptive smoothing momentum
Model mode	Batch	Online	Batch

From previous analyses and Table 6.5, we can see that the three proposed models are built based on the BP algorithm, which mainly used for short-term level prediction. The focus is to determine optimal learning rates and momentum factors. For example, optimal learning rates and optimal momentum rates are derived from basic BP learning algorithm in terms of optimization technique, while adaptive smoothing momentum is derived by adaptive smoothing technique. Of the three proposed models, the models proposed by Chapters 4 and 6 are batch learning model and the proposed model in Chapter 5 is an online learning model. Accordingly, we can select different neural network model for foreign exchange rates prediction in terms of their characteristics.

6.5 Conclusions

In this study, an improved BPNN learning algorithm with optimal learning rate and adaptive smoothing momentum is proposed to predict the foreign exchange rate index. In this new algorithm, adaptive smoothing techniques are used to adjust the momentum parameters automatically by tracking error signals in terms of "3σ limits theory". The effectiveness of the proposed algorithm applied to trade-weighted US dollar index prediction is demonstrated by a series of simulation experiments. At the same time, the effect of some important parameter setting on the performance of BPNN is also extensively investigated. The results obtained reveal that the proposed improved BPNN algorithm is a fast and efficient learning algorithm, which is very suitable for foreign exchange rate prediction. This implies that the proposed BPNN algorithm with optimal learning rate and adaptive smoothing momentum can be used as an alternative tool for foreign exchange rates forecasting in the future.

Part IV: Hybridizing ANN with Other Forecasting Techniques for Foreign Exchange Rates Forecasting

7 Hybridizing BPNN and Exponential Smoothing for Foreign Exchange Rate Prediction

7.1 Introduction

A challenging task in financial market such as stock market and foreign exchange market is to predict the movement direction of financial markets so as to provide valuable decision information for investors (Lai et al., 2006b). Thus, many researchers and business practitioners have developed various kinds of forecasting methods. Of the various forecasting models, the exponential smoothing model has been found to be an effective forecasting method. Since Brown (1959) began to use simple exponential smoothing to forecast demand for inventories, the exponential smoothing models have been widely used in business and finance (Winters, 1960; Gardner, 1985; Alon, 1997; Leung, 2000). For example, Winters (1960) proposed exponentially weighed moving averages for sales forecasting and obtained good results. Gardner (1985) introduced exponential smoothing methods into supply chain management for forecasting demand, and achieved satisfactory results. Similarly, Alon (1997) found that Winters' model forecasts aggregate retail sales more accurately than the simple exponential model. Leung et al. (2000) used an adaptive exponential smoothing model to predict stock indices, and achieved good forecasting performance for Nikkei 225.

However, the exponential smoothing method is only a class of linear model and thus it can only capture linear features of financial time series. But financial time series are often full of nonlinearity and irregularity. Furthermore, as the smoothing constant decreases exponentially, the disadvantage of the exponential smoothing model is that it gives simplistic models that only use several previous values to forecast the future. The exponential smoothing model is, therefore, unable to find subtle nonlinear patterns in the financial time series data. Obviously, the approximation of linear models to complex real-world problems is not always sufficient. Hence, it is necessary to consider other nonlinear methods to complement the exponential smoothing model.

Recently, artificial neural network (ANN) models have shown their promise in financial time series forecasting with their nonlinear modeling capability. Since Lapedes and Farker (1987) used ANN to predict the chaotic time series, the ANN models are widely used in the time series forecasting. For example, Refenes et al. (1993) applied multilayer feedforward neural network (MLFNN) models to forecast foreign exchange prices and obtained good results. More examples for foreign exchange rates forecasting can be referred to Chapter 1.

Although the ANN models achieve success in financial time series forecasting including foreign exchange rates forecasting, they have some disadvantages. Since the real world is highly complex, there exist some linear and nonlinear patterns in the financial time series simultaneously. It is not sufficient to use only a nonlinear model for time series because the nonlinear model might miss some linear features of the time series data. Furthermore, previous works (Denton, 1995; Markham and Rakes, 1998; Zhang, 2003) are shown that using ANN only to model linear problems may produce mixed results. In addition, the ANN models often suffer the overfitting problem and thus weaken the generalization capability of the ANN models.

Under such backgrounds, it is necessary to hybridize the linear model and nonlinear model for financial time series forecasting. This is because the ANN model and exponential smoothing model are complementary. On one hand, the ANN model can find subtle nonlinear features in the time series data, but may miss some linear patterns when forecasting. On the other hand, the exponential smoothing model can give good results in the linear patterns of time series, but cannot capture the nonlinear patterns, which might result in inaccurate forecasts. Motivated by the previous findings, this chapter proposes a hybrid synergy model to financial time series prediction integrating an exponential smoothing (ES) model and an ANN model via linear programming technique.

The exponential smoothing (ES) model rather than other linear models such as ARIMA is chosen as neural network model's complement for several reasons. First of all, the major advantage of exponential smoothing methods is that they are simple, intuitive, and easily understood. These methods have been found to be quite useful for short-term forecasting of large numbers of time series. At the same time, exponential smoothing techniques have also been found to be appropriate in such applications because of their simplicity. Second, the exponential smoothing model has less technical modeling complexity than the ARIMA model and thus makes it more popular in practice. As Lilien and Kotler (1983) reported, exponential smoothing models have been widely used by approximately 13% of industry. Third, Mills (1990) found little difference in forecast accuracy between exponential smoothing techniques and ARIMA models. In some examples, exponential smoothing models can even obtain better results than neural network model. Foster et al. (1992) once argued that the exponential smoothing is superior to neural networks in forecasting yearly data. Generally, the exponential smoothing model is regarded as an inexpensive technique that gives forecasts that is "good enough" in a wide variety of applications.

In this chapter, a hybrid synergy model integrating both an exponential smoothing (ES) model and a BPNN model is proposed to take advantage of the unique strength of exponential smoothing and BPNN models in linear and nonlinear modeling. For testing purposes, two main exchange rates, EUR/USD and JPY/USD, are used. For comparison, individual exponential smoothing model and individual BPNN model are used as benchmark models.

The remainder of the chapter is organized as follows. Section 7.2 provides basic backgrounds about the exponential smoothing and neural network approaches to financial time series forecasting. Then the hybrid methodology combining the exponential smoothing and neural network model is introduced in Section 7.3. Subsequently, some experimental results are reported in Section 7.4. Finally, Section 7.5 concludes the study.

7.2 Basic Backgrounds

In this section, we provide some basic knowledge about two forecasting techniques used in this chapter. First of all, exponential smoothing technique is introduced and then the ANN for financial time series forecasting is also presented.

7.2.1 Exponential Smoothing Forecasting Model

In the application of the exponential smoothing model, there are three types of models that are widely used on different time series. Simple exponential smoothing (Type I) is used when the time series has no trend. Double exponential smoothing (Type II) is an exponential smoothing method for handling a time series that displays a slowly changing linear trend. Two approaches are covered: one-parameter double exponential smoothing, which employs a single smoothing constant; and Holt-Winters' two-parameter double exponential smoothing, which employs two smoothing constants.

The third is Winters' method (Type III), which is an exponential smoothing approach to predicting seasonal data. This method also contains two approaches: multiplicative Winters' method, which is appropriate for increasing seasonal variation; and additive Winters' method, which is appropriate for constant seasonal variation (Winters, 1960; Gardner, 1985; Alon, 1997).

In financial time series, there is irregularity, randomicity and no trend. These features show that the simple exponential smoothing method is suitable for financial time series forecasting for the specified time period. Therefore, only the type I of the exponential smoothing model is described in detail. For the type II & III, interested readers can be referred to Winters (1960), Gardner (1985) and Alon (1997) for more details.

Suppose that the time series $y_1, y_2, \ldots, y_n$ is described by the model

$$y_t = \beta_0 + \varepsilon_t \tag{7.1}$$

where β_0 is the average of the time series and ε_t random error. Then the estimate S_t of β_0 made in time t is given by the smoothing equation:

$$S_t = \alpha y_t + (1-\alpha) S_{t-1} \tag{7.2}$$

where α is a smoothing constant between 0 and 1 and S_{t-1} is the estimate of β_0 in t-1.

Thus a point forecast made in time t for y_{t+1} is

$$\hat{y}_{t+1} = S_t = \alpha y_t + (1-\alpha)\hat{y}_t \tag{7.3}$$

From Equation (7.2), we have

$$S_{t-1} = \alpha y_{t-1} + (1-\alpha) S_{t-2} \tag{7.4}$$

Substituting Equation (7.4) to Equation (7.2), then

$$S_t = \alpha y_t + (1-\alpha)(\alpha y_{t-1} + (1-\alpha) S_{t-2}) = \alpha y_t + \alpha(1-\alpha) y_{t-1} + (1-\alpha)^2 S_{t-2} \tag{7.5}$$

Similarly, substituting recursively for S_{t-2}, S_{t-3}, …, S_1 and S_0, we obtain

$$S_t = \hat{y}_{t+1} = \alpha y_t + \alpha(1-\alpha) y_{t-1} + \alpha(1-\alpha)^2 y_{t-2} + \cdots + \alpha(1-\alpha)^{t-1} y_1 + (1-\alpha)^t S_0 \tag{7.6}$$

Here we see that S_t, the estimate made in time t of the average β_0 of the time series, can be expressed in term of observations y_t, y_{t-1}, …, y_1 and the initial estimate S_0. The coefficients measuring the contributions of the observations y_t, y_{t-1}, …, y_1 – that is, $\alpha, \alpha(1-\alpha), \cdots, \alpha(1-\alpha)^{t-1}$ – decrease exponentially with time. For this reason this method is referred as simple exponential smoothing.

In order to use Equation (7.6) to predict the time series, we need to determine the value of the smoothing constant α and the initial estimate S_0. For the smoothing constant, the ordinary least square (OLS) can be used to determine α , while for the initial value S_0, we can let S_0 be equal to y_1, i.e. $S_0 = y_1$, or let S_0 be equal to the simple arithmetic average of a few previous observations. For example, $S_0 = (y_1+y_2+y_3)/3$. When determining the α and S_0, the exponential smoothing model can be used for prediction.

The advantage of the exponential smoothing method is that it is capable of fitting the linear patterns of the time series well and easy to use. But the financial time series is often irregular and nonlinear, it is not sufficient to use exponential smoothing for financial time series modeling.

7.2.2 Neural Network Forecasting Model

Neural networks used in this chapter have three layers of neurons, i.e., the input layer, hidden layer and output layer. The input layer has one neuron for each input to the network. Each neuron in the input layer is connected to every neuron in a hidden layer. The hidden layer represents the interaction between the input variables and will be connected to the output layer, the dependent variable.

The strength of the connection between the various neurons varies with the purpose of the network. The strength of the connection between two neurons is referred to as weight. The stronger the connection is, the greater the weight will be. Also, important inputs are given large weighting values. It has been shown by Hornik et al. (1989) that this standard back-propagation neural network (BPNN) using an arbitrary transfer function can approximate any measurable function in a very precise and satisfactory manner, provided a sufficient number of hidden neurons are used.

It is widely accepted that a three-layer BP feed-forward network with an identity transfer function in the output unit and logistic functions in the middle-layer units can approximate any continuous function arbitrarily well given a sufficient amount of middle-layer units (White, 1990). Furthermore, more than 70 percent of all problems are usually trained on a back-propagation, three-layer network. A back-propagation network is one that compares the forecast with the actual and uses the difference to adjust the strength of the interconnections between neurons. The back-propagation learning algorithm, designed to train a feed-forward network, is an effective learning technique used to exploit the regularities and exceptions in the training sample.

In the BPNN learning algorithm, it consists of two phase, forward-propagation and error back-propagation phase, described in Chapter 2. The total squared error calculated by Equation (2.2) is propagated back, layer by layer, from the output units to the input units in the back-propagation phase. Weight adjustments are determined on the way of propagation at each level. The two phases are executed during each iteration of the back-propagation algorithm until error E converges. As Rumelhart et al. (1986) argued, the solutions obtained from this algorithm come close to the optimal ones in their experiments.

As to time series forecasting, according to the previous computation process the relationship between the output (y_t) and the inputs (y_{t-1}, y_{t-2}, …, y_{t-p}) has the following mathematical representation.

$$y_t = a_0 + \sum_{j=1}^{q} a_j f(w_{0j} + \sum_{i=1}^{p} w_{ij} y_{t-i}) + e_t \tag{7.7}$$

where a_j (j = 0, 1, 2, …, q) is a bias on the jth unit, and w_{ij} (i = 0, 1, 2, …, p; j = 0, 1, 2, …, q) is the connection weights between layers of the model, $f(\bullet)$ is the transfer function of the hidden layer, p is the number of input nodes and q is the number of hidden nodes. Actually, the BPNN model in (11) performs a nonlinear functional mapping from the past observation (y_{t-1}, y_{t-2}, …, y_{t-p}) to the future value (y_t), i.e.,

$$y_t = \varphi(y_{t-1}, y_{t-2}, \cdots, y_{t-p}, w) + e_t \tag{7.8}$$

where w is a vector of all parameters and φ is a function determined by the network structure and connection weights. Thus, in some senses, the BPNN model is equivalent to a nonlinear autoregressive (NAR) model (Yu et al., 2005c; Lai et al., 2006b).

A major advantage of neural networks is their ability to provide flexible nonlinear mapping between inputs and outputs. They can capture the nonlinear characteristics of time series well. However, using BPNN to model linear problems may produce mixed results (Denton, 1995; Markham and Rakes, 1998; Zhang, 2003). Therefore, we can conclude that the relationship between exponential smoothing and BPNN is complementary. To take full advantage of the individual strengths of two models, it is necessary to integrate the exponential smoothing and BPNN models, as mentioned earlier.

7.3 A Hybrid Model Integrating BPNN and Exponential Smoothing

In real life, a financial time series forecasting problem such as exchange rate prediction is far from simple due to high volatility, complexity, irregularity and noisy market environment. Furthermore, real-world time series are rarely pure linear or nonlinear. They often contain both linear and nonlinear patterns. If this is the case, there is no universal model that is suitable for all kinds of time series data. Both exponential smoothing models and BPNN models have achieved success in their own linear or nonlinear domains, but neither exponential smoothing nor BPNN can adequately model and predict time series since the linear models cannot deal with nonlinear relationships while the BPNN model alone is not able to handle both linear and nonlinear patterns equally well (Zhang, 2003). On the other hand, as previously mentioned, for time series forecasting the relationship between exponential smoothing and BPNN is complementary. Exponential smoothing is a class of linear models that can capture time series' linear characteristics, while BPNN models are a class of general function approximators capable of modeling nonlinearity and which can capture nonlinear patterns in time series. Hybridizing the two models may yield a robust method, and more satisfactory forecasting results may be obtained by incorporating an exponential smoothing model and a BPNN model. Therefore, we propose a hybrid methodology integrating the exponential smoothing and the BPNN for financial time series forecasting. Different from decomposition principle proposed by Zhang (2003), we adopt the "parliamentary" hybridization strategy (Wedding and Cios, 1996) to create a synergetic model, as shown in Fig. 7.1.

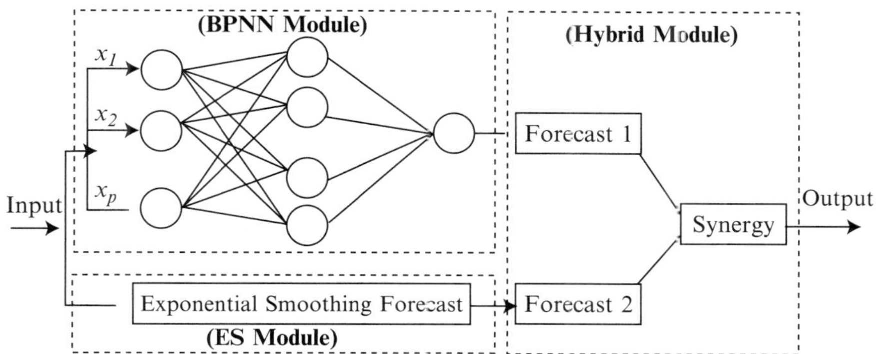

Fig. 7.1. The hybrid synergy forecasting model

From Fig. 7.1, the input is fed simultaneously into a BPNN model and an exponential smoothing forecasting model. The BPNN model generates a forecast result, while the exponential smoothing model also generates a time series forecast result. The two forecast results are entered into the hybrid forecast module and generate a synergetic forecast result as final output, as shown in Fig. 7.1. In the hybridization process, the "parliamentary" hybridization strategy (Wedding and Cios, 1996) is used, i.e.,

$$\hat{y}_t^{Hybird} = \alpha\hat{y}_t^{ES} + (1-\alpha)\hat{y}_t^{BPNN} \tag{7.9}$$

where $\hat{y}_t^{ES}$ is the forecast result obtained from exponential smoothing model, $\hat{y}_t^{BPNN}$ is the forecast result of the BPNN and α is the weight parameter.

Through integrating linear patterns and nonlinear patterns of financial time series, a synergetic effect will believed to be created to improve the prediction performance. In Equation (7.9), a critical problem is how to determine the weight parameter α. Generally, the value of α is estimated by the ordinary least square (*OLS*) method, i.e.,

$$MinQ = \sum_{t=1}^{n} (y_t - \hat{y}_t^{Hybird})^2 \tag{7.10}$$

However, its drawback of this approach is that the square treatment will move the fitted curve to some exceptional points and thus reducing the forecasting accuracy. One modification in this study is to minimize the sum of absolute error between estimated and the actual value, then

$$MinQ' = \sum_{t=1}^{n} \left| y_t - \hat{y}_t^{Hybird} \right| = \sum_{t=1}^{n} |e_t| \tag{7.11}$$

The Equation (7.11) can be solved by linear programming. First of all, we let

$$u_t = \frac{|e_t| + e_t}{2} = \begin{cases} e_t, e_t \geq 0 \\ 0, e_t < 0 \end{cases}, \tag{7.12}$$

$$v_t = \frac{|e_t| - e_t}{2} = \begin{cases} 0, e_t \geq 0 \\ -e_t, e_t < 0 \end{cases} \tag{7.13}$$

Clearly, $|e_t| = u_t + v_t, e_t = u_t - v_t$, then the linear programming (*LP*) model can be formulated below.

$$
(LP)\begin{cases} Min\ Q' = \sum_{t=1}^{n}(u_t + v_t) \\ \sum_{t=1}^{n}\left[y_t - \left(\alpha\hat{y}_t^{ES} + \beta\hat{y}_t^{BPNN}\right)\right] - \sum_{t=1}^{n}\left(u_t - v_t\right) = 0 \\ \alpha + \beta = 1 \\ \alpha \geq 0, u_t \geq 0, v_t \geq 0, t = 1,2,\cdots,n \end{cases} \tag{7.14}
$$

Using the simplex algorithm, an optimal hybridization parameter can be obtained from the *LP* problem in Equation (7.14). In order to verify the effectiveness of this hybridization approach, a simulated study about two typical exchange rates series is performed.

7.4 Experiments

The data set used in this paper are daily data from 1 January 2000 till 31 December 2004 (partial data sets excluding public holidays) and are obtained from Pacific Exchange Rate Service (http://fx.sauder.ubc.ca/), provided by Professor Werner Antweiler, University of British Columbia, Vancouver, Canada. The daily data cover three years of observations of two major international currency exchange rates — euros/US dollar (EUR/USD) and Japanese yen/US dollar (JPY/USD). In our empirical experiment, the data set is divided into two sample periods — the estimation (in-sample) and the test (out-of-sample) periods. The estimation period covers observations from 1 January 2000 till 31 December 2003 and is used to estimate and refine the forecast model parameters. Meantime we take the data from 1 January 2004 to 31 December 2004 as evaluation test sets, which are used to evaluate the good or bad performance of prediction based on some evaluation measurements. In order to save space, the original data are not listed in the paper, and detailed data can be obtained from the website.

In this study, the root mean square error (*RMSE*) and directional statistics (D_{stat}) (Yao and Tan, 2000; Yu et al., 2005c) are used as evaluation criteria. In addition, the individual exponential and BPNN models are selected as benchmark models for comparison purposes. Finally, only one-step-ahead forecasting is considered in this study.

Based on our analysis above, we examine the forecast performances of two major currency exchange rates in terms of *RMSE* and D_{stat}. The experimental results are shown in Table 7.1.

Table 7.1. The experiment results of EUR/USD and JPY/USD

Measure	EUR/USD			JPY/USD		
	ES	BPNN	Hybrid	ES	BPNN	Hybrid
RMSE	0.0847	0.0924	0.0531	0.6875	0.8146	0.5963
D_{stat} (%)	58.77	62.90	71.35	53.36	61.44	70.18

From the viewpoint of *RMSE*, the hybrid methodology is the best for both EUR/USD and JPY/USD, followed by the individual exponential smoothing and individual BPNN model. For example of EUR/USD, the *RMSE* of the individual BPNN model is 0.0924, and the individual exponential smoothing model is 0.0847, while the *RMSE* of the hybrid methodology is only 0.0531.

However, the direction prediction is more important than the accuracy prediction in the financial markets because the former can provide decision information for investors directly. Furthermore, the high *RMSE* can not lead to high D_{stat}, as the exponential smoothing and the BPNN reveals. Focusing on the D_{stat}, the hybrid method performs the best, followed by the individual BPNN model; the worst is the individual exponential smoothing. For example of JPY/USD, the D_{stat} of exponential smoothing is only 53.36%, the D_{stat} of the BPNN model is 61.44%, while that of the hybrid methodology arrives at 70.18%. The main reason is that the hybrid methodology integrating linear patterns and nonlinear patterns creates a synergetic effect and thus improves the prediction performance.

To summarize, several conclusions can be made. First of all, the individual exponential smoothing model outperform the individual BPNN model in terms of *RMSE*, but from the point of D_{stat}, the individual BPNN model is better than individual exponential smoothing model in forecasting the two currency exchange rates. Second, the prediction performance of EUR/USD is better than that of JPY/USD. The main reason may be that the JPY/USD is more volatile than the EUR/USD. Third, the hybrid methodology performs much better than the two benchmark models in terms of both *RMSE* and D_{stat}. This implies that the proposed hybrid methodology is a promise alternative solution to the financial time series forecasting.

7.5 Conclusions

In this chapter, we propose a hybrid synergy methodology incorporating exponential smoothing and neural network for foreign exchange rate forecasting and explore the forecasting capability of the proposed hybrid synergy methodology from the point of level prediction and direction prediction.

Experimental results obtained reveal that the hybrid methodology performs better than the two benchmark models, implying that the proposed hybrid synergy approach can be used as an alternative tool for foreign exchange rates prediction.

8 A Nonlinear Combined Model Hybridizing ANN and GLAR for Exchange Rates Forecasting

8.1 Introduction

Foreign exchange rates modeling and forecasting has been a common research stream in the last few decades. Over this time, the research stream has gained momentum with the advancement of computer technologies, which have made many elaborate computation methods available and practical (Yu et al., 2005c). However, it is not easy to predict exchange rates due to their high volatility and noise. But the difficulty in forecasting exchange rates is usually attributed to the limitation of many conventional forecasting models; this has encouraged academic researchers and business practitioners to develop more predictable forecasting models. As a result models using artificial intelligence such as artificial neural network (ANN) techniques have been recognized as more useful than conventional statistical forecasting models. Literature documenting the research shows this is quite diverse and involves different architectural designs. Some examples are presented. De Matos (1994) compared the strength of a multilayer feed-forward neural network (MLFNN) with that of a recurrent network based on the forecasting of Japanese yen futures. Kuan and Liu (1995) provided a comparative evaluation of the performance of MLFNN and a recurrent network on the prediction of an array of commonly traded exchange rates. Hsu et al. (1995) developed a clustering neural network model to predict the direction of movements in the USD/DEM exchange rate. Their experimental results suggested that their proposed model achieved better forecasting performance relative to other indicators.

Similarly, Tenti (1996) applied recurrent neural network models to forecast exchange rates. El Shazly and El Shazly (1999) designed a hybrid model combining neural networks and genetic training to the three-month spot rate of exchange for four currencies: the British pound, the German mark, the Japanese yen and the Swiss franc. The experimental results reported revealed that the networks' forecasts outperformed predictions made by both the forward and futures rates in terms of accuracy and correctness. In a more recent study by Leung et al. (2000), the forecasting accuracy of MLFNN was compared with the general regression neural network (GRNN). The study showed that the GRNN possessed a greater forecasting strength relative to MLFNN with respect to a variety of currency exchanges. Zhang and Berardi (2001) adopted a different approach. Instead of using single network architecture, their research investigated the use of ensemble methods in exchange rate forecasting. Essentially, the study proposed using systematic and serial partitioning methods to build ensemble models consisting of different neural network structures. Results indicated that the ensemble network could consistently outperform a single network design.

Recently, more hybrid forecasting models have been developed that integrate neural network techniques with many conventional and burgeoning forecasting methods such as econometrical models and time series models to improve prediction accuracy. There are a few examples in the existing literature combining neural network forecasting models with conventional time series forecasting techniques, especially auto-regression integrated moving average (ARIMA). Wedding II and Cios (1996) constructed a combination model incorporating radial basis function neural networks (RBFNN) and the univariant Box-Jenkins (UBJ) model to predict the Wolfer Sunspot, West German monthly unemployment figures and the number of monthly housing starts; the experiments proved that the hybrid model for time series forecasting is capable of giving significantly better results. Luxhoj et al. (1996) presented a hybrid econometrics and ANN approach for sales forecasting and obtained good prediction performance. Likewise, Voort et al. (1996) introduced a hybrid method called KARIMA using a Kohonen self-organizing map and ARIMA method to predict short-term traffic flow, and obtained relatively better results.

In the same way, Tseng et al. (2002) proposed using a hybrid model (SARIMABP) that combines the seasonal ARIMA (SARIMA) model and the back-propagation (BP) neural network model to predict seasonal time series data. The experimental results showed that SARIMABP was superior to the SARIMA model, to the back-propagation model with deseasonalized data, and to the back-propagation model with differenced data for the test cases of machinery production time series and soft drink time series. The values of the MSE, MAE and MAPE were all the lowest for the SARIMABP model. The SARIMABP also outperformed other models in terms of overall proposed criteria including MSE, MAE, MAPE and turning points forecasts. For the machinery production time series, the SARIMABP model remained stable even when the number of historical data was reduced from 60 to 36. Zhang (2003) proposed a hybrid methodology that combined both ARIMA and ANN models taking advantage of the unique strengths of these models in linear and nonlinear modeling for time series forecasting. Empirical results with real data sets indicated that the hybrid model could provide an effective way to improve the forecasting accuracy achieved by either of the models used separately.

In their pioneering work on combined forecasts (Reid, 1968; Bates and Granger, 1969), Bates and Granger showed that a linear combination of forecasts would give a smaller error variance than any of the individual methods. Since then, the studies on this topic (i.e., combined forecasts) have expanded dramatically. Makridakis et al. (1982) claimed that using a hybrid model or combining several models has become common practice in improving forecasting accuracy ever since the well-known M-competition in which a combination of forecasts from more than one model often leads to improved forecasting performance. Likewise, Pelikan et al. (1992) and Ginzburg and Horn (1994) proposed combining several feed-forward neural networks to improve time series forecasting accuracy. In addition, Clemen (1989) provided a comprehensive review and annotated bibliography in this area. The basic idea of the model combination in forecasting is to use each model's unique feature to capture different patterns in the data. Both theoretical and empirical findings suggest that combining different methods can be an effective and efficient way to improve forecast performances.

However, we find that there are two main problems in the existing hybrid models and combined models mentioned above. In this literature, both the hybrid forecasting models and combined forecasting models are limited to linear combination form. A linear combination approach is not necessarily appropriate for all the circumstances (Problem I). Furthermore, it is not easy to determine the number of individual forecasting models. It is well known to us that not all circumstances are satisfied with the rule of "the more, the better". Thus, it is necessary to choose an appropriate method to determine the number of individual models for combining forecasting (Problem II).

In view of the first problem (i.e., linear combination drawback) a novel nonlinear combined forecasting model utilizing the ANN technique is introduced in this paper. For the second problem (i.e., determining the number of individual models for combining forecasting), the principal component analysis (PCA) technique is introduced. We use the PCA technique to choose the number of individual forecasting models.

Considering the previous two main problems, this study proposes a novel nonlinear combined forecasting model for exchange rate forecasting. This model utilizes the ANN technique and PCA technique, integrates GLAR with ANN models, and takes full advantage of hybrid methods and existing combined techniques. The proposed novel nonlinear combined model is the PCA&ANN-based nonlinear combined forecasting approach, which integrates GLAR and ANN models as well as the single hybrid methodology. The aims of this study are two-fold: (1) to show how to predict exchange rates using the proposed nonlinear combined model; and (2) to display how various methods compare in their accuracy in forecasting foreign exchange rates. In view of the two aims, this study mainly describes the building process of the proposed nonlinear combined model and the application of the nonlinear combined forecasting approach in foreign exchange rate forecasting between the US dollar and three other major currencies — German marks, British pound and Japanese yen — incorporating GLAR and ANN, while comparing forecasting performance with all kinds of evaluation criteria.

The rest of the chapter is organized as follows. The next section describes the model building process in detail. In order to verify the effectiveness and efficiency of the proposed model, empirical analysis of the three main currencies' exchange rates is reported in Section 8.3. The conclusions are contained in Section 8.4.

8.2 Model Building Processes

In this section, we mainly describe the model building process step by step and in detail. First we present the GLAR and ANN modeling processes briefly. Then a hybrid forecasting method and some combined forecasting methodologies are introduced with the aid of existing literature. Finally, based on previous works, a nonlinear combined model is proposed and three forecasting evaluation criteria are presented.

8.2.1 Generalized Linear Auto-Regression (GLAR) model

The generalized linear auto-regression (GLAR) model, first introduced by Shephard (1995), is equivalent to a system of reduced form equations relating each endogenous variable to lag endogenous (predetermined) and pertinent exogenous variables. That is, in a GLAR model, the future value of a variable is assumed to be a linear function of several past observations and random errors. Generally, the form of the GLAR model is given in the following form:

$$y_t = \alpha + C(L)y_{t-1} + D(L)x_{t-1} + \varepsilon_t, \tag{8.1}$$

where y_t and ε_t are the actual value of endogenous variables and random disturbance at time t, x_{t-1} is the actual value of related exogenous variables, α is a constant term, $C(L)$ and $D(L)$ are the pth and q^{th} order lag polynomial of endogenous variable and exogenous variables respectively in the lag operator L, such that $C(L) = C_0 - C_1 L - C_2 L^2 - \cdots - C_p L^P$ and $D(L) = D_0 - D_1 L - D_2 L^2 - \cdots - D_q L^q$. L denotes the lag operator, e.g., $Ly_t = y_{t-1}, L^2 y_t = y_{t-2}$ and so on, and random disturbances, ε_t are assumed to be independently and identically distributed with a mean of zero and a constant variance of σ^2, i.e., $\varepsilon_t \sim IID(0, \sigma^2)$.

Actually, the GLAR is a special form of vector auto-regression (VAR) model (Gujarati, 1995) by comparison. The difference between GLAR and VAR is only in the representation of variables. That is, the former is a form of individual variable and the latter is the vector form. Similarly, the fact that we choose the GLAR model rather than autoregressive integrated moving average (ARIMA) model (Box and Jenkins, 1970) is based on the following idea: the ARIMA model does not contain the term of exogenous variables (x_t) and thus cannot capture the effect of other external factors. Furthermore, adding an explanatory variable may strengthen the model's forecasting capability. Therefore, from this perspective, because foreign exchange rates are often affected by many factors and all variables are time series, it seems that the GLAR model is simpler and easier to understand than the VAR model and ARIMA model, and hence the GLAR model is preferable to VAR and ARIMA (Yu et al., 2005c).

Based on earlier works (Shephard, 1995; Firth, 1990; McChllagh and Nelder, 1989), the GLAR model involves the following five-step iterative procedures:

(a) Stationary test of time series, such as unit root test;

(b) Identification of the GLAR structure, i.e., the model order is required to be determined in this step;

(c) Estimation of the unknown parameters;

(d) Model checks and diagnostics;

(e) Forecast future outcomes based on the known data.

This five-step model building process is typically repeated several times until a satisfactory model is finally selected. The final model selected can then be used for prediction purpose.

The advantage of GLAR is that it is capable of receiving external information from exogenous variables. For example, the foreign exchange rate is often influenced by many external factors (e.g., exports, inflation rate, GDP and interest rate, etc.) besides the effect of formation mechanism of itself. Furthermore, the GLAR model fits the linear characteristic of time series well. The disadvantage of the GLAR is that it cannot capture non-linear patterns of financial time series if nonlinearity exists.

8.2.2 Artificial Neural Network (ANN) Model

In this study, one of the widely used ANN models, the back-propagation neural network (BPNN), is used for time series forecasting (Zhang, 2003; Tang et al., 1991; Tang and Fishwick, 1993; Zhang et al., 1998; Hwang, 2001). Usually, the BPNN model consists of an input layer, an output layer and one or more intervening layers also referred to as hidden layers. The hidden layers can capture the non-linear relationship between variables. Each layer consists of multiple neurons that are connected to neurons in adjacent layers. Since these networks contain many interacting non-linear neurons in multiple layers, the networks can capture relatively complex phenomena. ANNs are already one of the models that are able to approximate various nonlinearities in the data series.

A neural network can be trained by the historical data of a time series in order to capture the non-linear characteristics of the specific time series. The model parameters (connection weights and node biases) will be adjusted iteratively by a process of minimizing the forecasting errors, as illustrated in Chapter 2.

It is worth noting that in this paper the ANN model is the typical back-propagation neural network; however, the core-training algorithm is not the gradient descent rule, but the Levenberg-Marquardt rule, as this algorithm can increase the network's training speed. The Levenberg-Marquardt rule is a kind of quick algorithm in which the network's matrix of weights is updated based on the Jacobian matrix, J, collecting the partial derivatives of the network error e with respect to the weights, as described in Chapter 4. In other words, the matrix ΔW collecting the corrections of the weights in matrix W is computed according to

$$\Delta W = (J^T J + \mu I)^{-1} J^T e \tag{8.2}$$

where μ is a constant. If μ is sufficiently large, the previous algorithm is similar to the gradient descent algorithm; if μ is equal to zero, the previous rule will be a Gauss-Newton algorithm. The advantage of the Levenberg-Marquardt algorithm lies in the fact that the Hessian matrix is not needed; moreover using the Levenberg-Marquardt algorithm can accelerate the network's training speed, save training time and achieve better training results simultaneously.

The processes for developing the back-propagation neural network model are as follows (these are proposed by Ginzburg and Horn (1994)):

(a) Normalize the learning set;

(b) Decide the architecture and parameters: i.e., learning rate, momentum, and architecture. There are no criteria in deciding the parameters other than a trial-and-error basis;

(c) Initialize all weights randomly;

(d) Training, where the stopping criterion is either the number of iterations reached or when the total sum of squares of error is lower than a predetermined value;

(e) Choose the network with the minimum error (i.e., model selection criterion);

(f) Forecast future outcome.

Readers interested in a more detailed introduction to ANN modeling are referred to the related literature (Ginzburg and Horn, 1994; Tang et al., 1991; Tang and Fishwick, 1993; Zhang et al., 1998; Hwang, 2001).

The major advantage of ANN models is their flexible nonlinear modeling capability. They can capture the nonlinear characteristics of time series well. However, using ANN to model linear problems may produce mixed results (Zhang, 2003; Denton, 1995; Markham and Rakes, 1998). Therefore, we can conclude that the relationship between GLAR and ANN is complementary. To take full advantage of the individual strengths of two models, it is necessary to integrate the GLAR and ANN models, as mentioned earlier.

8.2.3 A Hybrid Model Integrating GLAR with ANN

From the previous two subsections, we can find that the relationship between GLAR and ANN is complementary. GLAR is a class of linear models that can capture time series' linear characteristics, while ANN models trained by back-propagation with hidden layers are a class of general function approximators capable of modeling non-linearity and which can capture nonlinear patterns in time series. Commixing the two models may yield a robust method, and more satisfactory forecasting results may be obtained by incorporating a time series model and an ANN model. Therefore, we propose a hybrid model integrating GLAR and ANN for exchange rate forecasting. In view of the works of Zhang (2003) and Chen and Leung (2004), the proposed hybrid model is a two-phase forecasting procedure, which incorporating GLAR model with ANN model in an adaptive (sequential) manner. The motivation of hybrid methodology is to create a synergy effect that further improves the prediction power (Yu et al., 2005c; Lai et al., 2006b).

In the first phase, a GLAR model is used to fit the linear component of time series, which is assumed to be $\{y_t, t = 1,2,\cdots\}$, and generate a series of forecasts, which is defined as $\{\hat{L}_t\}$. However, since some time series, such as exchange rates, often contain more complex patterns than those of linear regression models in the practical application, it is not enough to fit only a linear component for a time series. Hence, we also need to discover the nonlinear relationship of the time series in order to improve prediction accuracy. By comparing the actual value y_t of the time series and forecast value $\hat{L}_t$ of the linear component, we can obtain a series of nonlinear components, which is assumed to be $\{e_t\}$. That is,

$$e_t = y_t - \hat{L}_t \tag{8.3}$$

Thus, a nonlinear time series is obtained. The next step is how to fit the nonlinear component of series.

In the second phase of the hybrid methodology, a neural network model is used to model the above nonlinear time series. By training the ANN model using previously generated nonlinear time series as inputs, the trained ANN model is then used to generate a series of forecasts of nonlinear components of time series, defined by $\{\hat{N}_t\}$.

In order to obtain the synergetic forecast results, the final forecasting results, defined by $\{\hat{y}_t\}$, are calculated as

$$\hat{y}_t = \hat{L}_t + \hat{N}_t \tag{8.4}$$

In summary, the proposed hybrid methodology consists of two phases or four steps. In the first step, a GLAR model should be identified and the corresponding parameters should be estimated, i.e., a GLAR model is constructed. In the second step, the nonlinear components are computed from the GLAR model. In the third step, a neural network model is developed to model the nonlinear components. In the final step, the combined forecast results are obtained from Equation (8.4).

Theoretically speaking, the hybrid model presented here is still a class of single forecasting model. However, many studies reveal that combining several different forecasts can further improve prediction performances (Reid, 1968; Bates and Granger, 1969; Makridakis et al., 1982; Kang, 1986; Batchelor an Dua, 1995; Shi et al., 1999). In the following, we introduce the combined forecasting model briefly and then point out some problems with them.

8.2.4 Combined Forecasting Models

The idea of using a combined forecasting model is not new. Since the pioneering works of Reid (1968) and of Bates and Granger (1969), a variety of studies have been conducted and showed that a combination of forecasts often outperforms the forecasts obtained from a single method. The main problem of combined forecasts can be described as follows. Suppose there are n forecasts such as $\hat{y}_1(t), \hat{y}_2(t), \cdots \hat{y}_n(t)$. The question is how to combine these different forecasts into a single forecast $\hat{y}(t)$, which is assumed to be a more accurate forecast. The general form of the model for such a combined forecast can be defined as:

$$\hat{y}(t) = \sum_{i=1}^{n} w_i \hat{y}_i(t) \tag{8.5}$$

where w_i denotes the assigned weight of $\hat{y}(t)$, and in general the sum of the weights is equal to 1, i.e. $\sum_i w_i = 1$. In some situations the weights may have negative values and the sum of them may be greater than one (Shi et al., 1999).

There are a variety of methods available to determine the weights used in the combined forecasts. First of all, the equal weights (EW) method, which uses an arithmetic average of the individual forecasts, is a relatively easy and robust method. In this method,

$$w_i = \frac{1}{n} \tag{8.6}$$

where n is the number of forecasts. However, since the weights in the combination are so unstable, in practice a simple average may not the best technique to use (Kang, 1986). Moreover, research indicates that this weighted method has not always performed well empirically (Shi et al., 1999). In the existing literature, an effective combined approach, minimizing error-variance (ME) (Perrone and Cooper, 1993), is often used.

The minimum-error (ME) method is an approach that minimizes the forecasting error-variance when combining individual forecasts into a single forecast, i.e.,

$$w_{opt,i} = \arg\min_{w_i} \left\{ \sum_{i=1}^{n} \left(w_i \sigma_i^2 \right)^2 \right\}, \quad (i = 1, 2, \cdots, n) \tag{8.7}$$

under the constraints $\sum_{i=1}^{n} w_i = 1$ and $w_i \geq 0$. Using the Lagrange multiplier, the optimal weights are:

$$w_{opt,i} = \frac{\left(\sigma_i^2\right)^{-1}}{\sum_{j=1}^{n}\left(\sigma_j^2\right)^{-1}}, \quad (i = 1, 2, \cdots, n) \tag{8.8}$$

The minimizing error-variance method is based on the assumption of error independence. However, many predictors are often strongly correlated for the same task. This indicates that this approach has serious drawbacks when the predictors with strong correlation are included within the combined members.

However, there are still two main problems in existing combined forecasting approaches. One is the number of individual forecasts. Theoretical proof of Yang et al. (1996) shows that the total forecasting errors of combined forecasts does not necessarily decrease with an increase of the number of individual forecasting models. If this is the case, there are redundant or repeated (even noisy) information among the individual forecasts. Furthermore, the redundant information often directly impacts on the effectiveness of the combined forecasting method. It is therefore necessary to eliminate some individual forecasts that are not required. In other words, a key problem of combined forecasts is to determine a reasonable number of individual forecasts. The other problem is that the relationships among individual forecasts are determined in a linear manner. The two methods mentioned above and many other weighted methods (Kang, 1986) are all developed under linear consideration. A combined forecast should merge the individual forecasts according to the natural relationships existing between the individual forecasts, including but not limited to linear relationships. Because a linear combination of existing information cannot represent the relationship among individual forecasting models in some situations, it is necessary to introduce nonlinear combination methodology in the combined forecasts. Subsequently, we set out to solve the two problems in the following subsection.

8.2.5 A Nonlinear Combined (NC) Forecasting Model

In view of the two main problems outlined in the previous subsection, a novel nonlinear combined model is proposed. The following is the process of building the proposed model for the two problems.

In the first problem, the question we face is how to extract effective information that reflects substantial characteristics of series from all selected individual forecasts and how to eliminate redundant information. The principal component analysis (PCA) technique (see Jolliffe (1986) and Karhunen and Joutsensalo (1995)) is used as an alternative tool to solve the problems.

The PCA technique, an effective feature extraction method, is widely used in signal processing, statistics and neural computing. The basic idea in PCA is to find the components (s_1, s_2, ⋯, s_p) that can explain the maximum amount of variance possible by p linearly transformed components from data vector with q dimensions. The mathematical technique used in PCA is called eigen analysis. In addition, the basic goal in PCA is to reduce the dimension of the data. Thus, one usually chooses $p \le q$. Indeed, it can be proven that the representation given by PCA is an optimal linear dimension reduction technique in the mean-square sense (Jolliffe, 1986). Such a reduction in dimension has important benefits. First, the computation of the subsequent processing is reduced. Second, noise may be reduced and the meaningful underlying information identified. The following presents the PCA process for combined forecasting.

Assuming that there are n individual forecasting models in the combined forecasts and that every forecasting model contains m forecasting results, then forecasting matrix (Y) can be represented as

$$Y = \begin{bmatrix} y_{11} & y_{12} & \cdots & y_{1m} \\ y_{21} & y_{22} & \cdots & y_{2m} \\ \vdots & \vdots & \ddots & \vdots \\ y_{n1} & y_{n2} & \cdots & y_{nm} \end{bmatrix} \tag{8.9}$$

where y_{ij} is the jth forecasting value with the ith forecasting model.

Next, we deal with the forecasting matrix using the PCA technique. First, eigenvalues (λ_1, λ_2, ⋯, λ_n) and corresponding eigenvectors A=(a_1, a_2, ⋯, a_n) can be solved from the forecasting matrix. Then the new principal components are calculated as

$$Z_i = a_i^T Y \quad (i = 1, 2, \cdots, n) \tag{8.10}$$

Subsequently, we choose m ($m \le n$) principal components from existing n components. If this is the case, the saved information content is judged by

$$\theta = (\lambda_1 + \lambda_2 + \cdots + \lambda_m)/(\lambda_1 + \lambda_2 + \cdots + \lambda_n) \tag{8.11}$$

If θ is sufficiently large (e.g., $\theta > 0.8$), enough information has been saved after the feature extraction process of PCA. Thus, re-combining the new information can further improve the prediction performance of combined forecasts.

For further explanation, a simple example is presented. Assuming that we predict the GBP/USD exchange rate using three different forecasting methods (e.g., GLAR, ANN and ARIMA) and obtain eight forecasts for every method, the corresponding forecasting matrix is shown below.

Original forecasting matrix is represented by

$$Y = \begin{bmatrix} 0.6723 & 0.6599 & 0.6474 & 0.6349 & 0.6224 & 0.6099 & 0.5974 & 0.5848 \\ 0.6712 & 0.6586 & 0.6471 & 0.6356 & 0.6251 & 0.6155 & 0.6064 & 0.5982 \\ 0.6697 & 0.6566 & 0.6436 & 0.6310 & 0.6186 & 0.6064 & 0.5946 & 0.5829 \end{bmatrix}$$

Before using combined forecasting methodology, the PCA is performed. Accordingly, its eigenvalues and eigenvectors are computed as

$$\lambda = \begin{pmatrix} 0.1762E-01 \\ 0.1288E-04 \\ 0.9019E-08 \end{pmatrix}, \ A = \begin{pmatrix} -0.5806 & 0.5390 & 0.6102 \\ 0.8084 & 0.2928 & 0.5106 \\ 0.0966 & -0.7898 & 0.6058 \end{pmatrix}.$$

According to the previous computation process, the new forecasting matrix (or new principal components) can be as

$$Y' = \begin{bmatrix} 0.0734 & 0.0514 & 0.0301 & 0.0089 & 0.0115 & 0.0315 & 0.0510 & 0.0699 \\ 0.0019 & 0.0002 & 0.0006 & 0.0014 & 0.0014 & 0.0008 & 0.0003 & 0.0020 \\ 0.0000 & 0.0000 & 0.0001 & 0.0000 & 0.0000 & 0.0000 & 0.0000 & 0.0000 \end{bmatrix}$$

From the information presented in the new forecasting matrix, we can find and conclude that the third row belongs to redundancy and should be eliminated, indicating that the third individual forecasting model provides repeated information and cannot improve combined forecasting performance: thus this forecasting method should be removed from the combined forecasts. Therefore, the reasonable number of individual forecasting methods is determined. Of course, the value of θ should be calculated for judgment if the new forecasting matrix is not obvious. In this example, θ = (0.1762E-01+0.1288E-04) / (0.1762E-01 + 0.1288E-04 + 0.9019E-08) = 0.999999489. This also shows that the former two methods contain enough information to combine forecasts.

In the case of the second problem, we propose a nonlinear combined forecasting model as a remedy. The detailed model is presented as follows.

A nonlinear combined forecasting model can be viewed as nonlinear information processing system, which can be represented as

$$y = f(I_1, I_2, \cdots, I_n) \tag{8.12}$$

where $f(\cdot)$ is a nonlinear function and I_i denotes the information provided by individual forecasting models. If individual forecasting model can provide an individual forecast $\hat{y}_i$, then Equation (8.12) can be represented as

$$\hat{y} = f(\hat{y}_1, \hat{y}_2, \cdots, \hat{y}_n) \tag{8.13}$$

Determining the function $f(\cdot)$ is quite challenging. In this study, ANN is employed to realize nonlinear mapping. The ability of back-propagation neural networks to represent nonlinear models has been tested by previous work.

In fact, the ANN training is a process of searching for optimal weights. That is, this training process made the sum of the square errors minimal, i.e.,

$$Min[(y - \hat{y})(y - \hat{y})^T] \tag{8.14}$$

Or

$$Min\{[y - f(\hat{y}_1, \hat{y}_2, \cdots, \hat{y}_n)][y - f(\hat{y}_1, \hat{y}_2, \cdots, \hat{y}_n)]^T\} \tag{8.15}$$

By solving the earlier two problems of linear combination models, a PCA&ANN-based nonlinear combined model is generated for foreign exchange rates forecasting.

To summarize, the proposed nonlinear combined model consists of four stages. Generally speaking, in the first stage we construct an original forecasting matrix according to the selected individual forecasts. In the second stage, the PCA technique is used to deal with the original forecasting matrix and a new forecasting matrix is obtained. In the third stage, based upon the judgments of PCA results, the number of individual forecasting models is determined. And in the final stage, an ANN model is developed to combine different individual forecasts; meantime the corresponding forecasting results are obtained.

After completing the proposed nonlinear combined model, we want to know whether the proposed model indeed improves forecasting accuracy or not. In this study, we use an individual GLAR model, an individual ANN model, the hybrid model, the linear combination models, and the PCA&ANN-based nonlinear combined model to predict exchange rates so as to compare forecasting performance. The basic flow diagram is shown in Fig. 8.1.

8.2.6 Forecasting Evaluation Criteria

As Weigend (1994) argued, a measure normally used to evaluate and compare the predictive power of the model is the normalized mean squared error (NMSE) (this was used to evaluate entries in the Santa Fe Time Series Competition).

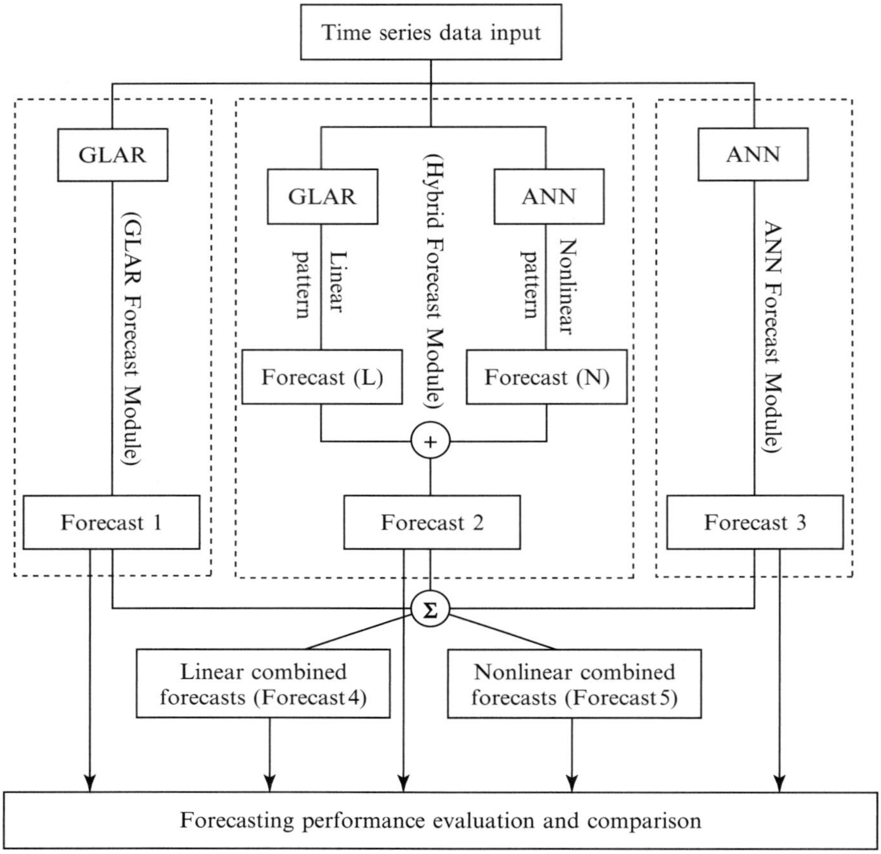

Fig. 8.1. A flow diagram of the hybrid and combined forecast

Given N pairs of the actual values (or targets, y_t) and predicted values ($\hat{y}_t$), the NMSE which normalizes the MSE by dividing it through the variance of respective series can be defined as

$$\text{NMSE} = \frac{\sum_{t=1}^{N}(y_t - \hat{y}_t)^2}{\sum_{t=1}^{N}(y_t - \bar{y}_t)^2} = \frac{1}{\sigma^2} \cdot \frac{1}{N} \cdot \sum_{t=1}^{N}(y_t - \hat{y}_t)^2 \tag{8.16}$$

where σ^2 is the estimated variance of the data and $\bar{y}_t$ the mean.

Clearly, accuracy is one of the most important criteria for forecasting models — the others being the cost savings and profit earnings generated from improved decisions. From the business point of view, the latter is more important than the former. For business practitioners, the aim of forecasting is to support or improve decisions so as to make more money. Thus profits or returns are more important than conventional fit measurements. But in exchange rate forecasting, improved decisions often depend on correct forecasting directions or turning points between the actual and predicted values, y_t and $\hat{y}_t$, respectively, in the testing set with respect to directional change of exchange rate movement (expressed in percentages). The ability to forecast movement direction or turning points can be measured by a statistic developed by Yao and Tan (2000). Directional change statistics (D_{stat}) can be expressed as

$$D_{stat} = \frac{1}{N}\sum_{t=1}^{N} a_t \times 100\% \tag{8.17}$$

where a_t=1 if$(y_{t+1} - y_t)(\hat{y}_{t+1} - y_t) \geq 0$, and a_t=0 otherwise.

However, the real aim of forecasting is to obtain profits based on prediction results. Here the annual return rate is introduced as another evaluation criterion. Without considering the friction costs, the annual return rate is calculated according to the compound interest principle.

$$R = \left(\left(\frac{M}{S} \right)^{12/N} - 1 \right) \times 100\,\% \tag{8.18}$$

where R is the annual return rate, M is the amount of the money obtained on the last month of testing, S is the amount of the money used for trading on the first month of testing, and N is the number of months in the testing periods. It is worth noting that the computation of R is based on the trading strategy in the following:

$$\text{If } (\hat{y}_{t+1} - y_t) > 0\text{, then “buy”, else “sell”} \tag{8.19}$$

where y_t is the actual value at time t, $\hat{y}_{t+1}$, is the prediction at time t+1. That is, we use the difference between the predicted value and the actual value to guide trading.

We want to use the latter two statistics as the main evaluation criteria because the normalized *MSE* measure predictions only in terms of levels. Hence, it is more reasonable to choose D_{stat} and annual return rate (R) as the measurements of forecast evaluation. Of course, *NMSE* is also taken into consideration in terms of a comparison of levels.

8.3 Empirical Analysis

In this section, we first describe the data used in this chapter and then report the experimental results.

8.3.1 Data Description

The foreign exchange data used in this paper are monthly and are obtained from Pacific Exchange Rate Service (http://fx.sauder.ubc.ca/), provided by Professor Werner Antweiler, University of British Columbia, Vancouver, Canada. They consist of the US dollar exchange rate against each of the currencies (DEM, GBP and JPY) studied in this paper. We take monthly data from January 1971 till December 2000 as in-sample (training periods) data sets (360 observations including 24 samples for validation). We also take the data from January 2001 to December 2003 as out-of-sample (testing periods) data sets (36 observations) which is used to evaluate the good or bad performance of prediction based on some evaluation measurement. For exogenous variables' consideration of GLAR, we can choose exports, inflation rate, interest rates and other factors related to exchange rates as exogenous variables. Unfortunately, monthly inflation rate and interest rate data is hard to obtain. Therefore, we only choose "exports" as the exogenous variable for the GLAR model. The monthly export data in this paper is taken directly from the CD-ROM of International Financial Statistics of IMF. In addition, all the evaluations in this paper (with the exception of specifications) are based on the whole testing sets. In order to save space, the original data are not listed here, and detailed data can be obtained from the website or from the authors.

8.3.2 Empirical Results

In this study, all GLAR models are implemented via the Eviews software package, which is produced by Quantitative Micro Software Corporation. The individual ANN model, the linear combination model with minimum error-variance method and nonlinear combined model are built using the Matlab software package, which is produced by Mathworks Laboratory Corporation. The single hybrid methodology involves the GLAR and ANN models and thus it use both the software packages. The ANN models use trial and error to determine the network architecture of 4-4-1 by minimizing the forecasting error with the exception of specifications. According to the idea shown in Fig. 8.1, the single GLAR model, the single ANN model, the single hybrid model, and the linear combination model with equal weights method, the linear combination model with minimum error-variance method, and the PCA&ANN–based nonlinear combined model are used to predict the exchange rates of three main currencies for comparison. For space reasons the computational processes are omitted but can be obtained from the authors if required.

Figs. 8.2 to 8.4 give graphical representations of the forecasting results for exchange rates of the three main currencies using different models. Tables 8.1 to 8.3 show the forecasting performance of different models from different perspectives. From the graphs and tables, we can generally see that the forecasting results are very promising for all currencies under study either where the measurement of forecasting performance is goodness of fit such as *NMSE* (refer to Table 8.1) or where the forecasting performance criterion is D_{stat} (refer to Table 8.2).

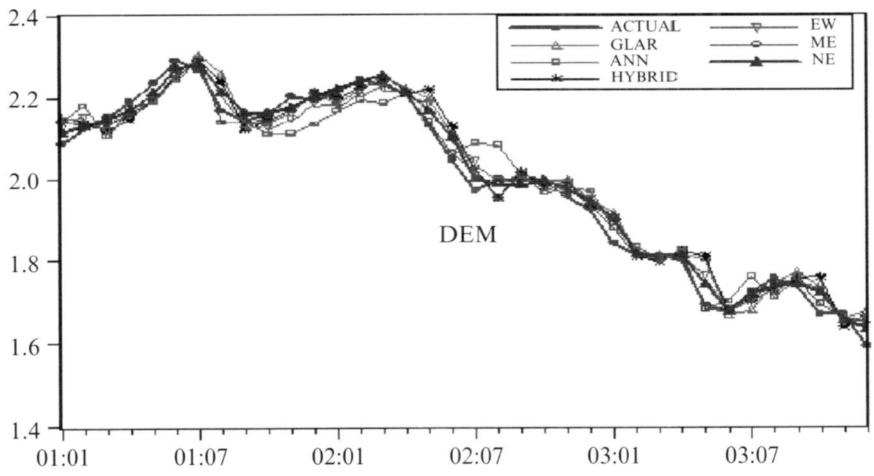

Fig. 8.2. A graphical comparison of DEM rate predictions using different models

In detail, Fig. 8.2 shows the comparison for the DEM rate of the nonlinear combined forecasting results versus the forecast results of individual models, the hybrid model, and linear combination model. Similarly, it can be seen from Fig. 8.3 that forecast accuracy for the GBP rate has significantly improved using the nonlinear combined forecasting models. Fig. 8.4 provides the comparison of the JPY prediction results using different models. The results indicate that the nonlinear combined forecasting model performs better than the other models presented here.

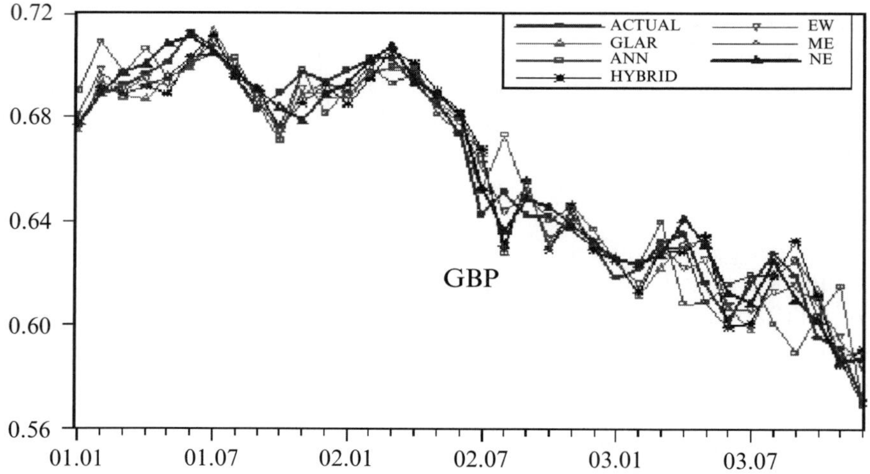

Fig. 8.3. A graphical comparison of GBP rate predictions using different models

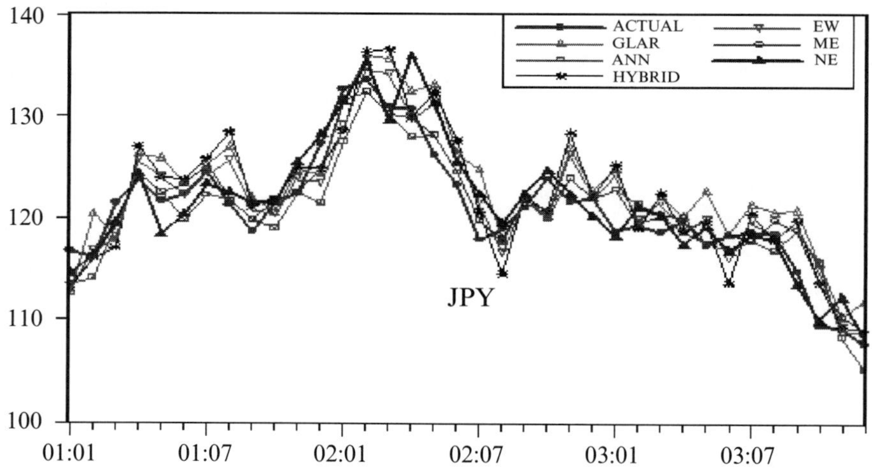

Fig. 8.4. A graphical comparison of JPY rate predictions using different models

Subsequently, the forecasting performance comparisons of various models for the three main currencies via *NMSE*, D_{stat} and return rate (*R*) are reported in Tables 8.1 to 8.3 respectively. Generally speaking, these tables provide comparisons of *NMSE*, D_{stat} and *R* between these different methods, indicating that the prediction performance of the nonlinear combined forecasting model is better than those of other models including these empirical analyses for the three main currencies.

Table 8.1. The *NMSE* comparison of different methods for the three currencies *

Model	DEM		GBP		JPY	
	NMSE	Rank	*NMSE*	Rank	*NMSE*	Rank
GLAR	0.0453	5	0.0737	4	0.4207	6
ANN	0.0433	4	0.1031	6	0.1982	3
Hybrid	0.0459	6	0.0758	5	0.3854	5
EW	0.0319	3	0.0469	3	0.2753	4
ME	0.0143	1	0.0410	2	0.1505	2
NE	0.0156	2	0.0357	1	0.1391	1

* GLAR: generalized linear auto-regression; ANN: artificial neural network; EW: equal weights; ME: minimum error; NE: nonlinear ensemble; NMSE: normalized mean square error.

Focusing on the *NMSE* indicator, the nonlinear combined model performs the best in all but the DEM case, followed by the linear combination model with ME method and EW method and other individual models. To summarize, the nonlinear combined model and the linear combination model with ME method outperform the other different models presented here in terms of *NMSE*. Interestingly, the *NMSE*s of the hybrid methodology are not better than those of the individual GLAR and ANN models. This result differs from that of Zhang (2003). This could be because the GLAR is different from ARIMA, or because the ANN architecture designs may differ.

Table 8.2. The D_{stat} comparison of different methods for the three currencies *

Model	DEM		GBP		JPY	
	D_{stat} (%)	Rank	D_{stat} (%)	Rank	D_{stat} (%)	Rank
GLAR	58.33	6	58.33	6	47.22	6
ANN	63.89	4	69.44	3	63.89	4
Hybrid	61.11	5	61.11	5	66.67	3
EW	69.44	3	66.67	4	58.33	5
ME	80.87	2	78.02	2	71.85	2
NE	83.33	1	86.11	1	75.00	1

* GLAR: generalized linear auto-regression; ANN: artificial neural network; EW: equal weights; ME: minimum error; NE: nonlinear ensemble; NMSE: normalized mean square error.

However, the low *NMSE* does not necessarily mean that there is a high hit rate of forecasting direction for foreign exchange movement direction prediction. Thus, the D_{stat} comparison is necessary. Focusing on D_{stat} of Table 8.2, we find the nonlinear combined model also performs much better than the other models according to the rank; furthermore, from the business practitioners' point of view, D_{stat} is more important than *NMSE* because the former is an important decision criterion. With reference to Table 8.2, the differences between the different models are very significant. For example, for the GBP test case, the D_{stat} for the individual GLAR model is 58.33%, for the individual ANN model it is 69.44%, and for the single hybrid model D_{stat} is only 61.11%; while for the nonlinear combined forecasting model, D_{stat} reaches 86.11%. For the linear combination model with EW method and the hybrid method, the rank of forecasting accuracy is always in the middle for any of the test currencies. The main cause of this phenomenon is that the bad performance of the individual models has an important effect on the holistic forecast efficiency. Similarly, the individual ANN can model nonlinear time series such as exchange rate series well, and the D_{stat} rank is also in the middle for any of the test currencies. In the same way, we also notice that the D_{stat} ranking for GLAR is the last. The main reason is that the high noise, nonlinearity and complex factors are contained in foreign exchange series, and unfortunately the GLAR is a class of linear model.

Table 8.3. The *R* comparison of different methods for the three currencies

Model	DEM		GBP		JPY	
	NMSE	Rank	NMSE	Rank	NMSE	Rank
GLAR	5.89	6	5.14	6	−1.76	6
ANN	7.53	4	8.69	3	6.23	4
Hybrid	7.38	5	6.47	5	7.58	3
EW	8.65	3	7.53	4	5.68	5
ME	11.12	2	11.05	2	9.87	2
NE	15.54	1	14.85	1	16.49	1

* GLAR: generalized linear auto-regression; ANN: artificial neural network; EW: equal weights; ME: minimum error; NE: nonlinear ensemble; NMSE: normalized mean square error.

In terms of the return or profit criterion, the empirical results show that the proposed nonlinear combined model could be applied to future forecasting. Compared with the other models presented in this paper, the nonlinear combined forecasting model performs the best. Likewise, we find that the rank of Table 8.3 is the similar to that of Table 8.2. It is not hard to understand such a rationale that right forecasting direction often leads to high return rates. As shown in Table 8.3, for the DEM test case the best return rate for the nonlinear combined model is 15.54%. Similarly, in the JPY test case, the annual return rate for GLAR is –1.76%, the rate for the ANN model is 6.23%, the rate for the hybrid model is 7.58%, and the rate for the linear combination model with ME method is also 9.34%;

however the return rate for the nonlinear combined model reaches 16.49% per year.

From the experiments presented in this study we can draw the following conclusions. (i) The experimental results show that the nonlinear combined forecasting model is superior to the individual GLAR model, the individual ANN model, the single hybrid model as well as the linear combination models with EW and ME methods for the test cases of three main currencies in terms of the measurement of directional change statistics (D_{stat}) and annual return rate (R), as can be seen from Tables 8.2 and 8.3. Likewise, the nonlinear combined model and the linear combination model with ME method also outperform other models in terms of goodness-of-fit or *NMSE*, as can be seen from Figs. 8.2 to 8.4 and Table 8.1. (ii) The nonlinear combined forecasts are able to improve forecasting accuracy significantly — in other words, the performance of the nonlinear combined forecasting model is better than those of other forecasting models in terms of *NMSE*, D_{stat} and R. This leads to the third conclusion. (iii) The nonlinear combined model can be used as an alternative tool for exchange rate forecasting to obtain greater forecasting accuracy and improve the prediction quality further in view of empirical results.

8.4 Conclusions

This study proposes using a nonlinear combined forecasting model that combines the time series GLAR model, the ANN model and the hybrid model to predict foreign exchange rates. In terms of the empirical results, we find that across different forecasting models for the test cases of three main currencies — German marks, British pounds and Japanese yen — on the basis of different criteria, the nonlinear combined model performs the best. In the nonlinear combined model test cases, the *NMSE* is the lowest and the D_{stat} and R is the highest, indicating that the nonlinear combined forecasting model can be used as a viable alternative solution for exchange rate prediction for financial managers and business practitioners.

9 A Hybrid GA-Based SVM Model for Foreign Exchange Market Tendency Exploration

9.1 Introduction

Forecasting foreign exchange rates has been regards as one of the most challenging application of modern time series forecasting (Yu et al., 2005f). Thus, numerous models have been developed to provide the investors with more precise predictions. Recently, artificial intelligence, such as artificial neural networks (ANN), has been widely applied to solve foreign exchange rates forecasting problems. Literature documenting the research shows the ANN has a powerful capability to predict foreign exchange rates. The previous chapters provide some literature reviews. Interested readers can be referred to the Chapter 1 and other chapters for more details about foreign exchange rates prediction. Two recently good surveys about foreign exchange rates prediction with ANN can be referred to Huang et al. (2004a, 2006) and Yu et al. (2005e) for more literature review.

Although a large number of successful applications have shown that ANN can be a very useful tool for time series modeling and forecasting (Zhang and Patuwo, 1998), but some studies have also pointed out that ANN often exhibits inconsistent results on noisy data (Hann and Steurer, 1996) because foreign exchange market is a typical noisy market with high volatility. With the exception of noisy market factor, the limitations of ANN itself are another important reason leading to some inconsistent results. As claimed by Yu (1992) and Lawrence et al. (1997), ANN often suffers from much difficulty in trapping into local minima, overfitting and selecting relevant input variables.

Furthermore, the problem may become more intractable when the model interpretability is important. For example, in foreign exchange market tendency judgment, it is critical for decision-makers to understand the key drivers of affecting foreign exchange price movement. However, a predictive ANN model that is essentially a "black box" is not helpful for developing comprehensive forecasting models (Kim and Street, 2004).

In order to provide a good solution to these problems, a hybrid intelligent data mining approach (i.e., GA-based SVM (GASVM)) that uses support vector machines (SVM) (Vapnik, 1995) guided by genetic algorithms (GA) (Holland, 1992; Goldberg, 1989) is proposed to explore foreign exchange market tendency. The SVM, as a novel neural network algorithm developed by Vapnik (1995) is used to avoid local minima and overfitting of traditional neural network models. Except for local minima and overfitting, GA attempts to solve input variable selection problem in this chapter. In this hybrid GASVM model, GA is first used for feature selection of SVM input variables in order to reduce the model complexity and increase the model interpretability, and then the SVM is used to predict foreign exchange market movement direction in order to improve the prediction performance based on the historical data. Note that two principal goals, forecasting performance (or predictive accuracy) and model complexity (or model interpretability), are often in conflict (Yu et al., 2005f). In our study, we try to arrive at a trade-off between performance and complexity from the following two aspects.

On the one hand, we will build predictive models that integrating SVM with GA to improve the prediction performance. Furthermore, we use GA-based SVM to identify the key factor of affecting foreign exchange rate movement. This can be done by learning linear or possibly nonlinear relationships between given input variables and the dependent variable. Relative to traditional predictive approaches, we go one step further. Because we are also interested in identifying the key determinants of foreign exchange market price movement, we select different subsets of feature variables using GA and use only those selected variables to train SVM.

The reasons of selecting SVM rather than ANN mainly include the following several aspects. First of all, unlike ANN, which is implemented by the principle of empirical risk minimization (ERM), SVM is implemented by the structural risk minimization (SRM) principle, which searches to minimize an upper bound of generalization error. Therefore, the solution of SVM may be global optimum rather than local optimum. Second, SVM models are also a class of data-driven, non-parametric models, and they can let "the data speak for themselves". Furthermore, overfitting is unlikely to occur with SVM (Vapnik, 1995; Kim, 2003). In addition, some empirical studies (Kim, 2003; Huang et al., 2005; Yu et al., 2005f) also confirmed the efficiency of SVM.

On the other hand, we enhance the interpretability of our model by reducing the dimensionality of data sets using GA. That is, the GA is used to select a subset of original features thus simplifying the SVM model and increasing the model interpretability. Usually, data dimensionality reduction is performed via feature selection. Feature selection is defined as the process of selecting a subset of the original features by eliminating redundant features or some features with little information (Kim and Street, 2004). Generally, feature selection algorithms such as principal component analysis (PCA) (Yu et al., 2005c) have been often used for this purpose. However, in our study, the PCA is not appropriate because our goal is not only to reduce the data dimensionality, but also to obtain highly accurate predictive models. Furthermore, PCA does not consider the relationship between the response variable and other input variables in the process of data reduction. In addition, the resulting principal components from PCA can be difficult to interpret when the dimensionality of input variables is huge. On the contrary, the GA has been proved to have superior performance to other algorithms for data set with high dimensionality (Kudo and Sklansky, 2000).

In the process of feature selection, if we extract as much information as possible from a given data set while using the smallest number of features, we cannot only save a great amount of computing time and cost, but also build a model with better generalization. Furthermore, feature selection can also significantly improve the comprehensibility of the resulting models. Even a complicated model, such as a SVM model, can be more easily understood if constructed from only a few variables.

In our study, the proposed GASVM approach makes full use of the desirable characteristics of GA and SVM models to achieve two principal goals: model interpretability and predictive accuracy. The detailed process is as follows. A standard GA is used to search through the possible combination of features. The input features selected by GA are used to train SVM. The trained SVM is tested on an evaluation set, and a proposed model is evaluated in terms of two evaluation criteria: hit ratio (which is maximized) and complexity (which is minimized). This process is repeated many times as the algorithm searches for a desirable trade-off between predictive accuracy and model complexity. The final results obtained is a highly accurate predictive model that uses only a subset of initial features, thus simplifying the model and providing some useful information on future data collection work. Thus the proposed GASVM model has two distinct advantages. One is that the computations of GASVM are reduced by the decrease of model inputs and running speed will be accelerated. Another is that GASVM can avoid some defects of traditional neural network models, such as local minima and overfitting as well as input variable selection.

The main motivation of this chapter is to propose a new hybrid intelligent data mining approach integrating SVM with GA for exploring foreign exchange market tendency and to test the predictability of the proposed hybrid intelligent model by comparing it with statistical models and neural network models. The rest of the chapter is organized as follows. The next section will describe the formulation process of the proposed hybrid intelligent data mining model in detail. In Section 9.3, we give an experiment scheme and report the experimental results. For comparison, the similarities and difference of three hybrids model proposed by this part are presented in Section 9.4. And Section 9.5 concludes this chapter.

9.2 Formulation of the Hybrid GA-SVM Model

In this section, a formulation process of the hybrid intelligent GASVM model is presented in detail. First of all, a basic theory of the SVM in classification is described. Then the GA for feature selection will be proposed to reduce the model complexity of SVM. Finally, a hybrid intelligent GASVM integrating GA and SVM is constructed.

9.2.1 Basic Theory of SVM

SVM was originally introduced by Vapnik (1995) in the 1990s. The basic idea of SVM is to use linear model to implement nonlinear class boundaries through some nonlinear mapping the input vector into the high-dimensional feature space. A linear model constructed in the new space can represent a nonlinear decision boundary in the original space. In the new space, an optimal separating hyperplane is constructed. Thus the SVM is known as the algorithm that finds a special kind of linear model, the maximum margin hyperplane. The maximum margin hyperplane gives the maximum separation between the decision classes. The training examples that are closest to the maximum margin hyperplane are called support vectors. All other training examples are irrelevant for defining the binary class boundaries. In this study, we use SVM to identify the movement direction of foreign exchange market price. Thu, support vector classification (SVC) is used.

Consider a set of training data $D=\{(x_1,y_1),\cdots,(x_i,y_i),\cdots,(x_d,y_d)\}, x\in R^n, y\in\{-1,1\}$, where x_i is the training data, d is the number of training data, y_i is the class label for x_i.

First of all, a nonlinear function is employed to map the original input space R^n to N-dimensional feature space: $\psi(x)=\varphi_1(x),\varphi_2(x),\ldots,\varphi_N(x)$. Then the separating hyperplane is constructed in this high dimension feature space. The classifier takes the form as

$$y(x) = \text{sgn}(w \cdot \psi(x) + b) \tag{9.1}$$

where w is the weight vector and b is a bias. To obtain the optimal classifier, $\|w\|$ should be minimized under the following constraints

$$y_i[\varphi(x_i) \cdot w + b] \geq 1 - \xi_i, i = 1,2,..., d \tag{9.2}$$

where ξ_i is positive slack variable, which is necessary to allow misclassification.

Thus, according to principle of structural risk minimization (SRM), the optimal problem can be formulated as minimization of the following problem:

$$\begin{cases} \text{Min} & \phi(w,\xi) = (1/2)\|w\|^2 + C\sum_{i=1}^{d}\xi_i \\ \text{s.t.} & y_i[\varphi(x_i) \cdot w + b] \geq 1 - \xi_i \\ & \xi_i \geq 0, i = 1,2,..., d \end{cases} \tag{9.3}$$

where C is a margin parameter, representing an upper bound. According to the Lagrangian principle, the above problem can be transformed into the following form:

$$\begin{aligned} L(w,b,\xi,\alpha,\beta) = {} & (1/2)\|w\|^2 + C\sum_{i=1}^{d}\xi_i \\ & - \sum_{i=1}^{d}\alpha_i(y_i[\varphi(x_i) \cdot w + b] - 1 + \xi_i) - \sum_{i=1}^{d}\beta_i\xi_i \end{aligned} \tag{9.4}$$

where α_i, β_i are the non-negative Lagrange multipliers. Using the Karush-Kuhn-Tucker (KKT) conditions (Fletcher, 1987) for the Equation (9.4), then the optimal conditions are

$$\frac{\partial L}{\partial w} = 0, \frac{\partial L}{\partial b} = 0, \frac{\partial L}{\partial \xi} = 0 \tag{9.5}$$

We have the following equations from Equation (9.5):

$$\sum_{i=1}^{d}\alpha_i y_i = 0, w = \sum_{i=1}^{d}\alpha_i y_i \varphi(x_i), C - \alpha_i - \beta_i = 0 \tag{9.6}$$

Hence, the dual problem can be obtained from Equations (4) to (6), i.e.,

$$\begin{cases} \text{Max} & J(\alpha) = -(1/2)\sum_{i,j=1}^{d}\alpha_i\alpha_j y_i y_j(\varphi(x_i), \varphi(x_j)) + \sum_{i=1}^{d}\alpha_i \\ \text{s.t.} & \sum_{i=1}^{d}\alpha_i y_i = 0 \\ & 0 \leq \alpha_i \leq C, i = 1, 2, ..., d \end{cases} \tag{9.7}$$

If there exist a kernel function such that $K(x_i, x_j) = (\varphi(x_i), \varphi(x_j))$, it is usually unnecessary to explicitly know what $\varphi(x)$ is, and we only need to work with a kernel function in the training algorithm. Any function satisfying Mercy condition (Vapnik, 1995) can be used as the kernel function. Common examples of the kernel function are the polynomial kernel $K(x,y)=(xy+1)^d$ and the Gaussian function $K(x,y)=\exp\left(-(x-y)^2/2\sigma^2\right)$. The construction of kernel function is important to SVM, but in practice the kernel function is often given directly. Therefore, the Equation (9.7) can be rewritten into Equation (9.8), i.e.,

$$\begin{cases} \text{Max} \quad J(\alpha) = -(1/2)\sum_{i,j=1}^{d} \alpha_i \alpha_j y_i y_j K(x_i, x_j) + \sum_{i=1}^{d} \alpha_i \\ \text{s.t.} \quad \sum_{i=1}^{d} \alpha_i y_i = 0 \\ \qquad 0 \le \alpha_i \le C, i = 1, 2, ..., d \end{cases} \tag{9.8}$$

From Equation (9.8), if $\alpha_i > 0$, then the corresponding x_i is called support vector. In general, support vectors are only a small part of the training samples. Equation (9.8) is a quadratic programming problem constrained in a convex set, and the solution is unique.

Finally, the optimal separating hyperplane is obtained as follows:

$$\sum_{SV} \alpha_i y_i K(x_i, x) + b = 0 \tag{9.9}$$

where *SV* are the support vectors. Then the nonlinear classifier is

$$y = \text{sgn}\left(\sum_{SV} \alpha_i y_i K(x_i, x) + b\right) \tag{9.10}$$

Because SVM is a form of kernel-based neural networks, the basic architecture of SVM model is similar to that of neural networks, as illustrated in Fig. 9.1. In the same way, the SVM also consists of an input layer, middle layer or transformation layer and output layer. The difference is that every node output of the middle layer is a support vector transformed by the kernel function $K(x_i, x)$.

9.2.2 Feature Selection with GA for SVM Modeling

In this study, we use standard genetic algorithm (GA) to extract feature subset of model input variables for SVM modeling. To date, GA has become a popular optimization method as they often succeed in finding the best optimum in contrast to most common optimization algorithms. Genetic algorithm imitates the natural selection process in biological evolution with selection, mating reproduction and mutation, and the sequence of the different operations of a genetic algorithm is shown in the left part of Fig. 9.2 (Yu et al., 2006a). The parameters to be optimized are represented by a chromosome whereby each parameter is encoded in a binary string called gene. Thus, a chromosome consists of as many genes as parameters to be optimized. Interested readers can be referred to previous work (e.g., Goldberg, 1989; Holland, 1992) for more details. In the following GA for feature variable selection is discussed.

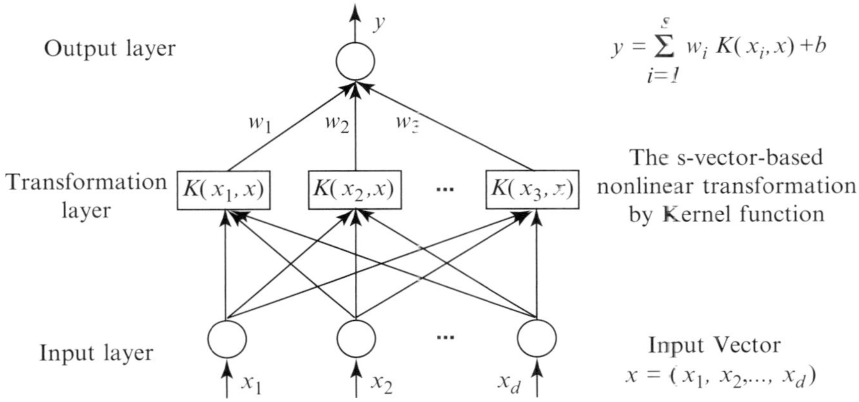

Fig. 9.1. The generic architecture of SVM model

First of all, a population, which consists of a given number of chromosomes, is initially created by randomly assigning "1" and "0" to all genes. In the case of variable selection, a gene contains only a single bit string for the presence and absence of a variable. The top right part of Fig. 9.2 shows a population of four chromosomes for a three-variable selection problem. In this study, the initial population of the GA is randomly generated except of one chromosome, which was set to use all variables. The binary string of the chromosomes has the same size as variables to select from whereby the presence of a variable is coded as "1" and the absence of a variable as "0". Consequently, the binary string of a gene consists of only one single bit.

The subsequent work is to evaluate the chromosomes generated by previous operation by a so-called fitness function, while the design of the fitness function is a crucial point in GA, which determines what a GA should optimize. In order to guide a GA toward more highly fit regions of the search space of chromosomes, one or more fitness values must be assigned to each string. Usually, a standard GA is designed for optimization problems with only one objective. Each fitness value can be determined by a fitness function that we want to optimize. In our work, the fitness value for a string is determined by a SVM model. Because our goal is to find a small subset of input variables from many candidate variables, the evaluation of the fitness starts with the encoding of the chromosomes into SVM model whereby "1" indicates that a specific variable is selected and "0" that a variable is not selected by the SVM model. Note that the SVM is used here for modeling the relationship between the input variables and the response variable. Then the SVM models are trained with a training data set and after that, the trained SVM is tested on a testing data set, and the proposed model is evaluated both on the hit ratio (which is maximized) and the complexity (number of selected feature variables, which is minimized) of the solution. Finally, these two evaluation criteria are combined into one so-called fitness function f and used to evaluate the quality of each string. For a classification problem, for example, our fitness function for the GA variable selection can use the following equation:

$$f = E_{accuracy} - \alpha E_{complexity} \tag{9.11}$$

where $E_{accuracy}$ is the hit ratio, representing the model predictive power, $E_{complexity}$ is the model complexity, α is a parameter. Parameter α is aimed at adjusting the number of variables used by the evolved SVM in terms of users' preference. Usually, a high value of the fitness function results in only few variables selected for each SVM model whereas a small value of fitness function results in more variables being selected. We expect that lower complexity will lead to easier interpretability of solutions as well as better generalization.

From Equation (9.11), it is not hard to find that two different goals, model accuracy and model complexity, are combined into one fitness value for candidate solutions. Usually, model accuracy of classification problems can be represented by hit ratio with the following form:

$$E_{accuracy} = \frac{\text{The number of correct classification}}{\text{The number of all evaluation sample}} \tag{9.12}$$

In Equation (9.11), the aim of the first part is to favor feature variable subsets with higher discriminative power to classification problems, while the second part is aimed at finding parsimonious solutions by minimizing the number of selected features as follows.

$$E_{complexity} = 1 - \frac{n_v}{N_{tot}} \tag{9.13}$$

where n_v is the number of variables used by the SVM models, N_{tot} is the total number of variables.

After evolving the fitness of the population, the best chromosomes with the highest fitness value are selected by means of the roulette wheel. Thereby, the chromosomes are allocated space on a roulette wheel proportional to their fitness and thus the fittest chromosomes are more likely selected. In the following mating step, offspring chromosomes are created by a crossover technique. A so-called one-point crossover technique is employed, which randomly selects a crossover point within the chromosome. Then two parent chromosomes are interchanged at this point to produce two new offspring. After that, the chromosomes are mutated with a probability of 0.005 per gene by randomly changing genes from "0" to "1" and vice versa. The mutation prevents the GA from converging too quickly in a small area of the search space. Finally, the final generation will be judged. If yes, then the optimized subsets are selected. If no, then the evaluation and reproduction steps are repeated until a certain number of generations, until a defined fitness or until a convergence criterion of the population are reached. In the ideal case, all chromosomes of the last generation have the same genes representing the optimal solution. More details can be referred to Chapter 3.

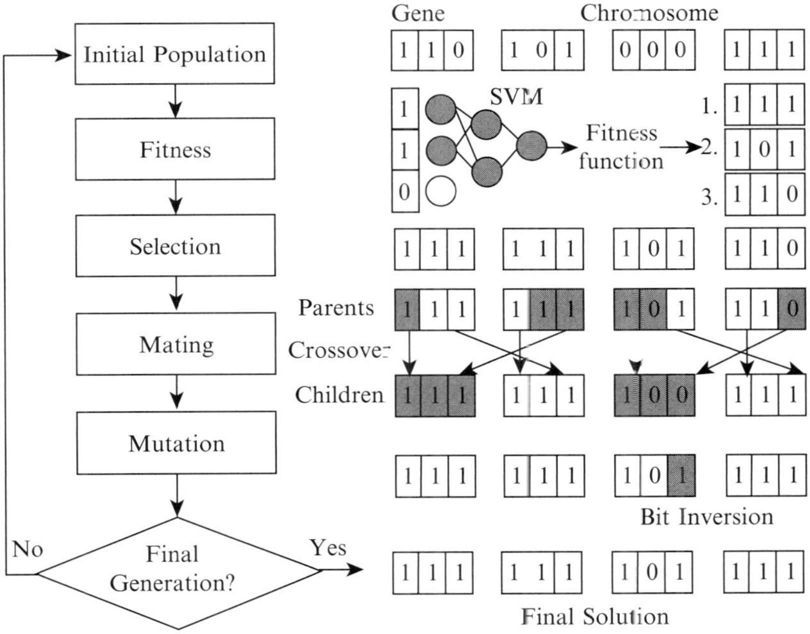

Fig. 9.2. The variable selection with GA for SVM modeling

9.2.3 A Hybrid GASVM Model

Generally, SVM itself cannot reduce the input variable. When the input space dimension is rather large, the problem-solving process with SVM models will require too much time. It is therefore necessary for SVM to preprocess the input feature variables. In this study, GA is first used for feature selection of SVM input variables in order to reduce the model complexity, and then the SVM is used to predict foreign exchange market movement direction based on the selected features. Thus, a hybrid intelligent approach integrating GA and SVM can be formulated for data mining purpose. Fig. 9.3 illustrates the basic architecture of the hybrid GASVM intelligent data mining approach.

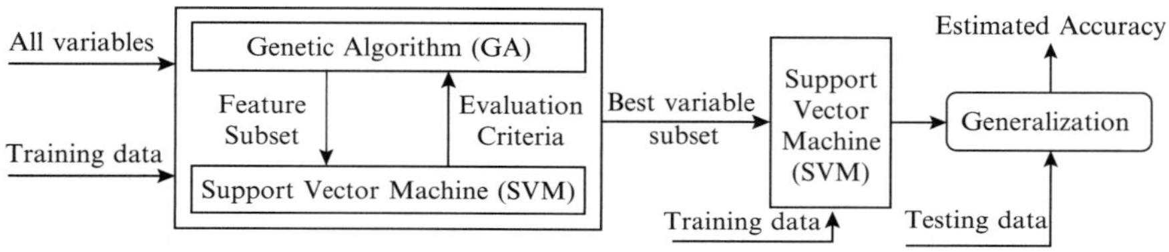

Fig. 9.3. The basic architecture of the hybrid GASVM model

As can be seen from Fig. 9.3, the hybrid GASVM intelligent data mining approach comprises two main phases: input variable selection and SVM generalization. In the first phase, the GA searches the exponential space of feature variable subsets and passes one subset of features to a SVM model. The SVM extracts predictive information from each subset and learns the patterns. Once a SVM learns the data patterns, the trained SVM is evaluated on a data set not used for training, and returns two evaluation criteria, $E_{accuracy}$ and $E_{complexity}$ to the GA. The two evaluation criteria are then combined into one fitness function. Based on the fitness value, the GA biases its search direction to maximize the combined objective. This routine continues for a fixed number of generalizations. Among all the evolved models over the generations, we select the best model in terms of fitness function value. It is worth noting that in the training and evaluation procedures, the SVM only uses the selected feature variables.

In the second phase, using the selected best feature variable subset, we can train the SVM with all the training data. The trained SVM model is then used to classify the unknown patterns (i.e., the records in the testing dataset). Based on the output results, we can evaluate the model's accuracy.

To verify the effectiveness of the proposed hybrid intelligent data mining approach, a typical foreign exchange rate index, trade-weighted US dollar against broad index currencies that circulate widely outside the country of issue, is used as a testing target. The simulation experiments are presented in the following section.

9.3 Empirical Study

In this section, we first describe the data used in this study, then present several comparable model, and finally the experimental results and corresponding analyses are reported.

9.3.1 Research Data

In this study, one typical foreign exchange rate index, trade-weighted US dollar against broad index currencies that circulate widely outside the country of issue, is used for testing purpose. The historical data are monthly and are obtained from the Wharton Research Data Service (WRDS), provided by Wharton School of the University of Pennsylvania. The entire data set covers the period from January 1973 to December 2005 with a total of 396 observations. The data sets are divided into two periods: the first period covers January 1973 to December 1997 with 300 observations, while the second period is from January 1998 to December 2005 with 96 observations. The first period, which is assigned to in-sample estimation, is used for network learning, i.e., training set. The second period, which is reserved for out-of-sample evaluation, is used for validation, i.e., testing set. For space limitation, the original data are not listed in this paper, and detailed data can be obtained from the WRDS.

The main aim of this study is to mine and explore the tendency of foreign exchange index movement. They are categorized as "1" and "-1" in the research data. "1" means that the next month's index is higher than this month's index, and "-1" means that the next month's index is lower than this month's index. Since we attempt to mine the foreign exchange index movement tendency, technical indicators and some factors affecting foreign exchange market price are used as input variables. In this study, we select 19 technical indicators and 4 macro variables to make up the initial attributes or features, as determined by the review of domain experts and prior studies (Kim, 2003; Huang et al., 2006). The descriptions of initially selected attributes or features are presented in Table 9.1. Note that the data of technical indicators can be obtained through computation with corresponding computational formula shown in Table 9.1 and the data of macro variables can also be obtained directly from the WRDS.

Table 9.1. Initially selected feature indicators and their formulae

ID	Feature description	Computational Formula
1	Price (P)	$x_t, (t = 1, 2, \ldots, n)$
2	Stochastic oscillator (SO)	$((x_t - x_l(m))/(x_h(m) - x_l(m)))$
3	Moving stochastic oscillator (MSO)	$\frac{1}{m}\sum_{i=t-m+1}^{t}(SO_{t-i})$
4	Slow stochastic oscillator (SSO)	$\frac{1}{m}\sum_{i=t-m+1}^{t}(MSO_{t-i})$
5	Rate of change (ROC)	x_t/x_{t-m}
6	Momentum (M)	x_t-x_{t-m}
7	Moving average (MA)	$\frac{1}{m}\sum_{i=t-m+1}^{t}x_i$
8	Moving variance (MV)	$\frac{1}{m}\sum_{i=t-m+1}^{t}(x_i - \bar{x}_t)^2$
9	Moving variance ratio (MVR)	MV_t^2 / MV_{t-m}^2
10	Exponential moving average (EMA)	$a \times x_t + (1-a) \times x_{t-m}$
11	Moving average convergence & divergence (MACD)	$\sum_{i=t-m+1}^{t} EMA_{20}(i) - \sum_{i=t-m+1}^{t} EMA_{40}(i)$
12	Accumulation/distribution oscillator (ADO)	$((x_l(m) - x_t)/(x_h(m) - x_l(m)))$
13	Disparity5 (D5)	x_t / MA_5
14	Disparity10 (D10)	x_t / MA_{10}
15	Price oscillator (OSCP)	$(MA_5 - MA_{10}) / MA_5$
16	Commodity channel index (CCI)	$(M_t - SM_t)/0.015D_t$ where $M_t = x_h(t) + x(t) + x_l(t)$, $SM_t = \sum_{i=t-m+i}^{t} M_i / m$, $D_t = \sum_{i=t-m+i}^{t} \lvert M_i - SM_t \rvert / m$.
17	Relative strength index (RSI)	$100 - \frac{100}{1 + RS}$ where $RS = \frac{\sum_{i=t-m+1}^{t}(x(i) - x(i-1))^{+}}{\sum_{i=t-m+1}^{t}(x(i) - x(i-1))^{1}}$
18	Linear regression line (LRL)	$\frac{m \times \sum_{i=t-m+1}^{t} i \times x(i) - \sum_{i=t-m+1}^{t} i \times \sum_{i=t-m+1}^{t} x(i)}{m \times \sum_{i=t-m+1}^{t} i^2 - \left(\sum_{i=t-m+1}^{t} i\right)^2}$
19	Monthly excess return (R)	$R_t = \ln(P_t / P_{t-1}) - r_{t-1}$, where r_t is risk-free interest rate
20	One-month T-bill (T)	N.A.
21	Consumer price index (CPI)	N.A.
22	Gross domestic product (GDP)	N.A.
23	Monthly trading volume (Q)	N.A.

9.3.2 Descriptions of Other Comparable Forecasting Models

In order to evaluate the forecasting ability of the proposed hybrid GASVM model, we compare its performance with those of conventional methods, such as statistical and time series model, and neural network model, as well as individual SVM model without GA preprocessing. Typically, we select random walk (RW) model, auto-regressive integrated moving average (ARIMA) model, linear discriminant analysis (LDA) model, individual back-propagation neural network (BPNN) model and individual SVM model with full feature variables as the benchmarks. We do not compare our approach to a standard logit regression model because a logit regression model is a special case of single BPNN model with one hidden node.

For RW and ARIMA models, only the foreign exchange index price P is used. RW is a one-step-ahead forecasting model, since it uses the current actual value to predict the future value as follows:

$$\hat{y}_{t+1} = y_t \tag{9.14}$$

where y_t is the actual value in current period t and $\hat{y}_{t+1}$ is the predicted value in the next period.

In an ARIMA model (Box and Jenkins, 1970), the future value of a variable is assumed to be a linear function of several past observations and random errors. That is, the underlying process that generates the time series takes the form

$$\phi(B)y_t = \theta(B)e_t \tag{9.15}$$

where y_t and e_t are the actual value and random error at time t respectively; B denotes the backward shift operator, i.e., $By_t = y_{t-1}$, $B^2y_t = y_{t-2}$ and so on; and $\phi(B) = 1 - \phi_1 B - \cdots - \phi_p B^p$, $\theta(B) = 1 - \theta_1 B - \cdots - \theta_q B^q$, where p, q are integers and often referred to as orders of the model. Random errors, e_t, are assumed to be independently and identically distributed with a mean of zero and a constant variance of σ^2, i.e., $e_t \sim IID(0, \sigma^2)$. If the dth difference of $\{y_t\}$ is an ARIMA process of order p and q, then y_t is called an ARIMA (p, d, q) process.

LDA (Huang et al., 2005) can handle the case in which the within-class frequencies are unequal and its performance has been examined on randomly generated test data. This method maximizes the ratio of between-class variance to the within-class variance in any particular data set, thereby guaranteeing maximal separability. Usually, a LDA model with d-dimension inputs takes the form

$$\hat{y} = a_0 + \sum_{i=1}^{d} a_i x_i \tag{9.16}$$

where a_0 is the intercept, x_i are various factors affecting foreign exchange price movement, and a_i are the coefficients of related factors. Each of the ARIMA and LDA models is estimated by in-sample data. The model selection process is then followed by using an empirical evaluation, e.g., RMSE, which is based on the out-of-sample data.

The BPNN model is widely used and produces successful learning and generalization results in various research areas. Usually, a BPNN can be trained by the historical data. The model parameters (connection weights and node biases) will be adjusted iteratively by a process of minimizing the forecasting errors. For prediction purposes, the final computational form of the BPNN model is as

$$\hat{y} = a_0 + \sum_{j=1}^{q} w_j f(a_j + \sum_{i=1}^{p} w_{ij} x_i) + \xi_t \tag{9.17}$$

where a_j (j = 0, 1, 2, …, q) is a bias on the jth unit, w_{ij} (i = 1, 2, …, p; j = 1, 2, …, q) is the connection weight between layers of the model, x_i (i = 1, 2,…, p) are the input variable factors, $f(\bullet)$ is the transfer function of the hidden layer, p is the number of input nodes and q is the number of hidden nodes. In our study, the BPNN has 23 input nodes because 23 input variables are employed. By trial and error, we set the number of training epochs is 500 and heuristically determine the number of hidden nodes using the formula ($2\times node_{in}\pm 1$) where $node_{in}$ represents the number of input nodes. The learning rate is 0.25 and the momentum factor is 0.35. The hidden nodes use sigmoid transfer function and the output node uses the linear transfer function.

For individual SVM model with all input features, the SVM model has also 23 input variables, the Gaussian radial basis function are used as the kernel function of SVM. In SVM model, there are two parameters, i.e., upper bound C and kernel parameter σ, to tune. By trial and error, the kernel parameter σ is 15 and the upper bound C is 60. More details about SVM, please refer to Section 2.1 and Vapnik (1995).

9.3.3 Experiment Results

In this subsection, we first show how to determine the key determinants of affecting exchange rate index movement using the hybrid GASVM approach integrating GA with SVM. Then we compare the performance of the hybrid GASVM model with some comparable forecasting models.

9.3.3.1 The analysis of GA-based feature selection

In order to show the robustness of the GA-based feature selection method and determine which features are the key driver of foreign exchange price movement, we use k-fold cross-validation estimation procedure to perform eight independent experiments. In the k-fold cross-validation estimation procedure, the training data is divided into k non-overlapping groups. We train a SVM using the first (k-1) groups of training data and test the trained SVM on the kth group. We repeat this procedure until each of the groups is used as a test set once. We then take the average of the performance measurements over the k folds. In our experiment, k is set to 2. This is a reasonable compromise considering the computational complexity of systems. Furthermore, an estimate from twofold cross validation is likely to be more reliable than an estimate from a common practice using a single testing set.

In the eight independent experiments, different training data partitions are performed. In this study, we randomly select i%(i = 20, 30, ..., 90) training data to perform the feature selection. Here we assume that the SVM training with less than 20% training data is inadequate for SVM learning. This testing process includes three steps. First of all, we randomly select some input variable using GA and pass them to SVM. Second, the dataset is randomly separated two parts, training samples and evaluation samples, in terms of i%, respectively. Third, we select i% training data to train SVM model and then test the trained SVM with (100–i)% evaluation data in terms of twofold cross-validation procedure. Accordingly, the selected features using GASVM model are shown in Table 9.2.

Table 9.2. The selected features by GASVM model

Data partition (i%)	Selected feature ID	Data partition (i%)	Selected feature ID
20	1, 7, 11, 17, 19, 20	60	1, 4, 11, 19, 20
30	1, 5, 11, 19, 20, 21	70	8, 11, 19, 20, 21
40	11, 16, 19, 20, 21	80	1, 7, 15, 19, 20
50	1, 5, 13, 19, 20	90	1, 11, 19, 20, 23

From Table 9.2, it is easy to see which features are key determinants of foreign exchange price movement. For each data partition, different predictive features are clearly highlighted. We partially attribute this finding to the strong correlation among price movement related features. However, note that one feature (19, monthly excess return) is selected by all data partitions. This makes considerable sense because the monthly excess return is significantly related to future price movement from this analysis. Therefore the investment decision-makers will add more weight to this feature for future predictions.

In Table 9.2, other features, such as One-month T-bill (20), MACD (11) and price (1), are also considered to be some key driver of foreign exchange price movement because they appear many times in eight experiments. In terms of these key determinants, we can reduce data collection and storage requirements and thus reducing labor and data transmission costs. Generally, the results obtained are consistent with previous studies, such as Campbell et al. (1993), Hiemstra and Jones (1994), Yu (1999), Kanas (2001), Kanas and Yannopoulos (2001).

In terms of data reduction, most data partitions choose at most six features except one that selects eight features at data partition i = 20%. On average, the GASVM method selects five features. This reduces the data dimensionality by (23-5)/23 ≈ 78.26%. This implies that we can reduce data collection, computation and storage costs considerably.

Although we only select several typical features to predict foreign exchange price movement tendency, the prediction performance of the model is still promising, as illustrated in Fig. 9.4.

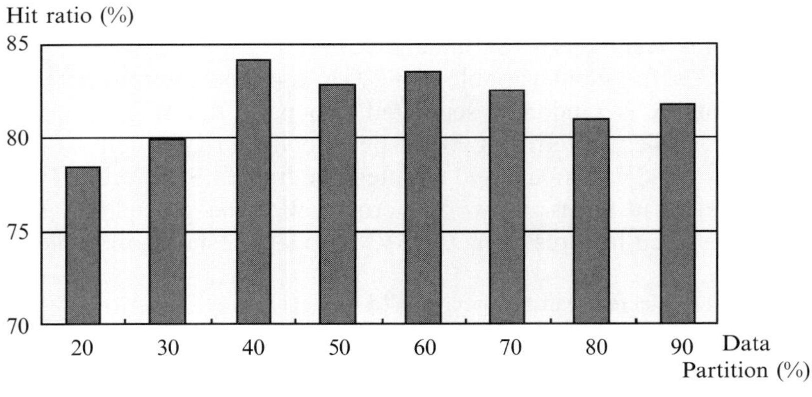

Fig. 9.4. The hit ratio of GA-SVM model with different data partitions

As can be seen form Fig. 9.4, the GA-based SVM model is rather robust. The hit ratios of almost all data partition are above 80% except that the data partitions are 20% and 30%, implying that the GASVM model provides a feasible solution to foreign exchange market tendency exploration. However, in order to measure the prediction power of the proposed hybrid intelligent model, the GASVM model is required to further compare with other forecasting model, which is performed in the following subsection.

9.3.3.2 Comparisons with other forecasting models

Each of the models described in the previous sections is estimated and validated by in-sample data. The model estimation selection process is then followed by an empirical evaluation based on the out-of-sample data. At this stage, the prediction performance of the models is measured by hit ratio. Fig. 9.5 illustrates the experimental results.

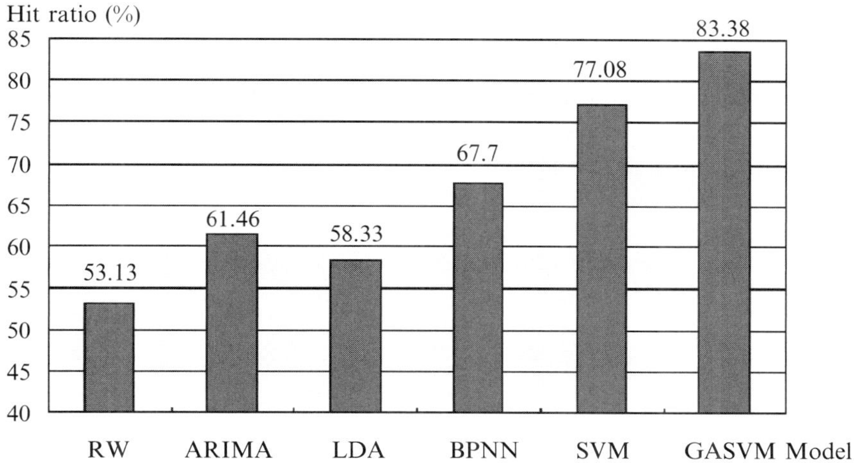

Fig. 9.5. The performance comparison of different models

From Fig. 9.5, we can see the following several findings. First of all, the differences between the different models are very significant. For example, the hit ratio for the RW model is only 53.13%, for the ARIMA model it is 61.46%, and for the BPNN model the hit ratio is only 67.70%; while for the proposed GASVM forecasting model, the hit ratio reaches 83.33%, which is significantly higher than the individual SVM, implying that the GA-based variable selection has a significant impact on SVM model in forecasting foreign exchange market tendency.

Second, comparing three conventional statistical and time series models (i.e., RW, ARIMA and LDA) with three intelligent mining models (i.e., BPNN, SVM and GASVM), it is clear that the intelligent mining models consistently outperform the conventional models. The main reasons include two aspects. On the one hand, the intelligent mining models can easily capture the nonlinear patterns in the foreign exchange market because they adopt nonlinear mapping functions, while the three conventional models are all linear models. On the other hand, in the intelligent mining models, more factors may be included. In our study, RW and ARIMA only use historical time series. That is, they only use one foreign exchange price factor.

Third, in the three conventional models, we can find that the performance of the ARIMA model is significantly better than that of the RW and LDA model. The possible reason is needed to further address in the future. In the three intelligent models, the GASVM is much better than the BPNN and the individual SVM. The main reason is that the GA-based variable selection procedure reduces the model input space and thus increases model interpretability and effectiveness.

In addition, the proposed GASVM model can have some comparative advantages relative to individual SVM and BPNN. First of all, there is less parameter to tune for GASVM than for BPNN. Second, the GASVM can overcome some shortcomings of BPNN, such as overfitting, local minima and input variable selection problem. Third, the input space of the GASVM is smaller, and the learning speed of GASVM is faster than the individual SVM model. Furthermore, the GASVM model with small input features has better generalization performance than individual SVM model.

9.4 Comparisons of Three Hybrid Neural Network Models

As earlier noted, the hybrid neural networks usually achieve better prediction performance than individual ANN model (Yu et al., 2005c). In this part (i.e., Part IV), three hybrid neural network models are proposed. Table 9.3 summarizes the comparisons of the three hybrid neural network models.

Table 9.3. Comparisons of the three hybrid neural network models

Chapter	Chapter 7	Chapter 8	Chapter 9
Model used	ANN, ES	ANN, GLAR	GA, ANN
Prediction range	Short-term	Short or medium-term	Medium or long-term
Prediction mode	Level prediction	Level prediction	Direction prediction

From Table 9.3, we can conclude that (1) in the three hybrid models, exponential smoothing (ES) model, generalized linear auto-regression (GLAR) model and genetic algorithm (GA) are hybridized into ANN models to formulate some synergetic model for foreign exchange rates forecasting. (2) Of the three hybrid models, the former two hybrid models proposed in Chapters 7 and 8 are level prediction models, which are often used for short- or medium-term prediction, while the hybrid model in

Chapter 9 is direction prediction model, which is often used for medium- or long-term prediction and foreign exchange market tendency exploration. According to these characteristics, we can select different hybrid models for foreign exchange rates forecasting.

9.5 Conclusions

This chapter proposes using a hybrid intelligent data mining model integrating GA and SVM to predict foreign exchange market tendency. In this hybrid GASVM approach, a standard genetic algorithm is used to search for possible feature combination and a support vector machine is used to predict the foreign exchange market tendency with selected feature variables. One of the distinct strengths of the hybrid intelligent GASVM model is its ability to build an interpretable prediction model because smaller numbers of features are used. Particularly, we show that our proposed hybrid intelligent approach not only maximizes the hit ratio but also selects a most parsimonious model.

To evaluate the predictability of the proposed hybrid GASVM model, we compare its performance with several parametric methods (e.g., statistical models) and nonparametric methods (e.g., neural network models). In terms of the empirical results, we find that across different forecasting models for the test cases of trade-weighted US dollar index on the basis of same criteria, the proposed GASVM model performs the best, indicating that the hybrid intelligent data mining approach can be used as an effective solution to foreign exchange market tendency exploration.

Part V: Neural Network Ensemble for Foreign Exchange Rates Forecasting

10 Forecasting Foreign Exchange Rates with a Multistage Neural Network Ensemble Model

10.1 Introduction

Some studies revealed that foreign exchange market is one of the most volatile markets. Due to its high volatility, foreign exchange rates forecasting is regarded as a rather challenging task (Yu et al., 2005c). For traditional statistical methods, it is hard to capture the volatility. In the last decades, many emerging artificial intelligent techniques, such as artificial neural networks (ANN), were widely used in foreign exchange rates forecasting and obtained good prediction performance.

However, neural networks are a kind of unstable learning techniques, i.e., small changes in the training set and/or parameter selection can produce large changes in the predicted output. Many experiments have shown that the generalization of single neural network is not unique, that is, the neural network's solutions are not stable. Even for some simple problems, different structures of neural networks (e.g., different number of hidden layers, different number of hidden nodes and different initial conditions) result in different patterns of network generalization. In addition, even the most powerful neural network model still cannot cope well when dealing with complex data sets containing some random errors or insufficient training data. Thus, the performance for these data sets may not be as good as expected (Naftaly et al., 1997; Carney and Cunningham, 2000).

Instability of the single neural network has hampered the development of better neural network models. Why can the same training data applied to different neural network models or the same neural models with different initialization lead to different performance? What are the major factors affecting this difference? Through the analysis of error distributions, it has been found that the ways neural networks have of getting to the global minima vary, and some networks just settle into local minima instead of global minima. In any case, it is hard to justify which neural network's error reaches the global minima if the error rate is not zero. Since the number of neural models and their potential initialization is unlimited, the possible number of results generated for any training data set applied to those models is theoretical infinite. The best performance we get is typically only the best one selected from a limited number of neural networks, i.e., the single model with the best generalization to a testing set. One interesting point is that, in a prediction case, other less accurate predictors may generate a more accurate forecast that the most accurate predictor. Thus it is clear that simply selecting the best predictors according to their performance is not the optimal choice. More and more researchers have realized that merely selecting the predictor that gives the best performance on the testing set will lead to losses of potentially valuable information contained by other less successful predictors. The limitation of individual neural network suggested a different approach to solving these problems that considers those discarded neural networks whose performance is less accurate as potential candidates for new network models: a neural network ensemble model (Hansen and Salamon, 1990).

Initially, the motivation for neural network ensemble forecasting procedure is based upon the intuitive idea that by combining the outputs of several individual predictors one might improve on the performance of a single generic one (Krogh and Vedelsby, 1995). However, this idea has been proved to be true only when the ensemble members are simultaneously accurate and diverse enough, which requires an adequate trade-off between these two conflicting conditions (Sharkey, 1996). As Tumer and Ghosh (1996a) pointed out, the increase in accuracy depend on error-independence far more than on the particular ensemble method used. Accordingly, many ensemble methods, including linear methods, such as simple averaging (Hansen and Salamon, 1990; Breiman, 1994) and weighted averaging (Breiman, 1996b, Benediktsson, et al., 1997; Tresp and Taniguchi, 1995), and nonlinear methods (Huang et al., 1995, Yu et al., 2005c), are proposed. Meanwhile, some experimental results reported in the literature have shown that the forecasting accuracy provided by the combination of a set of neural networks can outperform the accuracy of the best single network.

Although neural network ensemble research has gained considerable successes, most existing work has focused on the research of ensemble methods and pattern classification problems (Hansen and Salamon, 1990; Hashem, 1997; Tumer and Ghosh, 1996b; Ueda, 2000; Sharkey and Sharkey, 1997; Tumer and Ghosh, 1995; Kittler et al., 1998; Huang and Suen, 1995). A full neural network ensemble forecasting procedure for time series forecasting has not yet been formulated. In such situations, we aim to construct a multistage neural network ensemble model for remedying this gap. Some methods used to formulate multistage neural network ensemble forecasting procedure are provided, which consists of the following steps: preprocessing time series data; generating a collection of individual neural predictors by varying training data or network type; selecting ensemble members from a set of individual neural predictors; and combining selected members into an aggregated predictors. At the same time, some related issues in every phase are also addressed.

The rest of this chapter is organized as follows. The next section explains the reasons of motivating neural network ensemble for prediction problem. In Section 10.3, we describe the building process of the multistage neural network ensemble forecasting model in detail. For further illustration, two foreign exchange rate series are used for testing in Section 10.4. Finally, some concluding remarks are drawn in Section 10.5.

10.2 Motivations for Neural Network Ensemble Model

Recently, some experiments have been proved that neural network ensemble forecasting model is an effective approach to the development of a high performance forecasting system (Bishop, 1995; Yu et al., 2005c). In this section, we mainly interpret the reasons why do we need using neural network ensemble model for prediction problems. That is, what is the benefit of neural network ensemble for prediction problems? Assume that a task is to learn a function $f : R^N \to R^M$ for which you have a sample of p examples (x^u, y^u) for a specific prediction problem, where $y^u = f(x^u)$ and $u = 1,2,\ldots, p$. These examples are assumed to be drawn randomly from the distribution $p(x)$. The neural network ensemble predictor comprises n individual neural predictors and the output of the ith individual network on input x is called $f_i(x)$. Then a weighted average ensemble forecasting model can be represented in the following form:

$$\hat{f}(x) = \sum\nolimits_{i=1}^{n} w_i f_i(x) \qquad (10.1)$$

where $\hat{f}(x)$ is the output of the ensemble predictor, w_i is the ith network's weight, satisfying

$$0 \le w_i \le 1 \qquad (10.2)$$

$$\sum\nolimits_{i=1}^{n} w_i = 1 \qquad (10.3)$$

Now suppose the desired output of x is $d(x)=\{d_1(x),d_2(x),\ldots,d_n(x)\}$, then the generalization error $e_i(x)$ of the ith individual neural network on input x

and the generalization error $\hat{e}(x)$ of the ensemble predictor on x are respectively:

$$e_i(x) = f_i(x) - d_i(x) \tag{10.4}$$

$$\hat{e}(x) = \hat{f}(x) - d(x) \tag{10.5}$$

From Equation (10.5), the generalization error $\hat{e}(x)$ of ensemble predictor can be also denoted as

$$\begin{aligned} \hat{e}(x) &= \hat{f}(x) - d(x) \\ &= (f^{(1)}(x) - d^{(1)}(x), f^{(2)}(x) - d^{(2)}(x), \cdots, f^{(m)}(x) - d^{(m)}(x)) \end{aligned} \tag{10.6}$$

where the jth component $\hat{f}^{(j)}(x)(j = 1, 2, \cdots, m)$ of the ensemble predictor is defined by a weighted sample mean of generalization error by n individual neural predictor as

$$\begin{aligned} \hat{e}^{(j)}(x) &= \hat{f}^{(j)}(x) - d^{(j)}(x) \\ &= \sum_{i=1}^{n} w_i f_i^{(j)}(x) - (w_1 + w_2 + \cdots + w_n) \cdot d_i^{(j)}(x) \\ &= \sum_{i=1}^{n} w_i [f_i^{(j)}(x) - d_i^{(j)}(x)] \\ &= \sum_{i=1}^{n} w_i e_i^{(j)}(x) = \frac{\sum_{i=1}^{n} w_i e_i^{(j)}(x)}{\sum_{i=1}^{n} w_i} \end{aligned} \tag{10.7}$$

Consider that each neural network is learned by a set of training samples and assume that each neural predictor in the ith neural network generalization error $e_i(x)$ is independent of others. Then the jth component of $e_i^{(j)}(x)$ produces random error $T_e = \{e_i^{(j)}(1), \cdots, e_i^{(j)}(T)\}$, where the mean of T_e becomes zero since the expectation $E[f_i^{(j)}(x)] = E[d_i^{(j)}(x)]$ and the variance of T_e can be known as $(\sigma_i^{(j)})^2$ by measurement. Here we can make all the neural network's variances equal such that $(\sigma_i^{(j)})^2 = (\sigma^{(j)})^2$ for $i = 1, 2, \ldots, n$ by an appropriate choice of weight constants $\{w_i \| i = 1, \ldots, n\}$. Thus, according to the central limit theorem (Scheaffer and McClave, 1990), the jth individual $e^{(j)}(x)$ (j=1,2, …, m) of the ensemble predictor's error has approximately a normal distribution $N(m(e^{(j)}), \sigma^2(e^{(j)}))$ whose mean and variance are calculated as

$$m(e^{(j)}) = \sum_{i=1}^{n} w_i \cdot m(e_i^{(j)}) = 0 \tag{10.8}$$

$$\sigma^2(e^{(j)}) = \sum_{i=1}^{n} w_i^2 \cdot \sigma^2(e_i^{(j)}) + 2\sum_{k=1}^{n}\sum_{l=k-1}^{n} Cov(e_k^{(j)}, e_l^{(j)}) = \sum_{i=1}^{n} w_i^2 \cdot (\sigma_i^{(j)})^2 \tag{10.9}$$

where $Cov(e_k^{(j)}, e_l^{(j)})$ represents the covariance of the jth generalization error between the kth neural network and the lth neural network, and its value becomes zero since each network in the ensemble committee is made independently and executes its desired task individually. The approximation to normal distribution is improved, as the number of neural network models in the ensemble is increased. From the above equation, it is noted that the variance of the generalization error of the ensemble predictor can be smaller than that of the individual neural predictor by $(1/n)$ when $w_i = 1/n$ for $i = 1, 2, \ldots, n$. Therefore, the advantage or benefit of using a neural ensemble predictor is that it can reduce the variance of the generalized error and increase the neural network's generalization ability.

Motivating by this benefit of neural network ensemble, in the following we propose a multistage neural network ensemble forecasting model.

10.3 Formulation of Neural Network Ensemble Model

In this section, we first describe the procedure of our proposed multistage neural ensemble predictor. Subsequently we review and propose some design methods for every phase in the process of formulating the neural network ensemble model in detail. In addition, some related key issues in every phase are also addressed in this section.

10.3.1 Framework of Multistage Neural Ensemble Model

In order to improve the performance of foreign exchange rates forecasting, we propose a multistage neural network ensemble procedure for foreign exchange rates forecasting, as illustrated in Fig. 10.1.

From Fig. 10.1, it is easy to see that the proposed procedure consists of four stages, i.e., (1) preprocessing original data; (2) generating different neural predictors; (3) selecting the appropriate number of single neural predictors from the considerable number of candidate predictors generated by the previous phase; and (4) combining selected neural predictors into an aggregated output.

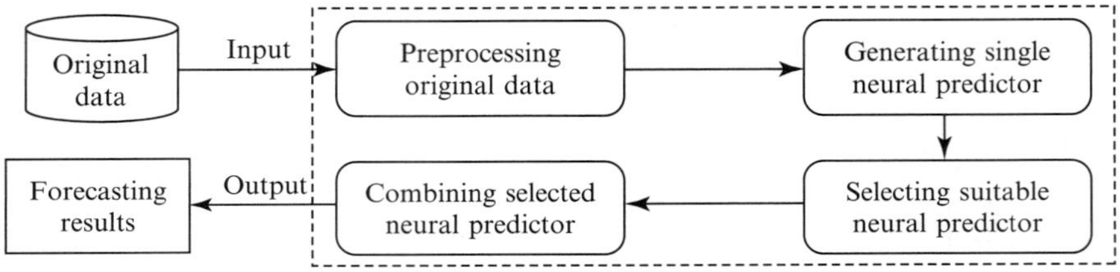

Fig. 10.1. The framework of neural network ensemble procedure

In the four phases, data preprocessing includes data reconciliation and data transformation. Individual neural predictors with weak error-correlation can be generated by varying the network type and network architecture, and using different initial conditions and different training data. In the individual neural predictors, some representative neural predictors are selected as ensemble members in terms of error-independence measurement. Finally, these selected members are combined into an aggregated predictor in a nonlinear way. Subsequently, we review and present some detailed approaches in formulating neural network ensemble predictors.

10.3.2 Preprocessing Original Data

As earlier mentioned, data preprocessing includes two tasks: data reconciliation and data transformation.

10.3.2.1 Data reconciliation

In neural network learning, using data with different scales often leads to the instability of neural networks (Weigend and Gershenfeld, 1994). At the very least, data must be scaled into the range used by the input neurons in the neural network. This is typically –1 to 1 or zero to 1 (Lou, 1993). Many commercially available generic neural network development programs, such as *BrainMaker* (http://www.calsci.com), automatically scale each input. Moreover, neural networks always require that the data range is neither too small nor too large, so that the computer's precision limits are not exceeded. Otherwise, the data should be scaled. Furthermore, data reconciliation helps to improve the performance of neural networks (Lou, 1993). Therefore, data reconciliation is necessary for treating multi-scale data. Note that we propose using the sigmoidal function reconciliation approach in this chapter.

A sigmoidal function can be used as a data reconciliation method, depending on the characteristics of the data. Usually, a sigmoidal function is represented by

$$I(x) = \frac{r_i}{1 + \exp[-p_i(x_i - q_i)]}, i = 1,2,\cdots,n. \tag{10.10}$$

where r_i is used to constrain the range for the ith transformed element of the n-element data set, q_i can be selected as the smallest in the ith element of data set, and p_i decides the sharpness of the transferred function. Also, Equation (10.10) can compress the abnormal data to a specific range. Note that different continuous and differentiable transformation functions can also be selected.

It is advisable to scale the data so that the different input signals have approximately the same numerical range. This is not necessary for feed-forward and Hopfield networks, but is recommended for all network models. The reason is that the other network models rely on Euclidean measures, so that un-scaled data could bias or interfere with the training process. Reconciling the data so that all inputs have the same range often speeds up the training and improves the resulting performance of the derived model.

10.3.2.2 Data transformation

This operation mainly aims at time series data. The main aim of data transformation is to transform time series data into a training matrix so as to satisfy the basic condition of neural network learning. The main reason for this is that reconciled data is still a string of simple numbers for time series, but neural networks are nonlinear dynamic systems with multiple inputs and outputs in general, and many parameters must be specified beforehand. Therefore, the question of how to transform a string of pure numbers into an input matrix vector or how to determine the size and structure of a training set has been one of the key problems in neural network time series forecasting. In existing literature, some studies (Azoff, 1994; Moody, 1995) presented simple transformation methods, for example, the moving window method (Moody, 1995). However, these methods will be trivial if large quantities of data are used. In order to overcome the drawbacks of previous methods, we therefore propose a data transformation method with interval variation to create a neural network training matrix.

For a univariate time series with a given size of training set, we can use the data transformation method with interval variation to obtain a training set matrix provided that an interval value is specified. Supposing that the size of the source data $\boldsymbol{D}$ is M, the size of new training data set is N, and sampling interval is K, we have the following algorithm:

(i) Input the source data.

(ii) The start point and end point of training, p_1 and p_2, are determined in terms of the pre-defined size of training data $N < M$: the lag period of every sub-sample is defined as L.

(iii) Formulate the training matrix according to the following program:

```
For i = 1 to N
    D_t (i) = D (p1 to p1+L-1)
    For j = 1 to (L/K) step -1
        N (j, i) = D_t (L, i)
        L = L-K
    Next j
        p1 = p1+1
Next i
```

(iv) Output the final training data sets generated by the previous basic algorithm: $N(j, i)$.

As an example, assume that we have 150 observations in time series data, i.e., $D = \{x_1, x_2, \ldots, x_{150}\}$, we assume the following parameters: the size of training set is 100, lag period of every training sample is 13, sample interval is 4, start point of training data is 1 and end point of training data is 100. Then with the above algorithm, the training data sets are:

$$\text{Training sets} = \begin{bmatrix} x_1 & x_2 & x_3 & \cdots & x_{88} \\ x_5 & x_6 & x_7 & \cdots & x_{92} \\ x_9 & x_{10} & x_{11} & \cdots & x_{96} \\ x_{13} & x_{14} & x_{15} & \cdots & x_{100} \end{bmatrix}$$

where every column in the matrix represents a training sample group. From the matrix, we find that there are 88 training groups from 100 observations. Thus, our proposed data transformation method with interval variation can effectively formulate a training matrix for network learning purposes. The advantages of such an approach are that: (1) it can overcome some drawbacks of the traditional moving window method; and (2) it can construct an effective training matrix even for little data only if a set of suitable parameters is specified.

10.3.3 Generating Individual Neural Predictors

10.3.3.1 The generation of individual neural predictors

With the work about bias-variance trade-off (Breiman, 1999; Yu et al., 2006a, 2006e), an ensemble model consisting of diverse models with much disagreement is more likely to have a good prediction performance. Therefore, how to generate the diverse model is a crucial factor. Usually, there are two different types for generating individual neural predictors: homogeneous models using the same network type, and heterogeneous models using the different network type.

For **homogeneous** neural network model generation, several methods have been investigated for the generation of ensemble members making different errors (Sharkey, 1996). Such methods basically rely on varying the parameters related to the design and to the training of neural network. In particular, the main methods in the literature include the following:

(i) Different network architecture: by changing the number of hidden layers and the number of nodes in every layer, different neural networks with different architectures can be created.

(ii) Different training data: by re-sampling and preprocessing time series data, we can obtain different training sets, thus making different network generations. There are six techniques that can be used to obtain diverse training data sets: bagging (Breiman, 1996a), noise injection (Raviv and Intrator, 1996), cross-validation (Krogh and Vedelsby, 1995), stacking (Wolpert, 1992; Breiman, 1996b), boosting (Schapire, 1990) and input decimation (Tumer and Ghosh, 1996a).

(iii) Different learning algorithm: by selecting different core learning algorithms, different neural networks can also be generated. For example, a multi-layer feed-forward network can use the steep-descent algorithm or Levenberg-Marquardt algorithm or other learning algorithms.

(iv) Different initial conditions: ensemble members can be created by varying the initial random weights, learning rate and momentum rate, from which each network is trained.

For **heterogeneous** neural network model generation, we can create neural network ensemble members by using different network types. For example, multi-layer perceptrons (MLPs), back-propagation networks (BPNs), radial basis function (RBF) neural networks, and probabilistic neural networks (PNNs) can be used to create the ensemble members. In addition, neural ensemble members could be created using a hybridization of two or more of the above methods, e.g., different network types plus different training data (Sharkey, 1996). In our study we adopt such a hybridization method to create ensemble members. Once some individual neural predictors are created, we need to select some representative members for ensemble purpose.

10.3.3.2 Key issue of this stage

In this phase, a key issue is whether homogeneous models or heterogeneous models should be created. That is, which is better of the two types of model? Here we use the bias-variance decomposition approach to explore this question.

Our approach is based on the observation that the generation error of a ensemble model could be improved if the predictors on which averaging is done disagree and their fluctuations are uncorrelated (Krogh, 1997). To see this, the generalization error of the ensemble predictor is decomposed into bias and variance in terms of Geman et al. (1992). For convenience of discussion, we still use the same denotations or symbols as Section 10.2 for consistency. The difference is that we use the squared error function as normal error function for further analysis in this section. According to German et al. (1992) and Equation (10.1), the bias/variance decomposition of the generalization error of ensemble predictor $\hat{e}(x)$ is

$$\begin{aligned}\hat{e}(x) &= (E[\hat{f}(x)] - E[d(x)])^2 + E[\hat{f}(x) - E[\hat{f}(x)]]^2 \\ &= (Bias(\hat{f}(x)))^2 + Variance(\hat{f}(x))\end{aligned} \tag{10.11}$$

where E represents the expectation and $E(d(x))$ is the deterministic term of the data because $d(x)$ is desired value. Now we can observe the effects concerning bias and variance from Equation (10.11).

From Equation (10.11), it is not difficult to find that the bias term is just the average of the bias of the individual models in the ensemble. Thus we should not expect a reduction in the bias term compared with single models. On the other hand, the variance term of the ensemble could be further decomposed as follows:

$$Variance(\hat{f}(x)) = E[(\hat{f}(x) - E(f(x)))^2] = E[(\sum_{i=1}^{n} w_i f_i(x))^2] - (E[\sum_{i=1}^{n} w_i f_i(x)])^2$$

$$= \sum_{i=1}^{n} w_i^2 (E[f_i^2(x)] - E^2[f_i(x)]) + 2\sum_{i<j} w_i w_j (E[f_i(x) f_j(x)] - E[f_i(x)]E[f_j(x)]) \tag{10.12}$$

In Equation (10.12), the first sum marks the bound of the ensemble variance and is the weighted mean of the variances of the ensemble members. The second sum contains the cross terms of ensemble members and disappears if the models are completely uncorrelated (Krogh and Sollich, 1997).

That shows that the reduction in the variance of the ensemble is related to the degree of independence of the single models (Naftaly et al., 1997). That is, if we introduce diverse models into the ensemble, the variance terms will be reduced greatly.

From our previous analysis, we find that homogeneous models can easily generate similar correlation errors although some different conditions are used. Thus it is hard to reduce the total generalization error by lowering bias using homogeneous models. On the contrary, heterogeneous models can generate some error-independence models, and thus the total generalization error will be reduced by using heterogeneous models. This leads to an interesting conclusion: in the ensemble forecasts, heterogeneous models with error-independence (i.e., diverse models) are often used to lower generalization error. Of course homogeneous models are also used to reduce the error, but the degree of error reduction is smaller than with heterogeneous models.

Based on the above analysis, in the case of neural ensemble forecasts, model diversity can be realized in two ways: one is to utilize different types of neural network; the other is to use different architecture, different training algorithms or different training subsets (Krogh and Vedelsby, 1995; Perrone and Cooper, 1993). For fixed architectures it is sufficient to use different initial conditions for training (Naftaly et al., 1997).

10.3.4 Selecting Appropriate Ensemble Members

10.3.4.1 Conventional methods for ensemble member selection

After training, each member of ensemble predictor has generated its own result. However, if there are a great number of individual members, we need to select a subset of representatives in order to improve ensemble efficiency. For heterogeneous neural models, we need to select a few typical representatives of the same type by way of some measurements and then combine selected models with different types. For homogeneous models, we select some models with error weak correlation. As earlier mentioned, we tend to use heterogeneous models, but for every different type of model we often generate a set of models with different initial conditions. Thus, in a certain model class with same type, we select a representative model for ensemble purposes. In the literature, three methods – principal component analysis (PCA) (Yu et al., 2005c), choose the best (CTB) (Partridge and Yates, 1996), and choose from subspace (CFS) (Partridge and Yates, 1996) – are used. The PCA is used to select a collection of members from candidate members using the maximizing eigenvalue of the error matrix. The CTB assumes a given size of the final ensemble E^*; then we select the networks with the highest forecasting performance from combined network members to formulate the ensemble E^*. The CFS is based on the idea that for each network type, it chooses the network exhibiting the best performance. It should be noted that the term "subspace" refers to the subset of network models related to a given network type. Here we propose a new approach (minimizing conditional generalized variance (CGV) method) to select the error-independent network model.

Supposed that there are p predictors with n forecast values. Then the error matrix $(e_1, e_2, \ldots, e_p)$ of p predictors is as

$$E = \begin{bmatrix} e_{11} & e_{12} & \cdots & e_{1p} \\ e_{21} & e_{22} & \cdots & e_{2p} \\ \vdots & \vdots & & \vdots \\ e_{n1} & e_{n2} & \cdots & e_{np} \end{bmatrix}_{n \times p}$$

From the matrix, the mean, variance and covariance of E can be calculated as

$$\text{Mean: } \bar{e}_i = \frac{1}{n}\sum_{k=1}^{n} e_{ki} \quad (i = 1, 2, \ldots, p) \tag{10.13}$$

$$\text{Variance: } V_{ii} = \frac{1}{n}\sum_{k=1}^{n} (e_{ki} - \bar{e}_i)^2 \quad (i = 1, 2, \ldots, p) \tag{10.14}$$

$$\text{Covariance: } V_{ij} = \frac{1}{n}\sum_{k=1}^{n} (e_{ki} - \bar{e}_i)(e_{kj} - \bar{e}_j) \quad (i,j = 1, 2, \ldots, p) \tag{10.15}$$

Considering Equations (10.14) and (10.15), we can obtain a variance-covariance matrix as follows:

$$V_{p \times p} = (V_{ij}) \tag{10.16}$$

Here we use the determinant of V, i.e., $|V|$ to represent the correlation among the p predictors. When p is equal to one, $|V| = |V_{11}| =$ the variance of e_1 (the first predictor). When p is larger than one, $|V|$ can be considered to be the generalization of variance: therefore, we call $|V|$ the generalized variance. Clearly when the p predictors are correlated, the generalization variance $|V|$ is equal to zero. On the other hand, when the p predictors are independent, the generalization variance $|V|$ reaches its maximum. Therefore, when the p predictors are neither independent nor correlated, the measurement of generalized variance $|V|$ reflects the correlation among the p predictors.

Now we introduce the concept of conditional generalized variance. The matrix V can be reformulated with the block matrix. The detailed process is as follows: $(e_1, e_2, \ldots, e_p)$ is divided into two parts: $(e_1, e_2, \ldots, e_{p1})$ and $(e_{p1+1}, e_{p1+2}, \ldots, e_p)$, denoted as $e_{(1)}$ and $e_{(2)}$, i.e.,

$$E = \begin{bmatrix} e_1 \\ e_2 \\ \vdots \\ e_p \end{bmatrix} = \begin{pmatrix} e_{(1)} \\ e_{(2)} \end{pmatrix}^{p_1 \times 1}_{p_2 \times 1}, p_1 + p_2 = p \qquad (10.17)$$

$$V = \begin{pmatrix} V_{11} & V_{12} \\ V_{21} & V_{22} \end{pmatrix}^{p_1}_{p_2} \quad \begin{matrix} p_1 & p_2 \end{matrix} \qquad (10.18)$$

where V_{11}, V_{22} represent the covariance matrix of $e_{(1)}$ and $e_{(2)}$.

Given $e_{(1)}$, the conditional generalized variance of $e_{(2)}$, $V(e_{(2)}|e_{(1)})$, can be represented as

$$V\left(e_{(2)}\middle|e_{(1)}\right) = V_{22} - V_{21}V_{11}^{-1}V_{12} \qquad (10.19)$$

The above equation shows the change in $e_{(2)}$ given that $e_{(1)}$ is known. If $e_{(2)}$ has a small change under $e_{(1)}$, then the predictors $e_{(2)}$ can be deleted. This implies that the predictors $e_{(1)}$ can obtain all the information that the predictors $e_{(2)}$ reflect. Now we can give an algorithm for minimizing the conditional generalized variance as follows:

(i) Considering that the p predictors, the errors can be divided into two parts: $(e_1, e_2, \ldots, e_{p-1})$ is seen as $e_{(1)}$, (e_p) is seen as $e_{(2)}$.

(ii) The conditional generalized variance $V(e_{(2)}|e_{(1)})$ can be calculated according to Equation (10.19). It is should be noted that here $V(e_{(2)}|e_{(1)})$ is a value, denoted as t_p.

(iii) Similarly, for the ith predictor (i =1, 2, …, p), we can use (e_i) as $e_{(2)}$, and other p-1 predictors are seen as $e_{(1)}$; then we can calculate conditional generalized variance of the ith predictor, t_i, with Equation (10.19).

(iv) For a pre-specified threshold θ, if $t_i < \theta$, then the ith predictor should be deleted from the p predictors. On the contrary, if $t_i > \theta$, then the ith predictor should be retained.

(v) For the retained predictors, we can perform the previous procedures (i)-(iv) iteratively until satisfactory results are obtained.

10.3.4.2 Key issue of this stage

From the previous stages, we know that neural network ensemble forecasts are effective only if the networks forming them make different errors. However, in this stage a key question is whether one should combine all the forecasts available? In other words, does the quantity of ensemble members satisfy the principle of "the more, the better"? In the following we answer the question through following analysis in terms of work of Zhou et al. (2002).

For convenience of discussion, we still use the same denotations or symbols of previous sections for consistency. Here the squared error function is used as the error function. In connection with the previous analysis, the generalization error $e_i(x)$ of the ith individual neural predictor and the generalization error $\hat{e}(x)$ of the ensemble predictor on input x are respectively:

$$e_i(x) = (f_i(x) - d_i(x))^2 \tag{10.20}$$

$$\hat{e}(x) = (\hat{f}(x) - d(x))^2 \tag{10.21}$$

Then the generalization error of the ith individual neural predictor and that of the ensemble predictor, i.e., e_i and $\hat{e}$, on the error distribution $p(x)$, are respectively:

$$e_i = \int e_i(x) p(x) dx \tag{10.22}$$

$$\hat{e} = \int \hat{e}(x) p(x) dx \tag{10.23}$$

Now we use Equation (10.15) to define the correlation between the ith and jth individual neural predictors based on the error distribution $p(x)$, then the correlation V_{ij} is:

$$V_{ij} = \int (f_i(x) - d_i(x))(f_j(x) - d_j(x)) p(x) dx \tag{10.24}$$

It is clear that V_{ij} satisfies the following Equations (10.25) and (10.26):

$$V_{ii} = e_i \tag{10.25}$$

$$V_{ij} = V_{ji} \tag{10.26}$$

Considering Equations (10.1) and (10.21) we can obtain the following:

$$\hat{e}(x) = \left(\sum\nolimits_{i=1}^{n} (w_i f_i(x) - d_i(x))\right)\left(\sum\nolimits_{j=1}^{n} (w_j f_j(x) - d_j(x))\right) \tag{10.27}$$

Then combining Equations (10.23), (10.24) and (10.27), we can get

$$\hat{e} = \sum\nolimits_{i=1}^{n} \sum\nolimits_{j=1}^{n} w_i w_j V_{ij} \tag{10.28}$$

For brevity, here we assume that all the individual neural predictors have equal weights, i.e., $w_i = 1/n$ ($i = 1, 2, \ldots, n$). In other words, the individual predictors are combined via simple averaging. Then Equation (10.28) becomes:

$$\hat{e} = \sum\nolimits_{i=1}^{n} \sum\nolimits_{j=1}^{n} V_{ij} \big/ n^2 \tag{10.29}$$

Suppose that the kth individual neural predictor is excluded from the ensemble members. Then the generalization error of the new ensemble predictors is:

$$\hat{e}' = \sum\nolimits_{\substack{i=1 \\ i \neq k}}^{n} \sum\nolimits_{\substack{j=1 \\ j \neq k}}^{n} V_{ij} \big/ (n-1)^2 \tag{10.30}$$

From Equations (10.29) and (10.30) we derive that if Equation (10.31) is satisfied then $\hat{e}$ is not smaller than $\hat{e}'$, which means that the ensemble predictor excluding the kth individual neural predictor is better than the one including the kth individual neural predictor.

$$\hat{e} \leq \left(2\sum\nolimits_{\substack{i=1 \\ i \neq k}}^{n} V_{ik} + e_k\right) \Big/ (2n-1) \tag{10.31}$$

Integrating Equation (10.29) with Equation (10.31), we can obtain the following constraint on the kth individual neural predictor that should be excluded from the ensemble predictor:

$$(2n-1)\sum\nolimits_{i=1}^{n} \sum\nolimits_{j=1}^{n} V_{ij} \leq 2n^2 \sum\nolimits_{\substack{i=1 \\ i \neq k}}^{n} V_{ik} + n^2 \cdot e_k \tag{10.32}$$

It is not hard to see that there are cases where Equation (10.32) is satisfied. For an extreme case, when all the individual neural predictors are the duplication of the same neural predictor, Equation (10.32) indicates that the size of the ensemble predictor can be reduced without sacrificing generalization ability (Zhou et al., 2002). This also leads to an interesting conclusion: for a large number of neural predictors available, the performance when combining many of them may be better than that of combining all of them. That is, the more does not necessarily mean the better. For a subsequent problem, how many individual forecasts should be combined? Please refer to the above approaches, such as PCA (Yu et al., 2005c), CTB (Partridge and Yates, 1996), CFS (Partridge and Yates, 1996) and CGV, which is proposed in this chapter for more details.

10.3.5 Ensembling the Selected Members

10.3.5.1 Existing ensemble strategies

Depended upon the work done in previous phases, a collection of appropriate ensemble members can be collected. The subsequent task is to combine these selected members into an aggregated predictor in an appropriate ensemble strategy (i.e., how to combine the selected members into an aggregate model for prediction purposes). Generally, there are two ensemble strategies: linear ensemble and nonlinear ensemble strategies.

A. Linear ensemble strategy

Typically, linear ensemble strategies include two approaches: the simple averaging (Tumer and Ghosh, 1995; Lincoln and Skrzypek, 1990) approach and the weighted averaging (Burges, 1998) approach. There are three types of weighted averaging: the simple mean squared error (MSE) approach (Benediktsson et al., 1997), stacked regression (modified MSE) approach (Breiman, 1996a) and variance-based weighted approach (Tresp and Taniguchi, 1995).

Simple averaging is one of the most frequently used ensemble approaches. After selecting the members of the ensemble, the final prediction can be obtained by averaging the sum of each forecaster's prediction of ensemble members. Some experiments (Hansen and Salamon, 1990; Breiman, 1994) have shown that simple averaging is an effective approach to improve neural network performance. It is more useful when the local minima of ensemble members are different, i.e., when the local minima of ensemble networks are different. Different local minima mean that ensemble members are diverse. Thus averaging can reduce the ensemble variance. However, this approach treats each member equally, i.e., it does not stress ensemble members that can make more contribution to the final generalization. If the variances of ensemble networks are very different, we do not expect to obtain a better result using simple averaging (Ueda, 2000).

Weighted averaging is where the final ensemble prediction result is calculated based upon individual members' performances with a weight attached to each individual member's prediction. The gross weight is one and each member of a ensemble is entitled to a portion of this gross weight according to their performance or diversity. There are three methods used to calculate weights: the simple MSE approach (Benediktsson et al., 1997), stacked regression approach (Breiman, 1996a) and variance-based weighted approach (Tresp and Taniguchi, 1995).

The simple MSE method estimates the linear weight parameter w_i in Equation (10.1) by minimizing the squared error (Benediktsson et al., 1997), that is, for $i = 1, 2, \ldots, n$,

$$\begin{aligned} w_{opt,i} &= \arg\min_{w_i}\left\{\sum_{j=1}^{m}\left(w_i^T f_i(x_j) - d_{ji}(x_j)\right)^2\right\} \\ &= \left(\sum_{j=1}^{m} f_i(x_j) f_i^T(x_j)\right)^{-1} \sum_{j=1}^{m} d_{ji}(x) f_i(x_j) \end{aligned} \tag{10.33}$$

The MSE solution seems to be reasonable, but, as Breiman (1996a) has pointed out, this approach has two serious problems in practice: a) the data are used both in the training of each predictor and in the estimation of w_i, and b) individual predictors are often strongly correlated since they try to predict the same task. Due to these problems, this approach's generalization ability will be poor.

The stacked regression method was proposed by Breiman (1996a) in order to solve the problems associated with the simple MSE method. Thus, the stacked regression method is also called the modified MSE method. This approach utilizes cross-validation data to modify the simple MSE solution, i.e.,

$$w_{opt,i} = \arg\min_{w_i}\left\{\sum_{j=1}^{m}\left(w_i^T g_i(x_j) - d_{ji}(x_j)\right)^2\right\}, \quad i = 1,2,\ldots,n \tag{10.34}$$

where $g_i(x_j) = \left(f_i^{(1)}(x_j; D_{cv}), \cdots, f_i^{(m)}(x_j; D_{cv})\right)^T \in \Re^M$ is a cross-validated version $f_i(x_j) \in \Re^M$ and D_{cv} is the cross-validated data.

Although this approach overcomes the limitations of the simple MSE method, the solution is based on the assumption that the error distribution of each validated set is normal (Ueda, 2000). In practice, however, this normal assumption does not always hold and thus this approach does not generally lead to the optimal solution in the Bayes sense(Ueda, 2000).

The variance-based weighting approach estimates the weight parameter w_i by minimizing error variance (Perrone and Cooper, 1993); all predictors are error-independent networks, that is, for $i = 1, 2, \ldots, n$,

$$w_{opt,i} = \arg\min_{w_i}\left\{\sum_{i=1}^{n}\left(w_i \sigma_i^2\right)^2\right\} \tag{10.35}$$

under the constraints $\sum_{i=1}^{n} w_i = 1$ and $w_i \geq 0$. Using the Lagrange multiplier, the optimal weights are:

$$w_{opt,i} = \frac{\left(\sigma_i^2\right)^{-1}}{\sum_{j=1}^{n}\left(\sigma_j^2\right)^{-1}}, \quad (\mathrm{i}=1,2,\cdots,\mathrm{n}) \tag{10.36}$$

The variance-based weighting method is based on the assumption of error independence. Moreover, as earlier mentioned, individual predictors are often strongly correlated for the same task. This indicates that this approach has serious drawbacks for minimizing error-variance when neural predictors with strong correlation are included within the ensemble members.

B. Nonlinear ensemble strategy

The nonlinear ensemble method is a promising approach for determining the optimal weight of neural ensemble predictor. The literature only mentions one nonlinear ensemble approach: the neural network-based nonlinear ensemble method (Huang et al., 1995; Yu et al., 2005c). This approach uses "meta" neural networks for ensemble purposes (Lai et al., 2006a). For further details, readers are referred to (Huang et al., 1995; Yu et al., 2005c; Lai et al., 2006a). Experiment results obtained show that the neural network-based nonlinear ensemble approach consistently outperforms the other ensemble approach. However, there are several shortcomings to the neural network-based nonlinear ensemble approach. First, a neural network often traps into local minima due to its local optimal algorithm; second, a neural network can easily exhibit the over-fitting problem because of too many learning times or too many training examples; third, neural network architecture and type heavily depend upon the user's experience. To avoid these negative effects, a new nonlinear model, the support vector machine regression (SVMR) model (Yu et al., 2006d; Lai et al., 2006c), is proposed for neural network ensemble in this study.

The support vector machine (SVM) (Vapnik, 1995; Burges, 1998; Smola, 1998) is an elegant tool for solving pattern recognition and regression problems. Over the past few years, it has attracted much attention of many researchers from the neural network community. Here we focus on the SVM regression (SVMR) algorithm to solve neural network ensemble problem in a nonlinear form.

Simply speaking, in an SVMR model, one has to estimate the functional dependence of the dependent variable y on a set of independent variables x. The model assumes, like other regression problems, that the relationship between the independent and dependent variables is given by a deterministic function f plus some additive noise:

$$y = f(x) + \varepsilon \tag{10.37}$$

The task is then to find a functional form for f that can correctly predict new cases that the SVMR has not been presented with before. This can be achieved by training the SVMR model on a sample set, i.e., training set – a process that involves the sequential optimization of an error function. It is interesting that the structure of SVMR model is similar to that of neural networks, as illustrated in Fig.10.2. In the same way, the SVMR also consists of an input layer, middle layer or transformation layer and output layer. The difference is that every node output of the middle layer is a support vector transformed by the kernel function $k(x, \hat{x})$ (Yu et al., 2006d; Lai et al., 2006c). Usually, the Gaussian function is used as a kernel function. Note that the SVMR could overcome the important drawbacks of the neural network (Vapnik, 1995).

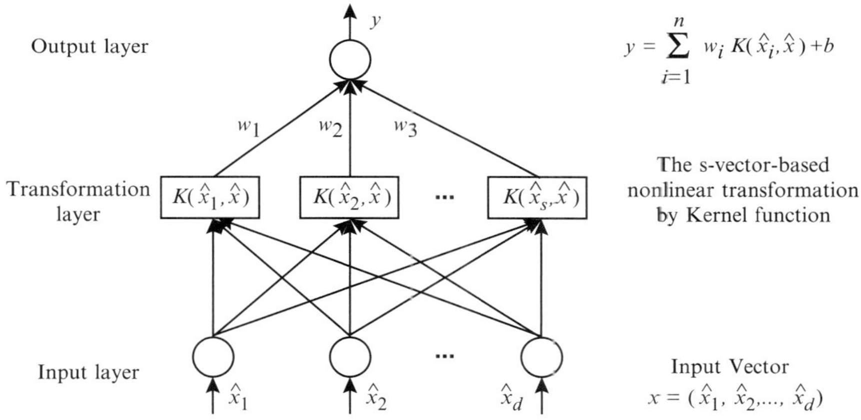

Fig. 10.2. The structure of SVMR-based neural network ensemble model

An SVMR-based nonlinear ensemble forecasting model can be viewed as a nonlinear information processing system that can be represented as:

$$\hat{y} = f(\hat{x}_1, \hat{x}_2, \cdots, \hat{x}_d) \tag{10.38}$$

where $(\hat{x}_1, \hat{x}_2, \cdots, \hat{x}_d)$ is the output of individual neural network predictors, $\hat{y}$ is the aggregated output, $f(\cdot)$ is a nonlinear function determined by SVMR. In this sense, SVMR-based ensemble is a nonlinear ensemble method (Yu et al., 2006d; Lai et al., 2006c).

10.4 Empirical Analysis

10.4.1 Experimental Data and Evaluation Criterion

In this study, two exchange rate series are presented to illustrate the proposed nonlinear ensemble forecasting model. The data set used for our experiment consists of two exchange rates: the euro against US dollar (EUR/USD for short) and the Japanese yen against US dollar (JPY/USD for short). The data used in this study are daily and are obtained from Pacific Exchange Rate Service (http://fx.sauder.ubc.ca/). The entire data set covers the period from January 1, 1991 to December 31, 2002. We take daily data from January 1, 1991 to December 31, 2000 as the in-sample data sets (including training set (January 1, 1991 to December 31, 1998) and validation set (January 1, 1999 to December 31, 2000)) and take the data from January 1, 2001 to December 31, 2002 as the out-of-sample data set (i.e., testing set), which are used to evaluate the prediction performance. In order to save space, the original data are not listed here, and detailed data can be obtained from the website or from the authors.

In order to compare the prediction performance, it is necessary to introduce the forecasting evaluation criterion. It is known that one of the most important forecasting evaluation criteria is the root mean square error (RMSE). Its computational form can be defined as:

$$RMSE = \sqrt{\frac{1}{N}\sum\nolimits_{t=1}^{N}(x_t - \hat{x}_t)^2} \tag{10.39}$$

where x_t represent the actual values, $\hat{x}_t$ are predicted values, and N is the number of evaluation periods. Here we use the RMSE to measure the forecasting performance of different models.

10.4.2 Experiment Design

The motivations of our experiments were mainly: (i) to evaluate the effective of the proposed ensemble forecasting approach; and (ii) to compare our proposed approach with other design approaches proposed in the literature.

Considering the first aim, we need to create different initial ensembles. For heterogeneous models, ensemble members were created using different neural network types, namely, multiplayer perceptrons (MLPs), back-propagation networks (BPNs) and nradial basis function (RBF) networks.

For homogeneous models (e.g., individual BPNs), ensemble members were generated by varying the network architecture, initial random weights, training data, etc. For simplicity, we report the results related to six initial ensembles, here referred to as ensembles E_1, E_2, …, E_6 created by the following stages:

(a) Ensemble E_1, E_2 and E_3 comprised 30 MLPs, BPNs and RBFs respectively. Three architectures with one or two hidden layers and various numbers of neurons per layer were used in the generation phase. For each architecture, 10 training processes with different training data and initial weights were performed. All the networks had six input nodes and one output node corresponding to the data feature and forecasting purpose respectively.

(b) Ensemble E_4 consisted of 20 MLPs and 10 RBF networks. Two different architectures with two hidden layers and a different number of neurons per layer were to create these MLPs (6-10-10-1 and 6-8-12-1). Ten RBF networks were created with 10 different trials of the K-means clustering algorithm used to define the network architecture.

(c) Ensemble E_5 comprised 10 BPNs and 20 RBFs. Ten different training processes with different initial random weights and training data were performed to create 10 BPNs. Twenty RBF networks were generated with 10 different training data and 10 different K-means clustering algorithms.

(d) Ensemble E_6 consisted of 10 MLPs (setting is same to E_4), 10 BPNs (setting is same to E_5) and 10 RBF networks (setting is same to E_4).

With regard to the second experimental aim, we compare our proposed approach with another approach. Accordingly, the comparison of the selection phase and ensemble phase are carried out. In particular, there are four selection strategies in selection phase: our proposed minimizing the conditional generalized variance (CGV for short), PCA (Yu et al., 2005c), "choose the best (Partridge and Yates, 1996)" (CTB for short), and "choose from subspace (Partridge and Yates, 1996)" (CFS for short). There are also six ensemble strategies used for comparative purposes in the ensemble phase: the simple averaging method (SAM), simple MSE method (SMSE), stacked regression method (SRM), variance-based weighting (VBW) method, artificial neural network-based ensemble method (ANN), and our proposed SVMR-based nonlinear ensemble (SVMR) method.

10.4.3 Experiment Results and Comparisons

10.4.3.1 Experiments results of ensemble E_1, E_2 and E_3

The main aim of these three experiments was to evaluate the effectiveness of our proposed approach in the design of neural network ensemble predictors. It is worth noting that this is a difficult task, since networks of the same type are poorly independent in terms of Partridge's results (Partridge, 1996). Our proposed approach selected ensemble E_1* made up of eight MLPs, E_2* made up of seven BPNs, and E_3* made up of 11 RBF networks, belonging to different network architectures. Table 10.1 shows the performance of the different neural network ensembles selected using our proposed approach. The performance was evaluated by *RMSE*. Note that all *RMSE* values reported in Table 10.1 are referred to the testing set.

Table 10.1. Performance for different neural network ensembles E_1 –E_3

Data Series	Ensemble	Selection Strategy	Ensemble Strategy					
			SAM	SMSE	SRM	VBW	ANN	SVMR
EUR/USD	E_1*	CGV	0.0115	0.0123	0.0098	0.0089	0.0078	0.0071
		PCA	0.0125	0.0156	0.0117	0.0101	0.0092	0.0085
		CTB	0.0118	0.0145	0.0109	0.0099	0.0089	0.0081
		CFS	0.0121	0.0133	0.0101	0.0091	0.0084	0.0077
	E_2*	CGV	0.0105	0.0112	0.0101	0.0089	0.0081	0.0067
		PCA	0.0125	0.0133	0.0121	0.0106	0.0098	0.0081
		CTB	0.0118	0.0128	0.0118	0.0098	0.0091	0.0075
		CFS	0.0111	0.0119	0.0112	0.0091	0.0085	0.0071
	E_3*	CGV	0.0088	0.0075	0.0071	0.0068	0.0054	0.0019
		PCA	0.0103	0.0096	0.0098	0.0077	0.0068	0.0054
		CTB	0.0093	0.0091	0.0085	0.0075	0.0061	0.0039
		CFS	0.0101	0.0085	0.0089	0.0072	0.0059	0.0028
JPY/USD	E_1*	CGV	0.0205	0.0224	0.0213	0.0189	0.0156	0.0144
		PCA	0.0263	0.0255	0.0274	0.0256	0.0182	0.0165
		CTB	0.0237	0.0242	0.0264	0.0245	0.0178	0.0161
		CFS	0.0219	0.0239	0.0248	0.0228	0.0166	0.0155
	E_2*	CGV	0.0221	0.0208	0.0196	0.0177	0.0145	0.0123
		PCA	0.0274	0.0256	0.0228	0.0209	0.0187	0.0165
		CTB	0.0261	0.0251	0.0211	0.0198	0.0189	0.0158
		CFS	0.0253	0.0244	0.0202	0.0188	0.0172	0.0140
	E_3*	CGV	0.0211	0.0221	0.0209	0.0201	0.0186	0.0158
		PCA	0.0254	0.0265	0.0258	0.0269	0.0209	0.0187
		CTB	0.0232	0.0251	0.0247	0.0254	0.0189	0.0175
		CFS	0.0219	0.0237	0.0216	0.0232	0.0191	0.0166

As can be seen from Table 10.1 we can conclude the following two conclusions.

Firs of all, by fixing a certain ensemble strategy, our proposed approach (i.e., CGV) is slight better, but the gap is small, indicating that the ensemble with the same type network does not contain error-independent networks that can be selected by a design method to improve performances.

Second, by fixing a certain selection strategy, our proposed ensemble strategy (SVMR-based nonlinear ensemble method) outperforms other ensemble strategies described in the literature, implying that our proposed ensemble approach is one of the effective and promising ensemble methods in foreign exchange rates forecasting.

10.4.3.2 Experiment results of ensemble E_4 and E_5

These two experiments were aimed to evaluate to what extent our design method can exploit error-independence to improve the performance of a set of "weak" neural networks (E_4) and "strong" neural networks (E_5). Therefore, ensemble E_4 consists of 20 MLPs and 10 RBFs whose performances were good (e.g., $RMSE$<0.02). In addition, introduction of the RBF aims to increase the independence of ensemble members. In two experiments, our proposed selection strategy extracts four MLPs (characterized by two different architectures and two different initial weights) and two RBF networks with two different kernels to formulate an optimal ensemble predictor E_4* from initial E_4. Similarly, two BPNs with different architectures and four RBF networks with four different kernels are extracted from E_5 to form another optimal ensemble predictor E_5*. Table 10.2 shows the performance of the different ensembles for the EUR/USD and the JPY/USD series. Similarly, all the reported results are referred to the testing data sets.

From Tables 10.1-10.2, five conclusions can be summarized.

(1) the performances of heterogeneous model ensemble are generally better than those of homogeneous model ensemble, comparing Table 10.1 and Table 10.2.

(2) The "strong" network ensemble is slightly better, but the difference is small, relative to the "weak" network ensemble, as shown in Table 10.2;

(3) The results show that our proposed selection strategy and ensemble strategy consistently outperform other strategies when the performance is as a measurement.

Table 10.2. Performance for different neural network ensembles E_4 and E_5

Data Series	Ensemble	Selection Strategy	Ensemble Strategy					
			SAM	SMSE	SRM	VBW	ANN	SVMR
EUR/ USD	E_4*	CGV	0.0081	0.0071	0.0068	0.0065	0.0056	0.0021
		PCA	0.0095	0.0088	0.0085	0.0081	0.0069	0.0048
		CTB	0.0089	0.0084	0.0078	0.0075	0.0065	0.0041
		CFS	0.0085	0.0077	0.0072	0.0077	0.0058	0.0035
	E_5*	CGV	0.0075	0.0070	0.0061	0.0058	0.0044	0.0017
		PCA	0.0093	0.0086	0.0078	0.0067	0.0058	0.0044
		CTB	0.0081	0.0081	0.0075	0.0062	0.0051	0.0036
		CFS	0.0078	0.0074	0.0069	0.0060	0.0047	0.0028
JPY/ USD	E_4*	CGV	0.0179	0.0156	0.0165	0.0177	0.0154	0.0123
		PCA	0.0199	0.0189	0.0187	0.0201	0.0175	0.0148
		CTB	0.0188	0.0180	0.0172	0.0196	0.0171	0.0133
		CFS	0.0181	0.0176	0.0169	0.0182	0.0165	0.0128
	E_5*	CGV	0.0166	0.0145	0.0162	0.0172	0.0148	0.0119
		PCA	0.0185	0.0174	0.0189	0.0196	0.0169	0.0138
		CTB	0.0179	0.0168	0.0178	0.0185	0.0160	0.0127
		CFS	0.0171	0.0152	0.0171	0.0177	0.0155	0.0121

(4) The performance of ensemble E_4* also performs better than the ensembles with the same network type, comparing Table 10.1 with Table 10.2 (this implies that neural network can also be combined effectively from a set of weak networks, but only by a detailed analysis of error independence).

(5) The performance of the ANN and SVMR approaches are better than that of the conventional linear ensemble approach. This also proves the conclusion of Granger and Ramanathan (1984) that the unconstrained least squares method can be applied to obtain a better forecasting performance than that obtained by using the ordinary least squares.

10.4.3.3 Experiment results of ensemble E_6

The goal of this experiment is basically similar to the previous subsection. Our proposed approach can formulate an effective ensemble E_6* extracting one MLP, one BPN and one RBF network from E_6. Table 10.3 shows the performance of the neural network ensembles. All values refer to the test data sets. It is not hard to see that since our proposed strategies clearly outperform the others, similar conclusions can be drawn as those made in the previous experiments.

Comparing the results in Tables 10.1-10.3, we can summarize the following conclusions.

Table 10.3. Performance for different neural network ensemble E_6

Data Series	Ensemble	Selection Strategy	Ensemble Strategy					
			SAM	SMSE	SRM	VBW	ANN	S VMR
EUR/ USD	E_6*	CGV	0.0063	0.0055	0.0047	0.0051	0.0039	0.0009
		PCA	0.0075	0.0067	0.0058	0.0065	0.0051	0.0028
		CTB	0.0070	0.0061	0.0051	0.0058	0.0048	0.0017
		CFS	0.0066	0.0056	0.0052	0.0057	0.0038	0.0014
JPY/ USD	E_6*	CGV	0.0105	0.0123	0.0133	0.0158	0.0118	0.0098
		PCA	0.0122	0.0158	0.0158	0.0177	0.0125	0.0117
		CTB	0.0118	0.0147	0.0149	0.0162	0.0119	0.0115
		CFS	0.0109	0.0135	0.0140	0.0165	0.0117	0.0111

(1) In the four selection strategies, the CGV is the best, followed by CFS, CTB and PCA, for of all the ensembles E_1*- E_6*.

(2) Of the six ensemble strategies, the nonlinear ensemble strategy is much better than the linear ensemble strategy; furthermore, the SVMR strategy is slightly better than the ANN strategy, indicating that the SVMR strategy is a very promising approach for ensemble forecasts.

(3) In the six optimal ensembles, E_6* shows the best results. The main reason for this is that the degree of error independence of the ensemble members in E_6* is better than other ensembles.

(4) Generally, the performance shown in Tables 10.2 and 10.3 is better than that in Table 10.1, implying that heterogeneous model ensemble performs better than the homogeneous model ensemble.

(5) the performance of the EUR/USD series is better than that of the JPY/USD. One main reason may be that the JPY/USD is more volatile than the EUR/USD in foreign exchange market.

10.5 Conclusions

Combining the prediction of several different neural predictors into an aggregated neural network prediction often gives improved performance over any individual prediction and is considered to be an effective technique for improving the generalization ability of single neural network predictors. In this study, we propose a novel multistage nonlinear ensemble predictor for foreign exchange rates forecasting. The experimental results reported in this paper demonstrate the effectiveness of the proposed design approach. The comparison shows that with our proposed approach it is possible to formulate an effective ensemble predictor by selecting some error-independent networks with minimizing conditional generalized variance algorithm and an SVMR-based nonlinear ensemble approach. Review of the experimental results yields the final conclusion: the proposed multi-stage SVM-based nonlinear ensemble model can be used as an alternative tool for foreign exchange rates forecasting.

11 Neural Networks Meta-Learning for Foreign Exchange Rates Ensemble Forecasting

11.1 Introduction

Artificial neural network (ANN), first introduced by McCulloch and Pitts (1943), is a system derived through neuropsychology models (Hertz, 1989). It attempts to emulate the biological system of the human brain in learning and identifying patterns. Moreover, ANNs can more aptly recognize poorly defined patterns. Instead of extracting explicit rules from sample data, the ANN employs a learning algorithm to automatically: (a) extract the functional relationship between input and output, which is embedded in a set of historical data (called training exemplars or learning samples), and (b) encode it in connection weights. Training exemplars that are readily available allow neural networks to capture a large volume of information in a rather short period of time and to continuously learn throughout their lifespan. Furthermore, neural networks have the ability to not only deal with noisy, incomplete, or previously unseen input patterns, but to also generate a reasonable response (Tsaih et al., 1998). However, ANN is far from being optimal learner. For example, the existing studies (e.g., Breiman (1999)) have found that the ways neural networks have of getting to the global minima vary and some networks just settle into local minima instead of global minima through the analysis of error distributions. In this case, it is hard to justify which neural network's error reaches the global minima if the error rate is not zero. Thus, it is not wise choice that only selecting a single neural network model with the best generalization from a limited number of neural networks if the error is larger than zero. For example, in foreign exchange rates forecasting, it is difficult to obtain a consistently good result by using a single neural network model due to high volatility and irregularity in foreign exchange markets. More and more researchers have realized that just selecting the neural network model that gives the best performance will result in losses of potentially valuable information contained by other neural network models with slightly weak performance relative to the best neural network model. Naturally, a different learning strategy to solving these problems that considers those discarded neural networks whose performance is less accurate as the best neural network model should be proposed. In such situations, a meta-learning strategy (Chan and Stolfo, 1993) based on the neural network is introduced.

Meta-learning (Chan and Stolfo, 1993), which is defined as learning from learned knowledge, provides a promising solution and a novel approach to the above challenges. The idea is to use neural network learning algorithms to extract knowledge from several different data sets and then use the knowledge from these individual learning algorithms to create a unified body of knowledge that well represents the entire data. Therefore meta-learning seeks to compute a meta-model (ensemble model) that integrates in some principled fashion the separately learned models to boost overall predictive accuracy. In this study, a four-stage neural-network-based meta-learning technique is proposed for foreign exchange rates forecasting (Lai et al., 2006a). In the first stage, an interval sampling technique is used to generate different training sets. Based on the different training sets, the different neural network models with different initial conditions are then trained to formulate different base models in the second stage. In the third stage, to improve the efficiency of meta-learning, a decorrelation maximization method is used as a pruning tool to generate an optimal set of base models. In the final stage, a neural-network-based meta-model or neural network ensemble model can be produced by meta-learning strategy from the selected base models (Lai et al., 2006a).

The main motivation of this study is to take full advantage of the inherent learning capability of meta-learning technique and to design a powerful neural network ensemble learning model. The rest of this chapter is organized as follows. Section 11.2 gives a brief introduction of neural network learning paradigm. In Section 11.3, a neural-network-based meta-learning process is provided in detail. To verify the effectiveness of the proposed meta-learning technique, an exchange rate prediction experiment is performed in Section 11.4. Finally, Section 11.5 concludes this chapter.

11.2 Introduction of Neural Network Learning Paradigm

In this section, we provide some basic knowledge about neural network learning technique. Similarly, the popular BPNN is used as a learning paradigm in this chapter. Usually, the back-propagation learning mechanism consists of two phases: training and testing phase.

During the training phase, the input and the output layer of the BPNN are set to represent a training pair (x, y) where x is the feature vector (i.e., attributes) and y is the target value. Commonly, $x \in R^n$ is an n-dimensional feature vector containing the independent variables or attributes, and $y \in \{0, 1\}$ is the dependent variable or truth. The goal is to construct a function or a model f,

$$y = f(x) = f_a(x) = f(x; a),\ a \in A, \tag{11.1}$$

where $f = f_a$ is defined by specifying parameters $a \in A$ from an explicitly parameterized family of models A. There are well-known parameterizations for a wide variety of statistical models, including various regression and classification models and neural network models. In this study, the BPNN learning algorithm (White, 1990) is performed with all training pairs to determine optimal parameters a. The BPNN learning algorithm repeatedly adjusts the link-weight matrices of BPNN in a way that minimizes the error for each training pair. When the average squared error computed over all training pairs is acceptably small, the BPNN learning procedure stops and produces the link-weight matrices, which is stored as the knowledge for the use of testing phase.

During the testing phase, the input layer of the BPNN is activated by the new feature vectors. This activation of the BPNN spreads from the input layer to the output layer using the link-weight matrices stored during the training phase. That is, the model $f = f_a$ determined by training phase is applied to previously unseen feature vectors x (i.e., feature vectors outside the training set) to produce the output of BPNN, $y = f(x)$. We measure success by using a testing set distinct from the training set.

A major advantage of BPNN is their ability to provide flexible mapping between inputs and outputs. The arrangement of the simple units into a multilayer framework produces a map between inputs and outputs that is consistent with any underlying functional relationship regardless of its "true" functional form. Having a general map between the input and output vectors eliminates the need for unjustified priori restrictions that are needed in conventional statistical and econometric modeling. Therefore, a neural network is often viewed as a "universal approximator", i.e., a flexible functional form that can approximate any arbitrary function arbitrarily well, given sufficient middle-layer units and properly adjusted weights (Hornik et al., 1989; White, 1990). Therefore, a three-layer BPNN model is used as a basic learning paradigm in this chapter.

11.3 Neural Network Meta-Learning Process for Ensemble

In this section, we first introduce the basic knowledge of meta-learning. Based on the meta-learning, a generic meta-modeling process with four phases is then provided.

11.3.1 Basic Background of Meta-Learning

As Section 11.1 mentioned, meta-learning (Chan and Stolfo, 1993), which is defined as learning from learned knowledge, is an emerging technique recently developed to construct a meta-model (i.e., ensemble model) that deals with the problem of computing a meta-model from multiple training data sets. Broadly speaking, learning is concerned with finding a model $f = f_a[j]$ from a single training set TR_j, while meta-learning is concerned with finding a meta-model $f = f_a$ from several training sets $\{TR_1, TR_2, \ldots, TR_n\}$, each of which has an associated model $f = f_a[j]$. The n individual models derived from the n training sets may be of the same or different types. Similarly, the meta-model may be of a different type than some or all of the single models. Also, the meta-model may use data from a meta-training set (MT), which are distinct from the data in the individual training set TR_j.

There are two types of meta-learning methods: different-training-set-based meta-learning and different-model-type-based meta-learning. For the first type, we are given a large data set DS and partition it into n different training subsets $\{TR_1, TR_2, \ldots, TR_n\}$. Assume that we build a separate model on each subset independently to produce n base models $\{f_1, f_2, \ldots, f_n\}$ with different parameters. Given a feature vector x, we can produce n scores $(f_1(x), f_2(x), \ldots, f_n(x))$, one for each model. Given a new training set MT, we can build a meta-model f on MT using the data $\{(f_1(x), f_2(x), \ldots, f_n(x), y) : (x, y) \text{ in } MT\}$.

For the second type, given a relative small training set $\{TR\}$, we replicate it n times to produce n training sets $\{TR_1, TR_2, \ldots, TR_n\}$ and create different models f_j on each training set TR_j, for example, by training the replicated data on n different-type models. For simplicity, assume that these models are binary classifiers so that each classifier takes a feature vector and produces a classification in $\{0, 1\}$. We can then produce a meta-model simply by using a majority vote of the n classifiers.

In the foreign exchange rates forecasting, data is abundant. To improve the prediction performance, we use the first type of meta-learning: different-training-set-based meta-learning. Generally, the maim aim of neural network meta-learning is to generate a number of independent models (i.e., base models) by applying neural network learning algorithms to a collection of data sets. The independent models are then selected and combined to obtain a global model or meta-model. Fig. 11.1 illustrates a neural network meta-learning process.

As can be seen from Fig. 11.1, the generic meta-learning process consists of four stages, which can be described as follows.

Stage 1: For an initial data set *DS*, the whole data set is first divided into training set *TR* and testing set *TS*. Then the different training subsets TR_1, TR_2, …, TR_n are created from *TR* with certain sampling algorithm.

Stage 2: For each training subset TR_i (i = 1, 2, …, n), the neural network model f_i (i = 1, 2, …, n) is trained by the specific learning algorithm to formulate n different base models. After training, the testing data was applied to these models to assess their performance.

Stage 3: For n different base models, the pruning techniques are used to generate some effective base models. After pruning, m ($m \leq n$) different base models are generated for the next stage's use.

Stage 4: Using the whole training data set to m different base models, different neural network's results can formulate a meta-training set (*MT*). Based on the meta-training set, another single neural network model is training to produce a meta-model or ensemble model.

From the generic neural network meta-learning process, there are **four main problems** to be further addressed, i.e., (a) how to create n different training subset from the original training data set *TR*; (b) how to create different base models f_i with different training subsets for the same-type model; (c) how to select some effective base model from many neural network base models in the previous stage; and (d) how to how to formulate a meta-model or ensemble model with different results produced by different base models (Lai et al., 2006a).

11.3.2 Data Sampling

For foreign exchange rates prediction task, our aim is to improve the prediction performance with historical time series data. For convenience of neural network learning, the original data set *DS* is first divided into two parts: training data set *TR* and testing data set *TS*. In order to capture the patterns of time series data, different data sampling is helpful for neural network learning. Here we propose an interval sampling algorithm to produce different training subsets from the original training data set.

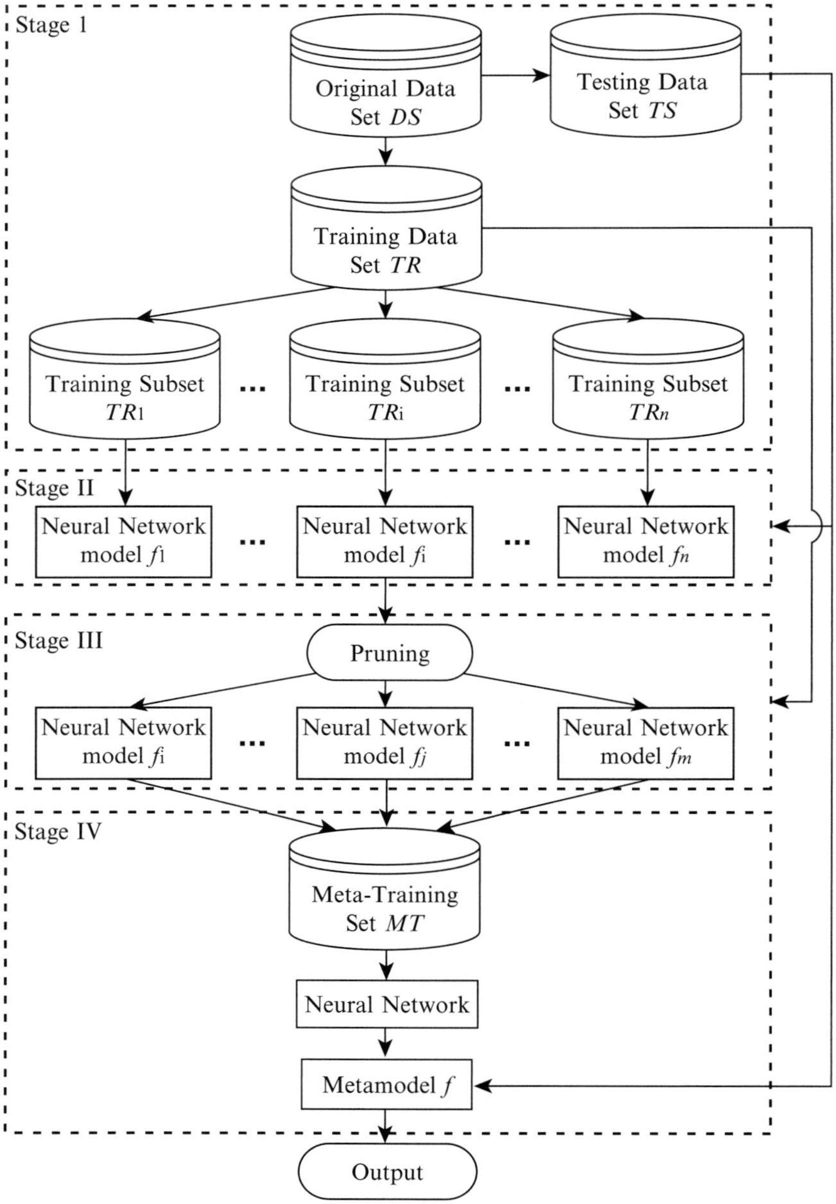

Fig. 11.1. The generic neural network meta-learning process

Given that the size of the original training data set *TR* is *M*, the size of new training data set is *N*, and sampling interval is *K*, we have the following algorithm, as illustrated in Fig. 11.2.

```
Input: original training data set TR
Output: The generated new training subsets
(1) Input the original training data set TR and rank the samples to be a sequence.
(2) The start point and end point of training, P1 and P2 are determined in terms of the
pre-defined size of training data N<M: the lag period of every sub-sample is defined as L.
(3) Formulate the training matrix according to the following program:
    For i = 1 to N
      TRt(i)=TR (p1 to p1+L-1)
      For j = 1 to (L/K)step -1
          N(j,i) = TRt(L, i)
          L = L-K
      Next j
      p1=p1+1
    Next i
(4) Output the final training data sets generated by the previous basic algorithm: N(j,i)
```

Fig. 11.2. The interval sampling algorithm

With the above interval sampling algorithm, we can obtain different training subsets only through varying the sampling interval or starting point of each time series (Lai et al., 2006a).

11.3.3 Individual Neural Network Base Model Creation

With the work about bias-variance trade-off (Breiman, 1999; Yu et al., 2006a, 2006e), an ensemble model (i.e., meta-model here) consisting of diverse models with much disagreement is more likely to have a good performance. Therefore, how to create the diverse base model is the key path to the creation of an effective meta-model. For neural network model, there are several methods for generating diverse models.

(1) Initializing different starting weights for each neural network models for different training subsets.

(2) Varying the architecture of neural network, e.g., changing the different numbers of layers or different numbers of nodes in each layer.

(3) Using different training algorithms, such as the back-propagation (Rumelhart et al., 1986), radial-basis function (Broomhead and Lowe, 1988), and Bayesian regression (Mackay, 1992) algorithms.

In this study, the individual neural network models with different initial conditions and different training subsets are therefore used as base models $f_1, f_2, \ldots, f_n$, as illustrated in Fig. 11.1.

When a large number of neural network base models are generated, it is necessary to select the appropriate number of component models for improving the efficiency of neural network meta-learning system. It is well known to us that not all circumstances are satisfied with the rule of "the more, the better". That is, some individual base models produced by this phase may be redundant, wasting resources and reducing system performance. Thus, it is necessary to prune some inappropriate individual base models for meta-model or ensemble model construction.

11.3.4 Neural Network Base Model Pruning

After training, each individual neural base model has generated its own result. However, if there are a great number of individual members, we need to select a subset of representatives in order to improve ensemble efficiency, as Chapter 10 indicated. Furthermore, in the neural network ensemble model, it does not follow the rule of "the more, the better", as proposed by Yu et al. (2005c). In this study, a decorrelation maximization method is used to select the appropriate number of neural network ensemble members.

The basic starting point of the decorrelation maximization algorithm is the principle of model diversity. That is, the correlations between the selected base models should be as small as possible, i.e., decorrelation maximization. Supposed that there are p neural predictors ($C_1, C_2, \ldots, C_p$) with n forecast values. Then the error matrix ($e_1, e_2, \ldots, e_p$) of p predictors is as

$$E = \begin{bmatrix} e_{11} & e_{12} & \cdots & e_{1p} \\ e_{21} & e_{22} & \cdots & e_{2p} \\ \vdots & \vdots & & \vdots \\ e_{n1} & e_{n2} & \cdots & e_{np} \end{bmatrix}_{n \times p} \tag{11.2}$$

From the matrix, the mean, variance and covariance of E can be calculated as

$$\text{Mean: } \bar{e}_i = \frac{1}{n}\sum_{k=1}^{n} e_{ki} \quad (i = 1, 2, \ldots, p) \tag{11.3}$$

$$\text{Variance: } V_{ii} = \frac{1}{n}\sum_{k=1}^{n} (e_{ki} - \bar{e}_i)^2 \quad (i = 1, 2, \ldots, p) \tag{11.4}$$

$$\text{Covariance: } V_{ij} = \frac{1}{n}\sum_{k=1}^{n}(e_{ki} - \bar{e}_i)(e_{kj} - \bar{e}_j) \quad (i,j = 1, 2, \ldots, p) \tag{11.5}$$

Considering Equations (11.4) and (11.5), we can obtain a variance-covariance matrix:

$$V_{p \times p} = (V_{ij}) \tag{11.6}$$

Based upon the variance-covariance matrix, correlation matrix R can be calculated using the following equations:

$$R = (r_{ij}) \tag{11.7}$$

$$r_{ij} = \frac{V_{ij}}{\sqrt{V_{ii}V_{jj}}} \tag{11.8}$$

where r_{ij} is correlation coefficient, representing the degree of correlation between base predictor C_i and base predictor C_j.

Subsequently, the plural-correlation coefficient $\rho_{C_i|(C_1,C_2,\cdots,C_{i-1},C_{i+1},\cdots,C_p)}$ between base predictor C_i and other p-1 base predictors can be computed based on the results of Equation (11.7) and Equation (11.8). For convenience, $\rho_{C_i|(C_1,C_2,\cdots,C_{i-1},C_{i+1},\cdots,C_p)}$ is abbreviated as ρ_i, representing the degree of correlation between C_i and (C_1, C_2, …, C_{i-1}, C_{i+1}, …, C_p). In order to calculate the plural-correlation coefficient, the correlation matrix R can be represented with block matrix, i.e.,

$$R \xrightarrow{\text{after transformation}} \begin{bmatrix} R_{-i} & r_i \\ r_i^T & 1 \end{bmatrix} \tag{11.9}$$

where R_{-i} denotes the deleted correlation matrix. It should be noted that r_{ii} = 1 (i = 1, 2, …, p). Then the plural-correlation coefficient can be calculated with the following equation:

$$\rho_i^2 = r_i^T R_{-i}^T r_i \quad (i = 1, 2, \ldots, p) \tag{11.10}$$

For a pre-specified threshold θ, if $\rho_i^2 > \theta$, then the base predictor C_i should be taken out from the p base predictors. On the contrary, the base predictor C_i should be retained. Generally, the decorrelation maximization algorithm can be summarized into the following steps:

(1) Computing the variance-covariance matrix V_{ij} and correlation matrix R with Equations (11.6) and (11.7);

(2) For the ith base model (i =1, 2, …, p), the plural-correlation coefficient ρ_i can be calculated with the Equation (11.10);
(3) For a pre-specified threshold θ, if $\rho_i < \theta$, then the ith base model should be deleted from the p base models. On the contrary, if $\rho_i > \theta$, then the ith base model should be retained.
(4) For the retained base models, we can also perform the procedure (1)-(3) iteratively until satisfactory results are obtained.

11.3.5 Neural-Network-Based Meta-Model Generation

As Fig. 11.1 illustrated, the initial data set is first divided into subsets, and then these subsets are input to the different individual models which could be executed concurrently. These individual models are called "base models'. In this phase, the main task is to generate a meta-model (i.e., ensemble model) to assimilate knowledge from different base models. Intuitively, the majority voting can produce a meta-model for classification problem. But majority voting ignores the fact that some networks that lie in a minority sometimes do produce the correct results. In meta-learning, it ignores the existence of diversity that can reduce error variance. Furthermore, it is unsuitable for regression problem, typically foreign exchange rates forecasting. In this study, another single neural network model different from base neural network model is used to perform regression task to generate a meta-model. The detailed process is described as follows.

Based upon different training sets, the base models can be generated in the previous phase. Using the testing set *TS*, the performance of the base models can be assessed. Afterwards, the whole training set is applied to these base models and these outputs of base neural network model can formulate a new training set called as "meta-training set (*MT*)". In order to reflect the principle of meta-learning, we utilize another neural network model to generate a meta-model by learning from the meta-training set (*MT*). That is, we use another neural network model to learn the relationship between base models by taking the outputs of the selected base models as input, combined with their targets or expected values. In this sense, neural network learning algorithm is used as a meta-learner (*ML*) shown in Fig. 11.1 for meta-model generation.

In this process of meta-model generation, another neural network model used as metal-learner (ML) can be viewed as a nonlinear ensemble forecasting system that can be represented as:

$$\hat{f} = f(\hat{f}_1, \hat{f}_2, \cdots, \hat{f}_n) \tag{11.11}$$

where $(\hat{f}_1, \hat{f}_2, \cdots, \hat{f}_n)$ is the output of base neural network predictors, $\hat{f}$ is the output of the neural-network-based meta-model by integrating the outputs of all individual neural network predictors, $f(\cdot)$ is nonlinear function determined by another neural network learning algorithm (Lai et al., 2006a).

In summary, the proposed meta-model can actually be seen as an embedded neural network system with two-layer neural network models, as illustrated in Fig. 11.1. Suppose that there is an original data set *DS* which is divided into two parts: training set (*TR*) and testing set (*TS*). The training set is usually preprocessed by various sampling methods (e.g., interval sampling here) in order to generate diverse training subsets $\{TR_1, TR_2, \ldots, TR_n\}$ before they are applied to the first layer's neural network learners: L_1, L_2, …, L_n. After training, the diverse neural network models (i.e., base models), $f_1, f_2, \ldots, f_n$ are generated. Through pruning of the decorrelation maximization algorithm, a set of selected base model are obtained. Afterwards the whole training set *TR* was applied and the corresponding results $(\hat{f}_1, \hat{f}_2, \cdots, \hat{f}_n)$ of each selected base model in the first layer were used as inputs of the second layer neural network model. This neural network model in the second layer can be seen as a meta-learner (*ML*). By training, the neural-network-based meta-model can be generated. Using the testing set *TS*, the performance of the neural-network-based meta-model can be assessed.

It is worth noting that the proposed meta-learning technique can be applied to both classification and regression problems. Due to the nature of foreign exchange rates prediction, the meta-learning for regression is adopted.

11.4 Empirical Study

11.4.1 Research Data and Experiment Design

The research data used in this study is euro against dollar (EUR/USD) exchange rate. This data are daily and are obtained from Pacific Exchange Rate Service (http://fx.sauder.ubc.ca/), provided by Professor Werner Antweiler, University of British Columbia, Vancouver, Canada. The entire data set covers the period from January 1 1993 to December 31 2004 with a total of 3016 observations. The data sets are divided into two periods: the first period covers from January 4 1993 to December 31 2002 while the second period is from January 1 2003 to December 31 2004. The first period, which is assigned to in-sample estimation, is used to network learning and training. The second period is reserved for out-of-sample evaluation, which is for the testing purposes. In addition, the training data covered from January 1 1993 to December 31 2002 are divided into ten training subsets by varying the sampling interval (K = 1, 2, …, 10) for this experiment. Using these different training subsets, different neural network base models with different initial weights are presented. For neural network base models, a three-layer back-propagation neural network with 10 TANSIG neurons in the hidden layer and one PURELIN neuron in the output layer is used. The network training function is the TRAINLM. For the neural-network-based meta-model, a similar three-layer back-propagation neural network (BPNN) with 10 inputs neurons, 8 TANSIG neural in the second layer and one PURELIN neuron in the final layer is adopted for meta-model generation. Besides, the learning rate and momentum rate is set to 0.1 and 0.15. The accepted average squared error is 0.05 and the training epochs are 1800. The above parameters are obtained by trial and error.

To evaluate the performance of the proposed neural-network-based meta-model, several typical financial time series prediction models, the auto-regressive integrated moving average (ARIMA), individual BPNN and support vector machine (SVM), are selected as benchmarks. In the individual BPNN model, a three-layer back-propagation neural network with 5 input nodes, 8 hidden nodes and 1 output nodes is used. The hidden nodes use sigmoid transfer function and the output node uses the linear transfer function. In the SVM, the kernel function is Gaussian function with regularization parameter $C = 50$ and $\sigma^2 = 5$. Similarly, the above parameters are obtained by trial and error.

For further comparison of the performance of neural network meta-model, simple averaging based meta-model and weighted averaging based meta-model are also used for time series prediction. Actually, these two meta-learning approaches are two neural network ensemble methods. For simple averaging approach, the final meta-model can be obtained by averaging the sum of each output of the neural network base models. Weighted averaging is where the final meta-model is calculated based on individual base model's performances and a weight attached to each base model's output. The gross weight is 1 and each base model is entitled to a portion of this gross weight according to their performance or diversity. For more details about these two meta-learning techniques, please refer to Hansen and Salamon (1990) and Benediktsson, et al. (1997). Finally, the root mean square error (*RMSE*) and direction change statistics (D_{stat}) (Yao and Tan, 2000) are used as performance evaluation criteria in terms of testing set.

11.4.2 Experiment Results

According to the above experiment design, different exchange rates prediction models with different parameters can be built. For comparison, the ARIMA model, individual BPNN, individual SVM, simple averaging based meta-model and weighted averaging based meta-learning approach, are also performed. The computational results are shown in Table 11.1.

Table 11.1. Performance comparison with different approaches

Model	Detail	*RMSE*	Rank	D_{stat} (%)	Rank
Individual model	ARIMA	0.2242	6	57.86	6
	BPNN	0.1856	5	72.78	4
	SVM	0.1124	4	74.75	2
Meta-model	Simple averaging	0.1058	3	73.35	3
(Ensemble model)	Weighted averaging	0.0986	2	72.56	5
	BP meta-learning	0.0813	1	87.44	1

As can be seen from Table 11.1, we can find the following conclusions from the general view. First of all, the neural network based meta-model perform the best in terms of both *RMSE* and D_{stat}, indicating that the proposed neural network based meta-model is a promising solution to the financial time series prediction. Second, the performances of the three meta-models are generally better than those of the individual forecasting models. The possible reason is that the meta-model can get more information from different base models and thus increase prediction accuracy. Third, the ARIMA model is the worst of the six models for both *RMSE* and D_{stat} in this case. The inherent reason is that the ARIMA model is difficult to capture the nonlinear patterns of financial time series since ARIMA is a class of linear model and the financial time series contain much nonlinearity and irregularity. Fourth, the results of D_{stat} are different from those of *RMSE* because two criteria are different. The former is a level estimation criterion, while the latter is a directional evaluation measurement.

Focusing on the *RMSE* indicator, it is difficult to find that the neural network based meta-model is the best, followed by weighted averaging based meta-model, simple averaging based meta-model, individual SVM, individual BPNN and the ARIMA model. To summarize, the meta-models outperform the individual model.

However, the low *RMSE* does not necessarily mean that there is a high hit ratio of forecasting direction for foreign exchange movement direction prediction. Thus, the D_{stat} comparison is necessary. Focusing on D_{stat} of Table 11.1, we find the neural network based meta-model also performs much better than the other models according to the rank. Furthermore, from the business practitioners' point of view, D_{stat} is more important than *RMSE* because the former is an important decision criterion. With reference to Table 11.1, the differences between the different models are very significant. In the testing case of EUR/USD, the D_{stat} for the individual ARIMA model is 57.86%, for the individual BPNN and SVM model, the D_{stat}s are 72.78% and 74.75%, respectively, and for the simple averaging based meta-model and the weighted averaging based meta-model, the D_{stat}s are only 73.35% and 7256%; while for the neural network based meta-model, D_{stat} reaches 87.44%. Interestingly, we can also find that (1) the individual BPNN model is better than the weight averaging meta-model, indicating that the weight averaging meta-model is only a linear combination of selected neural network base models. Sometimes, its performance is often worse than that of the individual base model or simple averaging meta-model; (2) the prediction performance of the individual SVM model is rather well, indicating that the SVM method is another feasible prediction model in exchange rates prediction. The main reason is that the SVM can overcome some shortcomings of neural networks, such as local minima and overfitting.

11.5 Conclusions

In this chapter, a neural-network-based meta-learning technique is proposed to formulate a meta-model or ensemble model for foreign exchange rates prediction. In terms of the empirical results, we find that across different forecasting models for the test cases of EUR/USD on the basis of different criteria, the proposed neural network based meta-model performs the best. In the proposed neural network based meta-model test cases, the *RMSE* is the lowest and the D_{stat} is the highest, indicating that the neural network meta-learning technique can be used as a viable alternative solution for exchange rate prediction for financial managers and business practitioners.

12 Predicting Foreign Exchange Market Movement Direction Using a Confidence-Based Neural Network Ensemble Model

12.1 Introduction

Neural network ensemble has been turned out to be an efficient strategy for achieving high classification performance, especially in fields where the development of a powerful single classifier system requires considerable efforts. Usually, neural network ensemble model outperforms the individual neural network models, whose performance is limited by the imperfection of feature extraction, learning/classification algorithms, and the inadequacy of training data. Due to these reasons, there is a growing research stream about neural network ensemble learning methods (Perrone and Cooper, 1993; Krogh and Vedelsby, 1995; Rosen, 1996; Tumer and Ghosh, 1996; Yang and Browne, 2004; Yu et al., 2005c, 2006d; Lai et al., 2006c, 2006d). For example, performance improvement can result from training the individual networks to be decorrelated with each other (Rosen, 1996) with respect to their errors. To achieve high classification performance, there are some essential requirements to the ensemble members and the ensemble strategy. First of all, a basic condition is that the individual neural network classifiers must have enough training data. Secondly, the ensemble members are diverse or complementary, i.e., classifiers show different classification properties. Thirdly, to obtain high performance, a wise ensemble strategy is also required on a set of complementary classifiers.

For a limited data set, some sampling approaches, such as bagging (Breiman, 1996a) have been used for creating different samples by varying the data subsets selected or perturbing training sets. Similarly, diverse ensemble members can be obtained by varying the initial conditions or using different training data. In the ensemble model, the most important point is to select an appropriate ensemble strategy. Generally, the variety of ensemble methods can be grouped into three categories according to the level of classifier outputs: abstract level (crisp class), rank level (rank order) and measurement level (class score) (Xu et al., 1992; Suen and Lam, 2000). In the existing studies, many ensemble systems still use empirical heuristics and ad hoc ensemble schemes at the abstract level. Typically, majority voting (Breiman, 1996a) uses the abstract level of output of ensemble members. An important drawback of this ensemble strategy is that it does not take confidence degree of neural network output into account. Actually, ensemble at the measurement level is advantageous in that the output measurements contain richer information of class measures. In a sense, an appropriate ensemble strategy is more crucial, especially for integrating the classifiers that output diverse measurements. Furthermore, the intensive investigation of neural network ensemble has not formulated a convincing theoretical foundation and an overall process model yet (Lai et al., 2006d).

In such situations, we propose a multistage confidence-based neural network ensemble approach that differs in that the final ensemble strategy is determined the confidence of neural network output at the measurement level. In this study, the proposed neural network ensemble model consists of five stages. In the first stage, a bagging sampling approach is used to generate different training datasets. In the second stage, the neural network model is trained by various training datasets from the previous stage. In the third stage, the trained neural network models are applied to testing datasets and some classification results and confidence values can be obtained. In the fourth stage, the confidence values are scaled into a unit interval by logistic transformation. In the final stage, the multiple neural network models are integrated to obtain final classification result by means of confidence measurement. For testing purpose, two typical foreign exchange rates (euro against US dollar (EUR/USD) and Japanese yen against US dollar (JPY/USD)) are used to verify the effectiveness of the proposed neural network ensemble model. In particular, our goal of this chapter is to predict the movement direction of foreign exchange rates.

The motivation of this chapter is to formulate a multistage confidence-based neural network ensemble model for predicting foreign exchange market movement direction and compare its performance with other existing approaches. The rest of the chapter is organized as follows. The next section presents a formulation process of the multistage neural network ensemble model in detail. To verify the effectiveness of the proposed method, two real examples are performed in Section 12.3. For comparison, Section 12.4 reports the comparisons of three ensemble neural network models. Finally, some conclusions are drawn in Section 12.5.

12.2 Formulation of Neural Network Ensemble Model

In this section, a five-stage confidence-based neural network ensemble model is proposed for classification. The basic idea of neural network ensemble originated from using all the valuable information hidden in neural network classifiers, where each can contribute to the improvement of generalization. In our proposed multistage neural network ensemble model, a bagging sampling approach is first used to generate different training sets for guaranteeing enough training data. According to various training set, multiple individual neural classifiers are trained. Then the trained neural classifiers are applied to testing set. Accordingly some classification results and confidence values are also obtained. Subsequently the confidence values are transformed into a unit interval for avoiding the situation that member classifier with large absolute value often dominate the final decisions of the ensemble. Finally the ensemble members are selected using some criteria, and their generated results are aggregated in terms of confidence measure. The final result is called the ensemble output. The general architecture of the multistage confidence-based neural network is illustrated in Fig. 12.1.

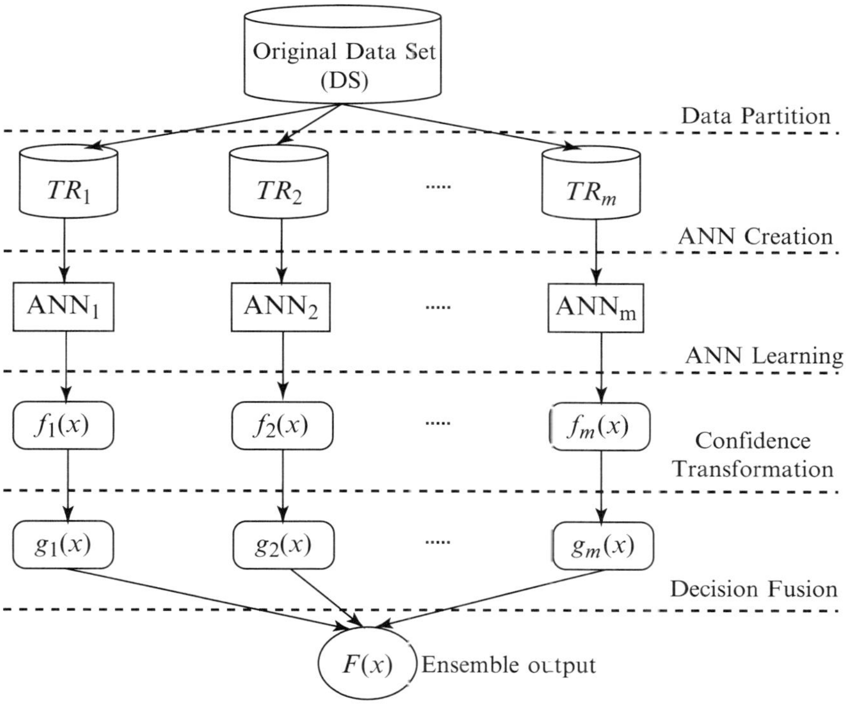

Fig. 12.1. The general architecture of multistage neural network ensemble model

12.2.1 Partitioning Original Data Set

In order to create different neural predictors, different training subsets should be generated. Typical methods for obtaining different training subsets include bagging (Breiman, 1996a) and cross-validation (Krogh and Vedelsby, 1995). Of course, other approaches such as noise injection (Raviv and Intrator, 1996), stacking (Wolpert, 1992), boosting (Schapire, 1990) and input decimation (Tumer and Ghosh, 1996) is also used.

In the situation of data shortage, bagging is an effective approach to creating samples by varying the data subsets selected or perturbing training sets (Yang and Browne, 2004). Due to the feature of its random sampling with replacement, bagging algorithm is very efficient in constructing a reasonable size of training set. Given that the size of the original data set *DS* is *P*, the size of new training data is *N*, and the number of new training data items is *m*, the bagging algorithm can be shown in Fig. 12.2.

```
Input: original data set DS
Output: The generated new training subsets (TR1, TR2,..., TRm)
For t = 1 to m
   For i = 1 to N
     RandRow P* rand ()
    If RandRow <=P
       Pt(i,All Columns) = DS(RandRow, AllColumns)
     End If
   Next i
Next t
Output the final training subsets {TR1, TR2, ...,TRm}
```

Fig. 12.2. The bagging algorithm for data partition

For most situations, cross-validation (Krogh and Vedelsby, 1995) which is borrowed from statistics is a effective tool to partition data set. First of all, the available data set is randomly divided into *m* disjoint subsets. By selecting one of these subsets as a testing set, the remainders are rejoined as its corresponding training set. For this case, *m* numbers of overlapping training set and *m* independent testing sets are obtained. As each training set is different somehow, the errors they generate after training are expected to fall in different local error minima and thus lead to different results. Model performance is measured on the corresponding testing set. This approach ideally requires that the number of partitions is the same as the number of data exemplars. In practice, ten-fold and twenty-fold cross-validations are adopted, i.e., m is equal to 10 and 20, respectively.

In view of the particularity of neural network predictors, cross-validation technique is selected as data set partitioning tool in this chapter.

12.2.2 Creating Individual Neural Network Classifiers

According to the definition of effective ensemble classifiers by Hansen and Salamon (1990), 'a necessary and sufficient condition for an ensemble of classifiers to be more accurate than any of its individual members is if the classifiers are accurate and diverse.' That is, an effective ensemble classifier consisting of diverse models with much disagreement is more likely to have a good generalization performance Therefore, how to generate the diverse model is a crucial factor. For neural network model, several methods have been investigated for the generation of ensemble members making different errors (Sharkey, 1996). Such methods basically rely on varying the parameters related to the design and to the training of neural networks. In particular, the main methods include the following:

(1) Different initial conditions: ensemble members can be created by varying the initial condition, such as starting random weights, learning rate and momentum rate, from which each network is trained.

(2) Different network architecture: by changing the number of hidden layers and the number of nodes in every layer, different neural networks with different architectures can be created.

(3) Different training data: by re-sampling and preprocessing data, we can obtain different training sets, thus making different network generations. There are six techniques that can be used to obtain diverse training data sets: bagging (Breiman, 1996a), noise injection (Raviv and Intrator, 1996), cross-validation (Krogh and Vedelsby, 1995), stacking (Wolpert, 1992), boosting (Schapire, 1990) and input decimation (Tumer and Ghosh, 1996).

(4) Different training algorithm: by selecting different core learning algorithms, different neural networks can also be created. For example, a multi-layer feed-forward network can use the steep-descent algorithm or Levenberg-Marquardt algorithm or other learning algorithms.

In our study, the third way is selected because the previous phase creates many different training datasets. In addition, the three-layer BP neural networks are selected because a three-layer BP neural network with an identity transfer function in the output unit and logistic transfer functions in the middle-layer units can approximate any continuous function arbitrarily well given a sufficient amount of middle-layer units (White, 1990). With these different training datasets, diverse neural network classifiers can be generated.

12.2.3 BP Network Learning and Confidence Value Generation

After selecting neural network type, the next step is to train the neural network with different training datasets. In our study, the selected neural network is a class of supervised error back-propagation learning mechanism in the form of the neural network associative memory. For classification task, a neural network can usually be trained by the in-sample dataset and applied to out-of-sample dataset for verification. The model parameters (connection weights and node biases) will be adjusted iteratively by a process of minimizing the error function *E*, as illustrated in Chapter 2. Basically, the final output of the BPNN model can be represented as

$$y = f(x) = a_0 + \sum_{j=1}^{q} a_j \varphi(w_{0j} + \sum_{i=1}^{p} w_{ij} x_i) \qquad (12.1)$$

where $a_j (j = 0, 1, 2, \ldots, q)$ is the connection weight on the *j*th unit between hidden and output layer, and w_{ij} $(i = 0, 1, 2, \ldots, p; j = 1, 2, \ldots, q)$ is the connection weight between input and hidden layer, $\varphi(\bullet)$ is the transfer function of the hidden layer, *p* is the number of input nodes and *q* is the number of hidden nodes.

By training neural network, model parameters in Equation (12.1) can be determined and accordingly the neural network classifier can be shown as

$$F(x) = sign\left(a_0 + \sum_{j=1}^{q} w_j \varphi(a_j + \sum_{i=1}^{p} w_{ij} x_i) \right) \qquad (12.2)$$

In our study, we mainly use neural network output value $f(x)$ as its classification score, instead of the classification results $F(x)$ directly. The main reason is that the neural network output value $f(x)$ is a good indicator of the confidence or reliability of ensemble classifiers. The larger the $f(x)$, the higher the neural network classifier for positive class is. Therefore the neural network output value $f(x)$ as a confidence value is used to integrate the ensemble members. By means of this treatment, we can realize the classification decision fusion at the measurement level.

12.2.4 Confidence Value Transformation

In the previous phase, the neural classifier outputs are used as confidence value. It is worth noting that the confidence value falls into the interval (-∞, +∞). The main drawback of this confidence value is that ensemble classifier with large absolute value often dominate the final decision of the ensemble model.

In order to overcome this weakness, one simple strategy is to re-scale the output values into zero mean and unit standard deviation, i.e.,

$$g_i^+(x) = \frac{f_i(x) - \mu}{\sigma} \tag{12.3}$$

where μ and σ are the mean and standard deviation of the pooled classifier outputs, respectively. However, for classifiers that output dissimilarity measures, the sign of the original outputs should be reversed before this transformation.

For convenience, transforming the confidence value into the unit interval [0, 1] is a good solution. In neural network, the logistic function behaves well in squashing neural output to approximate probability measures. Therefore we can take it as a scaling function for confidence transformation, i.e.,

$$g_i^+(x) = \frac{1}{1 + e^{-f_i(x)}} \tag{12.4}$$

In a binary classification problem, if the confidence degree for positive class is $g_i^+(x)$, the confidence degree for negative class can be represented as

$$g_i^-(x) = 1 - g_i^+(x) \tag{12.5}$$

According to the transformed confidence values, multiple classifiers can also be fused into an ensemble output, as illustrated in the following subsection.

12.2.5 Integrating Multiple Classifiers into an Ensemble Output

After neural network learning, each classifier of the ensemble has output its own results. Before integrating these ensemble members, strategies of selecting ensemble members must be noted. Generally, these strategies can be divided into two categories: (1) generating an exact number of ensemble members; and (2) overproducing ensemble members and then selected a subset of these (Yang and Browne, 2004).

For the first strategy, several common ensemble approaches, e.g., boosting (Schapire, 1990), can be employed to generate the exact number of diverse ensemble members for integration purpose. Therefore, no selection process will be used and all generated ensemble members will be combined into an aggregated output. For the second strategy, its aim is to create a large set of ensemble candidates and then choose some most diverse members for integration. The selection criterion is some error diversity measures, which is introduced in detail by Partridge and Yates (1996). Because the first strategy is based upon the idea of creating diverse neural networks at the early stage of design, it is better than the second strategy, especially for some situations where access to powerful computing resources is restricted. The main reason is that the second strategy cannot avoid occupying much computing time and storage while creating a large number of ensemble candidates, some of which are to be later discarded (Yang and Browne, 2004).

Actually, there are some fusion strategy in the literature at the abstract level and the rank level. Typically, majority voting, ranking and weighted averaging are three popular decision fusion approaches. Majority voting is the most widely used fusion strategy for classification problems due to its easy implementation. Ensemble members' voting determines the final decision. Usually, it takes over half the ensemble to agree a result for it to be accepted as the final output of the ensemble regardless of the diversity and accuracy of each network's generalization. Majority voting ignores the fact some neural network that lie in a minority sometimes do produce the correct results. At the stage of integration, it ignores the existence of diversity that is the motivation for ensembles (Yang and Browne, 2004). In addition, majority voting is only a class of integration strategy at the abstract level.

Ranking is where the members of an ensemble are called low level classifiers and they produce not only a single result but a list of choices ranked in terms of their likelihood. Then the high level classifier chooses from this set of classes using additional information that is not usually available to or well represented in a single low level classifier (Yang and Browne, 2004). However, ranking strategy is a class of fusion strategy at the rank level, as earlier mentioned.

Weighted averaging is where the final ensemble decision is calculated in terms of individual ensemble members' performances and a weight attached to each member's output. The gross weight is one and each ensemble member is entitled to a portion of this gross weight based on their performances or diversity (Yang and Browne, 2004). Although this approach is a class of ensemble strategy at the measurement level, but it is difficult for classification problem to obtain the appropriate weights for each ensemble member.

In such situations, this study proposes a confidence-based ensemble strategy to make the final decision of the ensemble at the measurement level. The following five strategies can be used to fuse the individual ensemble members (Lai et al., 2006d):

(1) Maximum strategy:

$$F(x)=\begin{cases}1, & \text{if } \max_{i=1,\cdots,m} g_i^+(x) \geq \max_{i=1,\cdots,m} g_i^-(x),\\ -1, & \text{otherwise.}\end{cases} \tag{12.6}$$

(2) Minimum strategy:

$$F(x)=\begin{cases}1, & \text{if } \min_{i=1,\cdots,m} g_i^+(x) \geq \min_{i=1,\cdots,m} g_i^-(x),\\ -1, & \text{otherwise.}\end{cases} \tag{12.7}$$

(3) Median strategy:

$$F(x)=\begin{cases}1, & \text{if } \underset{i=1,\cdots,m}{median}\left(g_i^+(x)\right) \geq \underset{i=1,\cdots,m}{median}\left(g_i^-(x)\right),\\ -1, & \text{otherwise.}\end{cases} \tag{12.8}$$

(4) Mean strategy:

$$F(x)=\begin{cases}1, & \text{if } \sum_{i=1}^{m} g_i^+(x) \geq \sum_{i=1}^{m} g_i^-(x),\\ -1, & \text{otherwise.}\end{cases} \tag{12.9}$$

(5) Product strategy:

$$F(x)=\begin{cases}1, & \text{if } \prod_{i=1,\cdots,m} g_i^-(x) \geq \prod_{i=1,\cdots,m} g_i^-(x),\\ -1, & \text{otherwise.}\end{cases} \tag{12.10}$$

To summarize, the multistage confidence-based neural network ensemble model can be concluded in the following steps:

(1) Partitioning original dataset into m training datasets, TR_1, TR_2, …, TR_m.

(2) Training m individual neural network models with the different training dataset TR_1, TR_2, …, TR_m and obtaining m individual neural network classifiers, i.e., ensemble members.

(3) Using Equation (12.1) to obtain the m neural classifiers' output values of new unlabeled sample x, $f_1(x)$, $f_2(x)$, …, $f_m(x)$.
(4) Using Equations (12.4) and (12.5) to transform output value to confidence degrees for positive class $g_1^+(x),\cdots,g_m^+(x)$ and for negative class $g_1^-(x),\cdots,g_m^-(x)$.
(5) Fusing the multiple neural classifiers into an aggregated output in terms of confidence value using Equations (12.6)-(12.10).

12.3 Empirical Study

In this section, two typical foreign exchange rates, EUR/USD and JPY/USD, are used for testing and verification purposes. The foreign exchange data used in this paper are daily and are obtained from Pacific Exchange Rate Service (http://fx.sauder.ubc.ca/), provided by Professor Werner Antweiler, University of British Columbia, Vancouver, Canada. We take daily data from January 1 2000 to December 31 2003 as in-sample (training periods) data sets with a total of 1003 observations for model building. We also take the data from January 1 2004 to December 31 2004 as out-of-sample (testing periods) data sets with a total of 252 observations, which is used to evaluate the good or bad performance of prediction based on some evaluation measurement.

The main goal of this chapter is to predict the movement direction of foreign exchange rates. The movement direction is categorized as "1" and "–1" in the research data. "1" means that the next day's price is higher than today's price, and "–1" means that the next day's price is lower than today's price. Since we attempt to mine the foreign exchange index movement tendency, technical indicators of foreign exchange market are used as input variables.

However, the research data collected in this study is time series of foreign exchange rates, some technical indicators and the direction of change in daily exchange rates are calculated as input variables for foreign exchange rate direction forecasting. This study calculates 19 technical indicators as the input variables, as determined by the review of domain experts and prior research (Murphy, 1986; Achelis, 1995; Choi, 1995; Gifford, 1995; Chang et al., 1996; Kim et al., 2004; Shen and Loh, 2004). The calculated indicators include the following items:

(1) Price (P)
(2) Daily excess return (R)
(3) Stochastic oscillator (SO)

(4) Moving stochastic oscillator (MSO)
(5) Slow stochastic oscillator (SSO)
(6) Rate of change (ROC)
(7) Momentum (M)
(8) Moving average (MA)
(9) Moving variance (MV)
(10) Moving variance ratio (MVR)
(11) Exponential moving average (EMA)
(12) Moving average convergence & divergence (MACD)
(13) Accumulation/distribution oscillator (ADO)
(14) Disparity5 (D5)
(15) Disparity10 (D10)
(16) Price oscillator (OSCP)
(17) Commodity channel index (CCI)
(18) Relative strength index (RSI)
(19) Linear regression line (LRL)

The meaning and computational formulae of these technical indicators can refer to Table 9.1 and the studies of Murphy (1986), Achelis (1995), Choi (1995), Gifford (1995), Chang et al. (1996), Kim et al. (2004), and Shen and Loh (2004) for more details.

For comparison purposes, linear regression (LinR), logit regression (LogR), single artificial neural network (ANN), support vector machine (SVM), single best neural network (BNN), majority voting ensemble are also conducted the experiments. Particularly, five ensemble strategies of the proposed confidence-based neural network ensemble model are also conducted in the experiment. The single ANN training an ANN classifier with the total training dataset and applies the classifier to predict the class of unknown samples. The single best neural network is constructed as following. Firstly, several training subsets of the total training dataset is created by some sampling algorithm, e.g., bagging. Then each subset is used to training neural classifier and several diverse neural classifiers can be obtained. Finally, the obtained classifiers are evaluated by validation dataset which is the collection of the total training sets minus the respective subsets, and the classifier with the best performance is chosen as a final classifier.

In the ANN model, a three-layer back-propagation neural network (BPNN) with 10 TANSIG neurons in the hidden layer and one PURELIN neuron in the output layer is used. The network training function is the TRAINLM. Besides, the learning rate and momentum rate is set to 0.25 and 0.35. The accepted average squared error is 0.05 and the training epochs are 2000. The above parameters are obtained by trial and error. In the SVM model, the kernel function is Gaussian function with regularization parameter $C = 50$ and $\sigma^2 = 8$. Similarly, the above parameters are obtained by trial and error.

In addition, the classification accuracy in testing set is used as performance evaluation criterion. Typically, three evaluation criteria are used to measure the prediction results.

$$\text{Type I accuracy} = \frac{\text{number of both observed up and classified as up}}{\text{number of observed up}} \tag{12.11}$$

$$\text{Type II accuracy} = \frac{\text{number of both observed down and classified as down}}{\text{number of observed down}} \tag{12.12}$$

$$\text{Total accuracy} = \frac{\text{number of correct prediction}}{\text{the number of evaluation sample}} \tag{12.13}$$

To reflect model robustness, each class of experiment is repeated 10 times and the final Type I, Type II and total accuracy is the average of the results of the 10 individual tests. According to the previous experiment design, the final computational results are shown in the Tables 12.1-12.2.

As can be seen from Tables 12.1-12.2, three main conclusions can be summarized as follows.

(1) Of the five single models, single support vector machine (SVM) performs the best for both EUR/USD and JPY/USD from the perspective of total accuracy. The main reason is that the SVM is a new kind of neural network model that can overcome the inherent local minima and overfitting problems of neural networks.

Table 12.1. Performance comparison of different models for EUR/USD*

Category	Model	Strategy	Type I (%)	Type II (%)	Total (%)
Single	LinR		53.44 [14.2]	49.63 [10.8]	51.49 [12.3]
	LogR		61.37 [10.3]	62.29 [11.4]	61.84 [10.9]
	ANN		70.61 [10.7]	71.36 [10.3]	71.00 [10.5]
	BNN		73.67 [8.78]	72.54 [8.64]	73.11 [8.71]
	SVM		74.36 [7.57]	73.52 [8.13]	73.94 [7.85]
Ensemble	Voting-based	Majority	76.43 [7.97]	77.67 [7.74]	77.05 [7.86]
	Confidence-based	Maximum	82.51 [7.48]	83.49 [7.33]	83.02 [7.41]
		Minimum	83.86 [7.41]	84.63 [7.74]	84.25 [7.59]
		Median	83.52 [7.47]	85.86 [8.03]	84.71 [7.76]
		Mean	**85.54 [6.72]**	**87.42 [6.43]**	**86.49 [6.58]**
		Product	83.48 [7.25]	84.78 [8.16]	84.14 [7.71]

* Standard deviations appear in brackets

Table 12.2. Performance comparison of different models for JPY/USD*

Category	Model	Strategy	Type I (%)	Type II (%)	Total (%)
Single	LinR		54.36 [12.8]	51.48 [11.6]	52.92 [12.2]
	LogR		62.42 [10.3]	60.27 [11.4]	61.34 [10.9]
	ANN		69.74 [9.75]	71.88 [10.0]	70.81 [9.88]
	BNN		71.69 [8.05]	73.46 [8.02]	72.58 [8.04]
	SVM		73.11 [7.57]	72.93 [8.18]	73.02 [7.88]
Ensemble	Voting-based	Majority	75.14 [7.11]	75.79 [8.04]	75.47 [7.58]
	Confidence-based	Maximum	80.51 [6.51]	81.36 [7.03]	80.94 [6.77]
		Minimum	81.83 [6.64]	80.35 [6.81]	81.09 [6.73]
		Median	81.15 [6.42]	82.82 [6.09]	81.98 [6.25]
		Mean	82.47 [6.50]	82.14 [6.43]	82.31 [6.47]
		Product	**83.28 [6.25]**	**84.01 [6.46]**	**83.64 [6.36]**

*Standard deviations appear in brackets

Basically, the performance of SVM is better than that of ANN, logit regression and linear regression for both two tested foreign exchange rates in terms of Type I accuracy and Type II accuracy as well as total accuracy. Using two tailed *t*-test, we find that the difference between performance of ANN and SVM is insignificant at five percent level of significance, while the difference between linear regression and SVM is significant at five percent level of significance. Interestedly, for JPY/USD, the Type II accuracy of single best neural network (BNN) is better than that of the SVM. The possible reason is that Japanese yen market is a more volatile market relative to the euro market.

(2) In the ensemble model, five confidence-based neural network ensemble models consistently outperform the majority voting based ensemble model, implying that the proposed confidence-based neural network ensemble model is a class of promising approach to predicting the foreign exchange market movement direction.

Among the five confidence-based neural network ensemble models, the neural network ensemble model with mean strategy perform the best for EUR/USD, while the neural network ensemble model with product strategy is the best of five confidence-based ensemble models for JPY/USD.

However, through two-tail paired *t*-test, the average performance difference of the five ensemble strategies is insignificant at 10 percent significant level. Furthermore, it is also obvious that the performances of the five ensemble strategies are quite close through observation of Tables 12.1-12.2. Although there is no significant difference in performance of the five confidence-based neural network ensemble models, the main reason resulting in such a small difference is still unknown, which is worth further exploring in the future.

(3) Generally speaking, the proposed confidence-based neural network ensemble model perform the best in terms of Type I accuracy, Type II accuracy, and total accuracy, indicating that the proposed confidence-based neural network ensemble learning technique is a feasible and effective solution to improve the accuracy of foreign exchange market movement direction prediction.

12.4 Comparisons of Three Ensemble Neural Networks

In this part (i.e., Part V), three ensemble neural network models are proposed. For comparison, Table 12.3 presents some similarities and differences of the three ensemble neural network models.

Table 12.3. Comparison of three ensemble neural network models

Chapter	Chapter 10	Chapter 11	Chapter 12
Ensemble method	SVM	Meta-Learning	Confidence
Prediction mode	Level prediction	Level prediction	Direction Prediction
Prediction range	Short-term	Short-term	Medium- or long-term

From Table 9.3, we can conclude that (1) in the three ensemble neural network models, different ensemble methods are used. For example, Chapter 10 uses an SVM-based nonlinear ensemble model for exchange rates prediction, Chapter 11 uses meta-learning method to construct a multi-stage neural network ensemble model, while Chapter 12 uses a confidence-based technique to fuse different neural network models for foreign exchange rates forecasting. (2) Of the three ensemble models, the former two ensemble models proposed in Chapters 10 and 11 are level prediction models, which are often used for short-term prediction, while the confidence-based neural network ensemble model in Chapter 12 is direction prediction model, which is often used for medium- or long-term prediction and foreign exchange market tendency exploration. According to these characteristics of the three proposed ensemble models, we can select different ensemble models for foreign exchange rates forecasting.

12.5 Conclusions

In this chapter, a multistage confidence-based neural network ensemble model is proposed to predict the foreign exchange market movement direction. Through the practical data experiments, we have obtained good prediction results and meantime demonstrated that the confidence-based neural ensemble model outperforms all the compared models listed in this study. These advantages imply that the confidence-based neural network ensemble model can provide a promising solution to foreign exchange rates movement direction prediction.

13 Foreign Exchange Rates Forecasting with Multiple Candidate Models: Selecting or Combining? A Further Discussion

13.1 Introduction

From Chapter 4 to Chapter 12, nine typical foreign exchange rates forecasting models, including three single neural network models, three neural network hybrid models and three neural network ensemble models, are proposed from the perspectives of level estimation and direction exploration. However, in the development process of forecasting models, we often have to come up against two important dilemmas (Yu et al., 2005g):

(1) Whether should we select an appropriate modeling approach for prediction purposes or should combine these different individual approaches into an ensemble forecast for the different/dissimilar models?

(2) Whether should we select the best candidate model for forecasting or to mix the various candidate models with different parameters into a new forecast for the same/similar modeling approaches?

In foreign exchange rates forecasting, various different models including linear and nonlinear methods has been utilized. For example, the autoregressive integrated moving average (ARIMA) model proposed by Box and Jenkins (1970), which is a typical linear model, has been proven to be effective in many time series forecasting problems including foreign exchange rates. Likewise, the artificial neural network (ANN) model, a typical nonlinear model and newly intelligent computational technique, has also been shown to be a very promising approach for foreign exchange rates modeling and forecasting. Due to the fact that different foreign exchange rates have different properties and features, the selected model based on a selection criterion (e.g., Akaike information criterion (AIC) (Akaike (1973))), hypothesis testing, and graphical inspections, is often mis-specified and thus may cause an unexpectedly high variability in the final prediction and affect forecasting accuracy and reliability. In practical applications, however, academic researchers and business practitioners also have to face the above two important dilemmas. For foreign exchange rates forecasting, the focus is on whether to select a model from multiple candidate models or to combine these different models into a single forecast. Generally speaking, the two issues are highly non-trivial and have received considerable attention with different approaches being studied. Here we briefly discuss some of these approaches that are closely related to our work.

In time series forecasting applications, the basic practice is to select one appropriate model from the multiple candidate models in terms of some selection criteria; thus final estimation, interpretation, and prediction are then based on the selected model. Generally speaking, there are three methods for multiple candidate model selection. The first is graphical inspection together with examination of simple summary statistics (such as autocorrelations (AC) and partial autocorrelations (PAC)), which is very useful for preliminary modeling analysis. The second is hypothesis testing, a formal technique for model selection in statistics. The third method is to use a well-defined and formal model selection criterion, such as Akaike information criterion (AIC) (Akaike (1973)) or Bayesian information criterion (BIC) (Schwartz (1978)). However, these methods exhibit some defects. For example, the first method is too subjective and too rough in general for model selection. Likewise, there are difficulties with the second approach due to the challenging issue of multiple testing (e.g., there is no objective guideline for the choice of the size of each individual test, and it is completely unclear how such a choice affects the forecasting accuracy). Furthermore, Breiman (1996b) argued that the estimators based on model selection criteria are instable. Yang (2001) also pointed out that a major drawback with model selection criteria is its instability. For example, with a small or moderate number of observations, as expected, it is usually hard to distinguish models that are close to each other (the model selection criterion values are generally quite close). In this case, the choice of the model with the smallest criterion value is unstable. That is, a slight change in the data may lead to a different model being chosen. As a consequence, the forecasts based on the selected model are highly variable (Yang (2000, 2001)). Furthermore, Yang (2001) claimed that identifying the true model is not necessarily optimal for forecasting. To reduce variability in model selection, researchers have turned to combining or hybridizing the different candidate models for prediction purpose. A variety of literature on combined forecasting is reported.

In the combination of the same/similar method, Draper (1995) and George and McCulloch (1997) proposed an "averaging" method to combine a stabilized estimator. Raftery (1995) suggested using a BIC approximation for Bayesian model averaging. Madigan and York (1995) used a Markov chain Monte Carlo approximation to obtain a stable forecast. Breiman (1996a) proposed a "bagging" method to generate multiple versions of the estimator and then average them into a stable estimator. In a similar manner, Buckland et al. (1995) proposed a plausible modeling weight method according to the value of a model selection criterion (e.g., AIC). Cross-validation and bootstrapping have also been used to linearly combine multiple similar models with the intention of improving accuracy by finding the best linear combination (see Wolpert (1992), Breiman (1996b), LeBlanc and Tibshirani (1996)). Juditsky and Nemirovski (2000) proposed a stochastic approximation method to combine k forecasts in the best linear combination. Yang (2000) applied the adaptive regression method (ARM) by mixing multiple candidates from the same models for regression. Similarly, Yang (2001) used the aggregated forecast through exponential re-weighting method (AFTER) to combine the forecasts from the different individual autoregressive moving average (ARMA) models. However, their works are only limited to linear combinations of the same methods with different parameters. Readers can refer to a review of this topic by Hoeting et al. (1999) for more details.

In the combination of different/dissimilar models, literature documenting the research shows it to be quite diverse. Combining forecasts with dissimilar methods has been extensively studied over three decades, starting with the pioneering work of Reid (1968, 1969) and Bates and Granger (1969). Various methods have been involved. For work pre-dating 1989, readers can refer to Clemen (1989) for a comprehensive review of this topic. After that time, with the increasing development and application of new computational technology, many artificial intelligent (AI) techniques such as artificial neural networks (ANNs) have been presented. There are quite a few examples in the existing literature hybridizing neural network forecasting models with conventional time series forecasting techniques. For example, Wedding II and Cios (1996) constructed a combination model integrated radial basis function neural networks (RBFNN) and the univariant Box-Jenkins (UBJ) model to predict three time series. Luxhoj et al. (1996) presented a hybrid enconometric and ANN approach for sales forecasting. Likewise, Voort et al. (1996) introduced a hybrid method called KARIMA using a Kohonen self-organizing map and ARIMA method to predict short-term traffic flow. Recently, Tseng et al. (2002) proposed a hybrid model (SARIMABP) that combines the seasonal ARIMA (SARIMA) model and the back-propagation (BP) neural network model to predict seasonal time series data. Zhang (2003) applied a hybrid methodology that combined both ARIMA and ANN models for time series forecasting.

Actually, the idea of hybrid or ensemble forecasts implicitly assumed that one could not identify the underlying process (i.e., one could not select an appropriate model for a specified time series), but that different forecasting models were able to capture different aspects of the information available for prediction. This is also one of starting points for this chapter. As Clemen (1989) concluded, "Using a combination of forecasts amounts to an admission that the forecaster is unable to build a properly specified model. To try ever more elaborate combining models seems to add insult to injury as the more complicated combinations does not generally perform all that well." Hence combining the multiple candidate models for prediction has been common practice in practical applications to reduce the variability and uncertainty of selecting an individual model.

However, the fact that we use the hybrid and ensemble (combined) method for time series prediction does not necessarily mean that we are against the practice of model selection. Generally, identifying the true model (when it makes good sense) is an important task in understanding relationships of time series variables. In deterministic linear time series, it is often observed that selection may outperform combining methods when one model is very strongly preferred, in which case there is little instability in selection. As Yang (2001) claimed, in the time series context, combining does not necessarily lead to prediction improvement when model selection is stable. That is to say, the use of hybrid or ensemble forecasts does not mean that the individual forecasts should be suppressed. In fact, these individual forecasts can provide evidence about the relative merits of their respective models, methods, or forecasters as well as about the relationships among the multiple forecasts. Furthermore, for purpose of reporting and comparison, it is helpful to give the individual forecast as well as the combined forecast in order to provide an indication of the variability among the forecasts. Decision makers using the forecasts may want to see if and how their decisions change when different forecasts are used (Winkler (1989)). In addition, it should be pointed out the hybrid or ensemble approach we take is related to (although different from) formal Bayesian consideration. In particular, no prior distribution will be specified for parameters in all developed models, as mentioned in previous chapters.

The motivation of this chapter is how to deal with the two dilemmas of selecting and combining for time series forecasting (in particular, foreign exchange rates forecasting problem) and meantime propose a solution to the two dilemmas. The rest of the chapter is organized as follows. The next section describes the procedure for dealing with the two dilemmas of selecting and combining in detail. To verify the effectiveness of the proposed procedures, a typical foreign exchange rate experiment is performed in Section 13.3. And Section 13.4 concludes this chapter and points out some future research directions.

13.2 Two Dilemmas and Their Solutions

Assume $\{y_1, y_2, \ldots, y_t\}$ to be a time series. At time t for $t > 1$, we are interested in forecasting the next value y_{t+1} based on the past observations of $y_1, y_2, \ldots, y_t$. Here we focus on one-step-ahead point forecasting.

As mentioned previously, two dilemmas are often encountered in time series modeling and forecasting. For a specific time series, the first dilemma is whether to select a "true" model or to combine multiple individual models for the different/dissimilar methods; another dilemma is whether to select a "true" model or to combine multiple models with different parameters for the same/similar method. That is, one faces two types of choice for a specified time series. If we do not decide on a determined method for the time series, then multiple different/dissimilar methods are tested by trial and error, thus one has to face the first dilemma. If a method is confirmed that is appropriate for the time series, multiple versions with different parameters of the determined method are generated: i.e., their model selection criterion values are rather close. Thus one has to face the second dilemma. This is graphically represented in Fig. 13.1. From Fig. 13.1, we can clearly see the two dilemmas for time series forecasting. In short, this is a problem of selecting or combining. Subsequently, we further explore the solution of the two dilemmas.

Many selection methods are often utilized in solving the selecting or combining problem for multiple candidate models. However, as mentioned earlier, the potentially large variability in the resulting forecast is common to all model selection methods. When there is substantial uncertainty in finding the best model, alternative methods such as hybrid or ensemble model should be taken into account. An interesting issue of selecting or combining is: how should we measure the model selection criteria or the resulting prediction? This is the key to solving the selecting or combining dilemma. It should be noted that two judgmental criteria are required to define dealing with two types of dilemma. In order to solve the dilemmas, a double-phase-processing procedure is proposed.

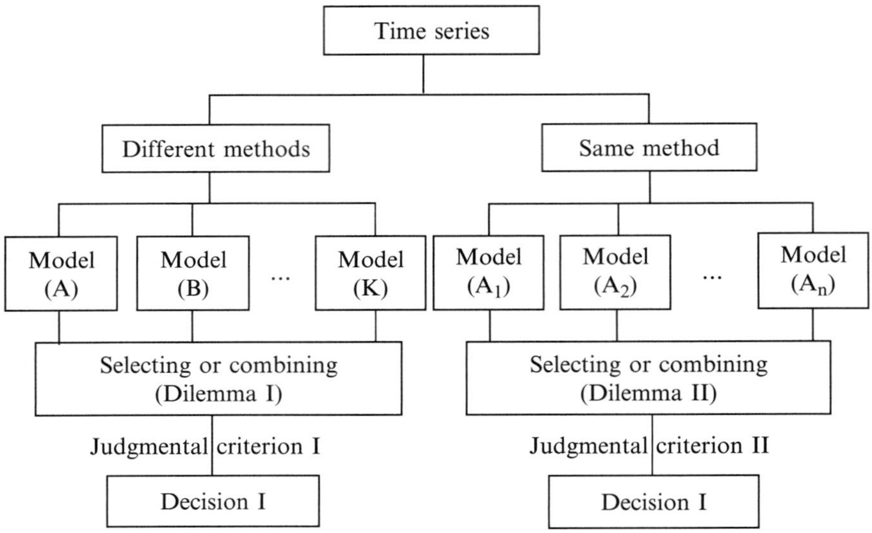

Fig. 13.1. Two dilemmas of time series forecasting

In the first phase, a solution to the first dilemma is provided. In the case of whether to select or combine the different/dissimilar methods, the relative prediction performance stability (S) is introduced. Here the conception of "prediction performance stability" is that time series prediction performance with different/dissimilar model is stable. In other words, for a specified time series, if the difference between the performance indicator for the best candidate model and the performance indicator for the remaining models is very large, then the prediction performance for the time series is unstable and another strategy, such as hybrid or ensemble method, should be considered. Otherwise, the time series prediction performance is stable and then selecting the "best" model from the different/dissimilar candidates may be a wise decision. To be concrete, assume that a different model $\{A, B, \ldots, K\}$ is determined based on all the observations $\{y_i\}$ (i =1, 2, …, k) and use them as predictors. Accordingly, in the same prediction period, the prediction performance indicator (e.g., root mean squared error ($RMSE$)), defined by $\{PI_i\}$ (i =1, 2, …, k) and its volatilities (σ_i) (i =1, 2, …, k) with the different evaluation sets, are calculated. Hence, the relative prediction performance stability (S) is calculated within the framework of the different evaluation sets.

$$S_i = \frac{\left(\frac{1}{m}\sum_{j=1}^{m} PI_{ij}\right)\Big/\sigma_i}{Min_i\{PI_i\}/\sigma_{PI_i}} \quad (i = 1, 2, \ldots, k) \tag{13.1}$$

where $PI_i = \sum_{j=1}^{m} PI_{ij}\Big/m$.

By comparing the calculated value S_i and the pre-defined threshold value S_θ, we can judge whether to select a "true" model or combine multiple different/dissimilar models, i.e.,

$$\text{Decision} = \begin{cases} \text{Selecting}, & \text{if } S_i \geq S_\theta, \\ \text{Combining}, & \text{if } S_i < S_\theta. \end{cases} \quad (i = 1, 2, \ldots, k) \tag{13.2}$$

From Equation (13.2), we can see that if the all the S_i are larger than the threshold value S_θ, a model with the smallest value of PI is selected. If the opposite is true, the combining strategy is recommended. In other words, once the value of S_θ is determined, we can make corresponding decisions for the first dilemma. It should be noted that it is hard to determine a rational value of S_θ. Furthermore, once S_θ is determined, we face a probable situation: some values of S_i larger than S_θ and others smaller than S_θ. If this is the case, how to deal with the situation is still a problem. These possible problems are worth exploring further in the future work.

When the selecting decision is made in the first phase, the second dilemma is generated. In the second phase, we use the "robustness" judgmental criterion. When the robustness of the selected model based on some model selection criteria is strong (i.e., when one model is strongly preferred in terms of the corresponding selection criterion), the model should be selected rather than combining from the multiple candidate models with different parameters for the same method. On the other hand, when the robustness of the selected model is weak, model combining should be considered in order to obtain good prediction performance. Here the definition of model robustness has been a focus issue.

In fact, model robustness can be defined in such a way as follows. The selected model should be robust in the sense that it is indifferent to a radical change to a small portion of the data or a slight change in all of the data (Yu et al., (2005b)). Simply speaking, if a model is robust, a minor change in the data should not change the outcome dramatically. The idea of measuring model robustness is to examine its consistency with different data sizes. Suppose that the model $\hat{M}_n$ is selected by the model selection criterion based on the observations $\{y_i\}$ (i =1, 2, …, n). Let k be an integer between 1 and n-1. For each j in $\{n\text{-}k, n\text{-}k\text{+}1, \ldots, n\text{-}1\}$, applying the model selection criteria to the data $\{y_i\}$ (i =1, 2, …, j) and let $\hat{M}_j$ denote the selected model. Then let R (robustness metric) be the percentage of time series for which the same model ($\hat{M}_n$) is selected, i.e.,

$$R = \sum_{j=n-k}^{n-1} I\{\hat{M}_j = \hat{M}_n\} \Big/ k \tag{13.3}$$

where $I\{\}$ denotes the indicator function: that is, if $\hat{M}_j = \hat{M}_n$, then $I\{\hat{M}_j = \hat{M}_n\}$ =1, otherwise $I\{\hat{M}_j = \hat{M}_n\}$ =0. The rationale behind the consideration of R is quite clear: removing a few observations should not cause much change for a stable model. In addition, the integer k should be chosen appropriately. On one hand, we want to choose small k so that the selection problems for j in $\{n\text{-}k, n\text{-}k+1, \ldots, n\text{-}1\}$ are similar to the real problem with full observations. On the other hand, we want k not to be too small so that R is reasonably stable (Yang (2000)). Once the rational threshold value R_θ is determined, we can make the decision of whether to select or combine by judging the robustness of the selected model in the second dilemma, i.e.,

$$\text{Decision} = \begin{cases} \text{Selecting,} & \text{if } R \geq R_\theta, \\ \text{Combining,} & \text{if } R < R_\theta. \end{cases} \tag{13.4}$$

That is, if the robustness value of the selected model is larger than the predefined threshold value, then the selecting strategy is preferred; else the combining strategy is preferred. Although Yang (2000) also pointed out the same question, his work was only limited to the selection of the same method. Thus, the dilemmas we present here are wider and the alternative solution is more general. Therefore, the proposed procedure is different from previous work.

To summarize, the proposed double-phase-processing procedure is executed as follows. In the first sub-procedure, we use the performance stability indictor to deal with the first dilemma (Equations (13.1)-(13.2)). In the second sub-procedure, we use the robustness metric to treat the second dilemma, as Equations (13.3)-(13.4) shows. With respect to the specified time series, common practice is to use multiple dissimilar methods to model the problem by trial and error and then introduce the judgmental criteria to make the corresponding decisions. That is, for many different/ dissimilar models, if the time series model has stable performance, then a selecting decision is made, otherwise the combining strategy is recommended. If the model is robust, then we select a "best" model from the multiple candidates with different parameters, otherwise we combine the candidates into a single time series forecast. A flow chart of proposed procedure is presented in Fig. 13.2.

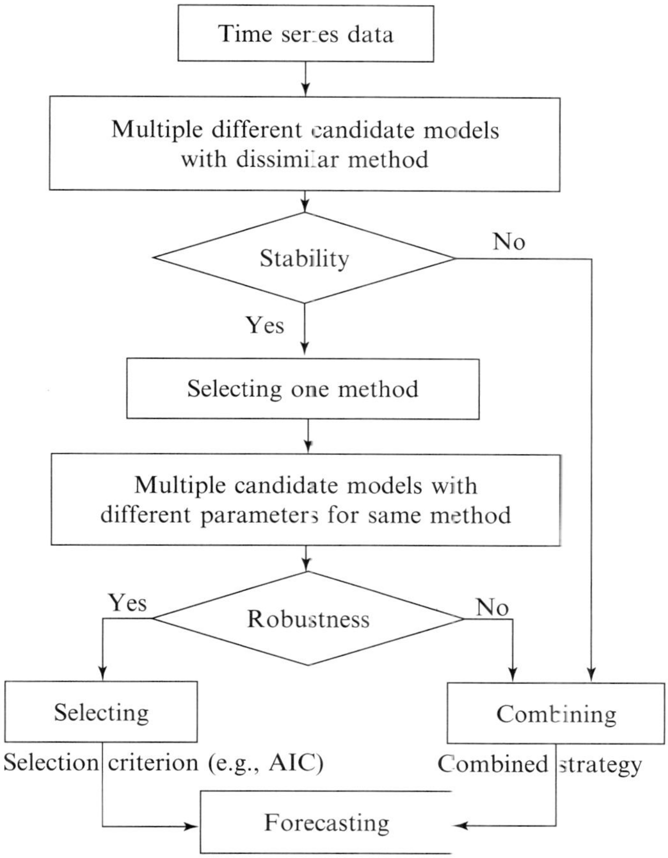

Fig. 13.2. The double-phase-processing procedure of time series forecasting

Relying on the proposed double-phase-processing procedure, the above two dilemmas can be solved. This gives a clear decision on whether to select an appropriate model from multiple candidates or to combine these models for a specified time series. In practical application, however, the combining strategy is often selected due to the uncertainty and volatility of time series formation.

In order to verify the effectiveness of this proposed procedure, a typical foreign exchange rate series, British pounds against US dollar (GBP/USD), are used as a testing target in the following section.

13.3 Empirical Analysis

In this section, many experiments about a typical foreign exchange rates series (GBP/USD) are conducted to verify the efficiency of the proposed procedure. The foreign exchange (GBP/USD) data used in this chapter are monthly and are obtained from Pacific Exchange Rate Service (http://fx.sauder.ubc.ca), provided by Professor Werner Antweiler, University of British Columbia, Vancouver, Canada. We take monthly data from January 1971 to December 2000 as in-sample data sets (360 observations) and take the data from January 2001 to December 2005 as out-of-sample data sets (60 observations), which are used to evaluate the good or bad performance of predictions based on evaluation measurements. In order to save space, the original data are not listed here, and detailed data can be obtained from the website or from the authors.

To perform these experiments, four forecasting methods – the ARIMA model, exponential smoothing (ES) model, simple moving average (SMA) model and artificial neural network (ANN) model with three-layer network structures – are used in the experiments. In addition, in order to compare the prediction performance, the root mean square error (RMSE) is used in this chapter.

In accordance with the procedure proposed in Section 13.2, we use the foreign exchange rate series to verify and test the proposed procedure. Some different forecasting methods are used to fit the specified time series, thus facing the first dilemma (whether to select one method from the alternatives or whether to combine these candidates into a single forecast). If the selection strategy is preferred, then the second dilemma is generated (whether to select one model from multiple candidate models with different parameters or whether to combine these candidate models with different parameters for prediction purposes).

As mentioned in experimental plans, the ARIMA, ES, SMA and ANN models are used to fit the time series and to select an appropriate model for every method. Then we use the selected model to predict the future value of time series. For convenience, the testing sets (or evaluation sets) are presented for verification. Accordingly, the performance indicators are calculated. In this experiment, two examples are presented. In view of the judgment rule of performance stability, the corresponding decision can be made. Here we assume the threshold value $S_\theta = 2.00$ and $R_\theta = 0.80$.

In the experiment, the results of foreign exchange rates forecasting performance indicators with different evaluation sets are shown in Table 13.1.

Table 13.1. RMSE of different methods with different evaluation sets*

Evaluation sets	PI_{ij}	ARIMA	ES	SMA	ANN
10 observations	RMSE	2.34×10^{-2}	1.86×10^{-2}	2.49×10^{-2}	1.86×10^{-2}
20 observations	RMSE	2.08×10^{-2}	2.01×10^{-2}	2.43×10^{-2}	1.34×10^{-2}
30 observations	RMSE	1.91×10^{-2}	1.89×10^{-2}	2.14×10^{-2}	1.28×10^{-2}
	$PI_{i\bullet}$	2.11×10^{-2}	1.92×10^{-2}	2.35×10^{-2}	1.49×10^{-2}
	σ_i	2.16×10^{-3}	0.79×10^{-3}	1.87×10^{-3}	3.19×10^{-3}

*PI: performance indicator; ARIMA: autoregressive integrated moving average; ES: exponential smoothing; SMA: simple moving average; ANN: artificial neural network; RMSE: root mean square error.

From Table 13.1, we can calculate the performance stability indicator in accordance with Equation (13.1). First of all, we find that the smallest value of PI_i is 2.6916, i.e., $\text{Min}_i\{PI_i\} = 0.0149$. Then, the values of S_i are calculated as follows:

$$S_1 = (2.11\times10^{-2}/2.16\times10^{-3})/(1.49\times10^{-2}/3.19\times10^{-3})=2.0914$$
$$S_2 = (1.92\times10^{-2}/0.79\times10^{-3})/(1.49\times10^{-2}/3.19\times10^{-3})=5.2033$$
$$S_3 = (2.35\times10^{-2}/1.87\times10^{-3})/(1.49\times10^{-2}/3.19\times10^{-3})=2.6905$$

Because all $S_i > S_\theta = 2.00$ (i =1, 2, 3) for different evaluation sets, the selection strategy is recommended. The ANN model is selected as a "true" model for prediction purpose because it has a minimal PI value. However, if we assume S_θ to be 6.00, then the combination strategy is preferred. In addition, if S_θ is equal to 4.00, it is hard to judge between the two strategies and further exploration is needed. This implies that determining an appropriate S_θ is extremely important.

When the selection strategy is preferred, as mentioned earlier, we face the second dilemma. Here we take the ANN model as an example for further explanations. First of all, we select an appropriate ANN model based on the full observations by training and learning. Then we change the sample sizes and fit the series again and again. Meanwhile, the robustness metric (R) is calculated. Finally, the corresponding decision can be made by comparing the value of R with the threshold value (R_θ). The experimental results are presented in Tables 13.2-13.3.

Table 13.2. The process of method selection

Methods	ANN (input node, hidden node, output node)							
	(3,2,1)	(3,3,1)	(3,4,1)	(3,5,1)	(4,2,1)	(4,3,1)	(4,4,1)	(4,5,1)
RMSE	0.0152	0.0158	0.0131	0.0169	0.0139	0.0153	0.0129	0.0148

Table 13.3. The robustness testing process ($k = 30$)*

Sample size	102	103	104	105	106	107	108	109	110	111
ANN(I, H, O)	(3,3,1)	(4,3,1)	(4,4,1)	(4,4,1)	(4,3,1)	(4,4,1)	(4,4,1)	(3,5,1)	(4,4,1)	(4,4,1)
$I_j\{\hat{M}_j = \hat{M}_n\}$	0	0	1	1	0	1	1	0	1	1
Sample size	112	113	114	115	116	117	118	119	120	121
ANN(I, H, O)	(4,4,1)	(4,5,1)	(4,4,1)	(3,4,1)	(4,4,1)	(4,4,1)	(4,5,1)	(4,4,1)	(4,4,1)	(4,4,1)
$I_j\{\hat{M}_j = \hat{M}_n\}$	0	1	1	0	1	1	0	1	1	1
Sample size	122	123	124	125	126	127	128	129	130	131
ANN(I, H, O)	(4,4,1)	(4,4,1)	(4,4,1)	(4,4,1)	(4,4,1)	(4,4,1)	(4,4,1)	(4,4,1)	(4,4,1)	(4,4,1)
$I_j\{\hat{M}_j = \hat{M}_n\}$	1	1	1	1	1	1	1	1	1	1

*If $\{\hat{M}_j = \hat{M}_n\}$, then $I_j\{\hat{M}_j = \hat{M}_n\}$=1, else $I_j\{\hat{M}_j = \hat{M}_n\}$=0; ANN (I, H, O): ANN (input node, hidden node, output node).

From Table 13.2, we select the ANN(4,4,1) as an appropriate model using the corresponding model selection criteria. By changing the sample size, we can make the corresponding decisions. From Table 13.3, we can find that different k values often lead to different robustness metric values. When k is equal to 20, the value of R is 0.85 ($R > R_\theta$), then the decision of selecting ANN(4,4,1) as a predictor is rational. However, when k is equal to 30, the value of R is 0.7667 ($R < R_\theta$), then ANN(4,4,1) should be rejected for the time series. Combining multiple models with different parameters into a new forecast is a good choice for prediction purpose. Thus, it is very important to choose k with an appropriate value because this value affects our final decisions to a great extent. Unfortunately, the determination of k still depends on the trial and error and is worth further exploring in the future research.

13.4 Conclusions and Future Directions

In this study, we propose a double-phase-processing procedure to treat the two dilemmas about model selection problem. Based on the proposed procedure, a typical foreign exchange rate, GBP/USD, is used as a testing target. Experimental results obtained reveal that the proposed procedure is useful for dealing with the problems of selecting or combining in the time series forecasting including foreign exchange rates forecasting problem.

However, some crucial issues are generated and need to be further addressed in the future. One important issue is the determination of the threshold values, such as the threshold value of performance stability and robustness metric, since they affect the final decision-making judgments. Another is the determination of the value of k in the computation of the robustness metric, and the size of the evaluation period in the calculation of performance stability. In addition, the other reasonable judgmental rules of selecting one method or combining multiple methods are worth exploring in further research.

Part VI: Developing an Intelligent Foreign Exchange Rates Forecasting and Trading Decision Support System

14 Developing an Intelligent Forex Rolling Forecasting and Trading Decision Support System I: Conceptual Framework, Modeling Techniques and System Implementations

14.1 Introduction

Decision support system (DSS) is a powerful tool which can be used to support decision makers in making strategic decisions (Quah et al., 1996). One key objective of DSS is to improve management judgment (Kean and Morton, 1978). From the viewpoint of management, all forecasts are considered as a prerequisite for effective decisions that are based upon planning (Armstrong, 1983). Effective foreign exchange (forex) trading decision is usually dependent upon effective forex forecasting. Tsoi et al. (1993) have shown that forex forecasting on direction is more important than the actual forecast itself in terms of determining the profitability of a forecasting and trading system. A more comprehensive survey on forex forecasting can refer to Yu et al. (2005e, Huang et al., 2004a, 2006) for more details. Financial market (e.g., stock market or foreign exchange market) is a rather complicated environment. Traders must predict market price movements in order to sell at high points and to buy at low points. Therefore, forecasting often plays an important role in the process of decision-making in financial market.

Forex market is often influenced by many related factors, including political events, general economic conditions, and traders' expectations. Forex forecasting is difficult because of its high volatility and complexity (Yu et al., 2005c). However, investors often have to make some hard decisions over forex investment. Therefore, predicting forex movement direction to provide valuable trading decision information for financial institutions and individual investors is very important and necessary in financial market. The real challenge for forex forecasting is always to beat the market by more accurate forecasting results. This chapter develops an advanced intelligent forex rolling forecasting and trading decision support system (FRFTDSS) integrating a back-propagation neural network (BPNN)-based forex rolling forecasting sub-system (BPNNFRFS) and a web-based forex trading decision support sub-system (WFTDSS) to predict forex movement direction and support forex trading decisions for investors in financial market (Lai et al., 2005; Yu et al., 2005h, 2006f). This study about this intelligent integrated system is divided into two parts: The first part describes BPNN-based forex forecasting method, the system architecture, main functions and operation of the developed DSS system. The BPNN-based forex forecasting method worked out for this system can outperform other commonly used forex forecasting systems. This part will formulate this chapter. The second part will do a comparative study between the currently developed forex forecasting and trading DSS system and other commonly used ones, as well as an empirical assessment on the overall forecasting and trading performance of the developed system. This part will formulate next chapter, which will be illustrated in Chapter 15.

The reminder of this chapter is organized as follows. Section 14.2 describes the general framework architecture and main functions of the advanced intelligent system. Modeling approach and quantitative measurements used in the system are presented in Section 14.3. Subsequently, the development, implementation and operation of the intelligent system are discussed in Section 14.4. Section 14.5 concludes this chapter.

14.2 System Framework and Main Functions

In this study, our developed forex rolling forecasting and trading decision support system (FRFTDSS) is implemented by integrating a BPNN-based forex rolling forecasting sub-system (BPNNFRFS) and a web-based forex trading decision support sub-system (WFTDSS). The main design goal of this integrated system is to generate accurate short-term forex forecasting results and support forex trading decisions for forex traders and investors. The system framework of FRFTDSS is shown in Fig. 14.1. FRFTDSS consists of eight main components, which are elaborated as follows.

The *man-machine interface* (MMI) is a window through which users interact with the integrated system. It handles all the inputs/outputs between users and the system. It also handles all the communications with knowledge engineers or domain experts during the development of the knowledge base.

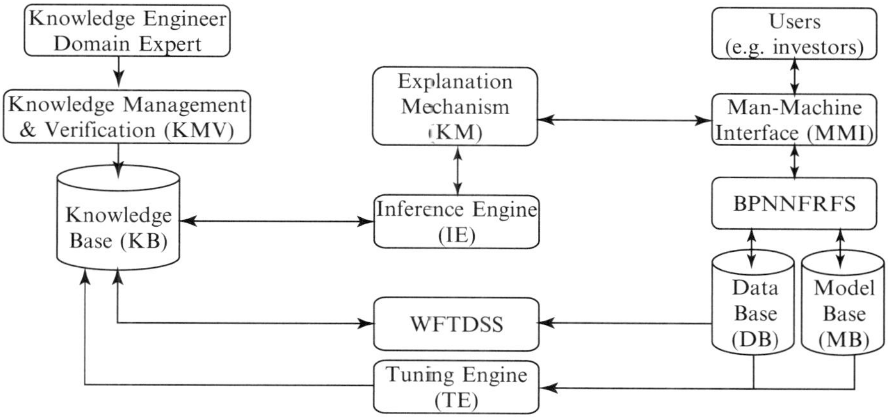

Fig. 14.1. The system framework of the FRFTDSS

The *knowledge base* (KB) is the aggregation of domain materials, algorithms and rules from knowledge engineers and domain experts, as well as from WFTDSS. This component is important in determining the quality of the system

The *database* (DB) consists of data collected from real forex price and forex prediction results from BPNNFRFS. It provides data support for BPNNFRFS and WFTDSS.

The *model base* (MB) mainly includes BPNN model from BPNNFRFS and various decision models from WFTDSS. These models are the cores of the system modules. In addition, the DB and MB can be used to fine-tune knowledge in order to adapt to a dynamic situation.

The *inference engine* (IE) and *explanation mechanism* (EM) provide automatic criteria comparison and interpretation of rules. With the help of this information, traders and investors can make a decision to buy, sell, or hold foreign currencies.

The *Knowledge management and verification* (KMV) has following functions: it deletes existing rules, adjusts existing rules, adds new rules to the knowledge base using a knowledge acquisition tool and verifies the knowledge base to check consistency, completeness and redundancy. There are hundreds of rules in the knowledge base which represent domain experts' heuristics and experience. Using a knowledge acquisition tool, domain experts specify their rules for the knowledge base and describe the rules with the format "IF…THEN…". Then the knowledge acquisition tool automatically converts the rules into an inner encoded format. After new rules are added, the knowledge base verifier checks for any inconsistency, incompleteness or redundancy that might have arisen as a result of adding rules (Nedovic and Devedzic, 2002).

The *Tuning engine* (TE) adjusts knowledge after analyzing recent forex trends from the database. In a dynamic situation, continuous knowledge tuning is needed because old knowledge can no longer be useful over the time.

The conceptual framework of intelligent forex rolling forecasting and trading decision support system (FRFTDSS) consists of two sub-systems: one is the BPNN-based forex rolling forecasting sub-system (BPNNFRFS), which is an online training and forecasting sub-system; the other is the web-based forex trading decision support sub-system (WFTDSS), which is the daily online trading decision sub-system, as shown in Fig. 14.1. In addition, BPNNFRFS, using burgeoning artificial intelligence (AI) technology (i.e., BPNN here), provides forecasting results of some main international currencies to the forex trading decision support sub-system (WFTDSS); while WFTDSS is developed using decision models and web technology to provide some specific trading suggestions for investors (Lai, et al., 2005; Yu et al., 2005h, 2006f). Therefore, the former can provide data support for the latter and the latter feeds back forecasting performance information to the former for further improving forecast.

The developed forex forecasting functions of the system include one-day-ahead forecasts, three-day-ahead forecasts, and five-day-ahead forecasts. Forex trading decision suggestions for the three types of forecasting are generated by WFTDSS sub-system. In addition, some auxiliary functions are also designed, which include forex real-time quotes, online forex e-service, financial tools and forex market information. The implementation of these functions will be illustrated in Section 14.4.

14.3 Modeling Approach and Quantitative Measurements

This section presents the modeling approach and quantitative measurement of FRFTDSS. First, it describes the modeling approach for the first sub-system — BPNNFRFS, and then the modeling approach and quantitative measurements for the second sub-system, WFTDSS, are presented. The development of FRFTDSS integrating BPNNFRFS and WFTDSS will be described in Section 14.4.

14.3.1 BPNN-Based Forex Rolling Forecasting Sub-System

As noted earlier, due to the high volatility, irregularity and noisy market environment, it is difficult for investors to accurately predict forex movement direction. According to prior academic studies, however, movements in market prices seemed not to be random, and rather, they behave in a highly nonlinear, dynamic manner (Tsaih et al., 1998). Yao and Tan (2000) argued that forex markets were fractal and not random processes. Furthermore, Yu et al. (2005e) showed that forex markets were predictable from an artificial neural network perspective. Thus exchange rate prediction seems possible. This study aims to develop a forex forecasting sub-system with a good user interface and system flexibility to accurately predict forex movement direction employing BPNN technique.

A BPNN (Rumelhart et al., 1986) generally consists of an input layer, an output layer and one or more intervening layers that are referred to hidden layers, as shown in Fig. 14.2. The hidden layers can capture non-linear relationships between variables. Each layer consists of multiple neurons that are connected to neurons in adjacent layers. Since these networks contain many interacting non-linear neurons in multiple layers, the networks can capture relatively complex phenomena. The basic principle is that it iteratively adjusts network's parameters (i.e., connection weights and node biases) to minimize error functions using certain training algorithm, and thus find satisfactory solutions for specified problems.

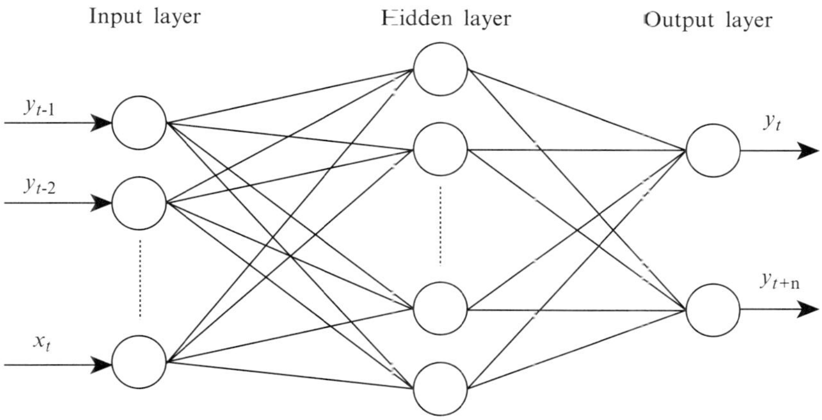

Fig. 14.2. A typical BPNN structure with internal and external effects

Developing a BPNN-based forex forecasting sub-system follows following phases, as illustrated in Fig. 14.3. Fig. 14.3 includes four phases and nine steps. Activities in each of these steps are described below.

14.3.1.1 Phase I: Data sampling

To develop a BPNN model for a forecasting scenario, training, validating and testing data need to be collected. However, data collected from various sources must be selected according to corresponding criteria. As Kuo et al. (2001) revealed, a BPNN model's success mostly depends upon the quality of training data.

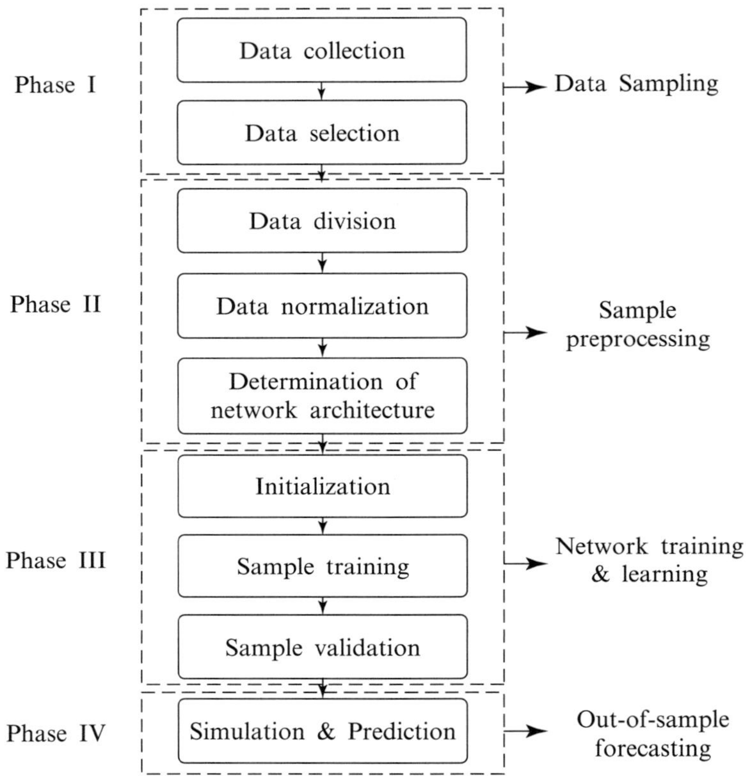

Fig. 14.3. A flow chart of BPNN-based forecasting sub-system

In this phase, the first step is data collection process. Forex markets are unstable markets with high volatilities. Furthermore, a number of qualitative factors, e.g., macro-economical or political events, may substantially influence the forex price movement. Even investors' psychology could result in forex price oscillations (Kuo et al., 2001). Thus, collecting high-quality and related training sample data for BPNN implementation is necessary. If unrelated or low-quality training data are selected, BPNN can only "memorize" specific patterns rather than extract essential features from training samples (Freeman and Skapura, 1991), and thus lower the ability of network generalization.

In the case of forex, many factors including economic development of related countries, international trade conditions, inflation rate and interest level, all influence forex fluctuation. Therefore these factors should be taken into consideration. In this study, forex time series data are collected as internal effects. In addition, GDP, exports and imports, CPI and interest rates of related nations are collected from IFS CD-ROM as external factors. As shown in Fig. 14.2, y denotes the internal factor (i.e., time series data), x represents the external effects (i.e., some related factors affecting forex price).

After finishing the data collection, sample selection, including input variable determination and sample size determination, is very important for the whole performance of neural networks. Generally, input variable selection is required to satisfy following three conditions: (1) input variable should adequately represent fundamental features that a network must detect in order to obtain correct outputs; (2) input data should provide sufficient variation to allow generalization and discourage "memorization"; and (3) input data should not contain contradictory examples, i.e., examples with identical inputs with different outputs (Pendharkar, 2001). In our study, forex price, GDP, exports and imports, CPI are selected based on the above three conditions. Interest rate is not included due to its incomplete data.

Having chosen appropriate input variables, the next step is the determination of appropriate lags or sample size for each of input variables (Maier and Dandy, 2000). In the same way, determination of the sample size is another factor that can affect a neural network's forecasting capability. Redundancy in training data is undesirable as it slows down training, particularly when the entire training set is presented to a network between weights updates (Rojas, 1996). Hence, a network needs an appropriate number of sample sizes. Refines et al. (1997) proposed a stepwise method with two phases. Narendra and Parthasarathy (1990) suggested an approach incorporating time structure into neural network inputs to determine appropriate lags. Unfortunately, although both methods are suitable, the determination of sample size remains an art that must rely on trial and error in practice. In this article, we use the stepwise method proposed by Refines et al. (1997) to determine an appropriate sample size.

14.3.1.2 Phase II: Sample preprocessing

The second phase is sample preprocessing. It includes three steps: data division, data normalization and determination of network architecture.

Data division. In neural network application, it is a common practice to split available sample data into two sub-sets: a training set and a testing set. The training set is used for neural network model development and the testing set is used to evaluate the forecasting capability. Sometimes a third set, called the validation set, is used to avoid the over-fitting problem or to determine the stopping point in training process.

There is no general solution to splitting the training set and the testing set. Brainmaker software randomly selects 10% of the facts from the data set and uses them for testing. Yao and Tan (2000) suggested that historical data are divided into three portions: training, validation and testing sets. The training set contains 70% of the collected data, while the validation and the testing sets contain 20% and 10% respectively. The division is based on a rule of thumb derived from the authors' experience. Many other researchers just give the division directly, not justifying it. In this study, we extended and used the auto-correlation (AC)-based data division method based on the work of Huang et al. (2004). Through simulation experiments, we found that the extended approach worked very well.

Data normalization. This is a very important step. In any model development process, familiarity with available data is of the utmost importance (Maier and Dandy, 2000). ANN models are no exception, and data normalization can have a significant effect on model performance (Kaastra and Boyd, 1995). It is important to note that available data need to be divided into their respective subsets (e.g. training, validation and testing) before any data normalization is carried out (Burden et al., 1997).

Furthermore, the degree of care invested in normalizing the data is of decisive importance to the network's learning speed and the quality of approximation it can attain. Every hour invested in preprocessing the data may save some days in training a network. Once all input samples have been selected, the data needs to be transformed into certain appropriate range for network learning. Various normalization methods are generally employed to this end. For example, Tenti (1996) normalized inputs to zero mean and two standard deviations. In the study by Hu et al. (1999), all inputs to BPNN were linearly normalized to [0, 1], but in the study conducted by Pendharkar and Rodger (2003), all inputs to BPNN were nonlinearly normalized to (0, 1) using logistic function $f(x) = 1/(1+e^{-x})$. El Shazly and El Shazly (1999) suggested that data should be manipulated and converted to the required format for further processing. In Lisi and Schiavo's study (1999), the log-differenced data were scaled linearly in the range of 0.2-0.8 in order to adapt them to the output range of the sigmoid activation function. Qi and Zhang (2001) applied natural logarithm transformation to raw data to stabilize the series. They considered that if an augmented Dick-Fuller (ADF) test showed that the transformed time series contained a unit root, thus the first order difference was applied.

In this article, we normalized the original data to (0, 1) using the logistic function proposed by Pendharkar and Rodger (2003). The motivation of this method comes from the following two perspectives. First of all, our sample data selection includes different variables and these variables span different ranges. In order to ensure that all variables receive equal attention during training process, standardization is necessary. On the other hand, the original data were transformed in the range of 0-1 in order to adapt them to the output range of the logistic transfer function, as revealed by Lisi and Schiavo (1999).

Determination of network architecture. This is also a key step. Network architecture determines the number of connection weights and the way information flows through a network. Determination of appropriate network architecture is one of the most important, but also the most difficult, tasks in the model building process. There are two problems that need to be taken into account at this step.

(i) *Type of connection and degree of connectivity.* There are two main types of network connections. Traditionally, feed-forward networks, where nodes in one layer are only connected to nodes in the next layers, have been used for prediction and forecasting applications. However, recurrent networks, where nodes in one layer can be connected to nodes in the next layer, the previous layer, the same layer and even to themselves, have been proposed as alternatives (Maier and Dandy, 2000).

There are also two degrees of connectivity between the nodes in a network: full connectivity and partial connectivity. Generally, the degree of connectivity has an impact on the number of connection weights that need to be optimized. In many instances, fully connected networks are utilized. However, for most studies, partial connectivity has been used (Maier and Dandy, 2000; Zhu, 1994). In our study, typical feed-forward networks (i.e., BPNN) with partial connectivity are adopted for this sub-system. The main reason is that the BPNN with partial connectivity is a class of the most stable neural network and thus is widely used and admitted in most applications.

(ii) *The number of nodes in every layer.* As Hornik et al. (1989) claimed, it has been shown that BPNNs with one hidden layer could approximate any function, given that sufficient degrees of freedom (i.e. the number of hidden nodes) were provided. In the practical application of networks, since the number of nodes in the input layer is fixed by the number of model inputs — whereas the number of nodes in the output layer equals the number of model outputs — the critical aspect is the choice of the number of hidden nodes. In the study by Pendharkar and Rodger (2003), the number of hidden nodes was twice the number of input nodes plus one, which is a more common heuristic for smaller sample sizes (Pendharkar, 2001; Bhattacharyya and Pendharkar, 1998). In the case of larger sample sizes, a higher number of hidden nodes are recommended. In most studies, a trial-and-error approach is often adopted to help find the best number of hidden nodes.

In this sub-system, we adopt two-phase robustness analysis approach (Yu et al., 2005b) to determine optimal network architecture. Simulation experiments revealed that the proposed approach worked well.

14.3.1.3 Phase III: Network training and learning

In this phase, it includes three main tasks: initialization, sample training and sample validation. It is the core process of a BPNN model.

Initialization. This step is the start-point of network learning. The main task of this step is to assign initial weights and set transfer (activation) functions for every layer, training epoch size, error function, learning rate, momentum, etc. The initialization process is of importance to the network's ability to learn and generalize.

(i) *Initial weights.* As Maier and Dandy (2000) argued, the results obtained when the back-propagation algorithm is used are sensitive to initial conditions. Generally, the weights are initialized to zero-mean random values. Care must be taken to choose adequate upper and lower bounds $\{-a, a\}$ for the weights. If the value of a is too small, training may be paralyzed; on the other hand, if a is too large, premature saturation of the nodes may occur, which in turn will slow down training and result in the cessation of training at suboptimal levels (Lee and Shen, 1991). The results of empirical studies indicate that there is no single optimum value of a, but a large range for which similar performance is obtained. Faraway and Chatfield (1998) suggested that a number of different sets of random starting values should be used to see whether consistent results are obtained. In our study, we use the method proposed by Faraway and Chatfield (1998) to set initial weights in order to obtain consistent forecasting results. Furthermore, Faraway and Chatfield (1998) also found that this approach is effective and feasible for achieving consistent results.

(ii) *Transfer (activation) function.* The transfer functions that are most commonly used are sigmoidal-type functions such as the logistic and hyperbolic tangent functions. Other transfer functions (e.g. radial basis function and polynomial function) may be used as long as they are continuous and differentiable. Generally, using sigmoidal-type transfer functions in the hidden layers and linear transfer functions in the output layer can be an advantage when it is necessary to extrapolate beyond the range of the training data (Kaastra and Boyd, 1995; Karunanithi et al., 1994). In our study, this latter strategy (i.e., sigmoidal-type transfer functions in the hidden layers and linear transfer functions in the output layer) is adopted. In like manner, most studies also proved that this strategy is appropriate for the BPNN forecasting.

(iii) *Training epoch size.* The training epoch size is equal to the number of training samples presented to the network between weight updates. If the epoch size is equal to 1, the network is said to operate in on-line mode. If the epoch size is equal to the size of the training set, the network is said to operate in batch mode. In many applications, batch mode is the preferred option, as it forces the search to move in the direction of the true gradient at each weight update. However, the larger the training epoch size, the longer the training time in general. So far, there is no optimal epoch size for network. Generally, trial and error is used as a usual way to select the epoch size. In this study, 100, 500, 1000, 1500 and 2000 epochs is simulated for the testing case of JPY, as revealed in Table 14.1. The experiment results show that 500 epochs is the good selection in terms of mean squared error (MSE) from Table 14.1.

Table 14.1. The MSE of different training epochs

Training epochs	100	500	1000	1500	2000
MSE	0.1053	0.0756	0.0869	0.0963	0.1125

(iv) *Error function.* The error function is the function that is minimized during training. The mean squared error (MSE) function is most commonly used, but other error functions have also been proposed (Solla et al., 1988). The advantage of using the MSE include that it is calculated easily, that it penalizes large errors, that its partial derivative with respect to the weights can be calculated easily and that it lies close to the heart of the normal distribution.

In our study, direction accuracy is introduced into the error function as our research not only considers the error level but also takes into account forex price movement direction. In general, the directional error function (E_d) (Yu et al., 2006c) is constructed as follows:

$$E_d = \frac{1}{N}\sum_{t=1}^{N} f(d) \times (y_t - \hat{y}_t)^2 \qquad (14.1)$$

where N is sample size, y_t and $\hat{y}_t$ are actual and prediction value respectively, and $f(d)$ is a indicator function, i.e.,

$$f(d) = \begin{cases} p_1, & (y_t - y_{t-1})(\hat{y}_t - y_{t-1}) \leq 0, \\ p_2, & \text{Otherwise.} \end{cases} \tag{14.2}$$

Comparing with the traditional mean square error (MSE), directional forecasting result is considered. That is, incorrectly predicted directions are penalized more heavily than the correct prediction. In this study, $p_1 = 5.0$ and $p_2 = 2.5$ are used in the directional error function.

(v) *Learning rate.* The learning rate is directly proportional to the size of the steps taken in weight space. However, it is worth re-iterating that the learning rate is only one of the parameters that affect the size of the steps taken in weight space. Traditionally, learning rates remain fixed during training (Warner and Misra, 1996) and optimal learning rates are determined by optimization techniques (Yu et al., 2005a). For simplification, learning rate is set to 0.25 in this study and remains fixed during training, as revealed by Warner and Misra (1996). Experimental results show the effect of this setting is very good. Furthermore, most studies also adopted some certain fixed values.

(vi) Momentum. The momentum term may be considered to increase the effective step size in shallow regions of the error surface and can speed up the training process by several orders of magnitude. In general, the momentum term is less than 1.0 for convergence (Maier and Dandy, 2000). In the same way, optimal momentum factor can be obtained by derivation, as shown in Chapter 4. For simplification, momentum factor in this study is set to 0.35 and remains fixed during training. Empirical simulations also show this value is appropriate for our study.

Sample training. After the initialization step, we can start sample learning or training. The sample learning or training step is equivalent to the parameter estimation phase in conventional statistical models. In fact, the process of sample learning is to find a global optimal solution to what is typically a highly non-linear optimization problem. This optimization problem can be carried out using local or global methods. Local methods fall into two major categories: first-order and second-order methods. First-order methods are based on a linear model (gradient descent) whereas second-order models are based on a quadratic model (e.g. Newton's method). However, local optimal methods are easy to trap into the local minima. Therefore, global methods, which have the ability to escape local minima in the error surface and thus are able to find optimal or near optimal weight configurations, are generated. The common global methods include stochastic gradient algorithms, simulated annealing algorithm and genetic algorithms (GAs) (Maier and Dandy, 2000). However, these algorithms are the time-consuming approaches in the network training and learning.

Generally speaking, steepest gradient descent algorithm is widely used. Furthermore, this algorithm has good performance while away from a local minimum. However, it requires a lot of iterations to converge when close to the minimum, and thus the speed of training and convergence is very slow. Thus, in our sub-system, the Levenberg-Marquardt method is introduced as a kind of quick and effective algorithm for improving the speed of convergence. One characteristic of the Levenberg-Marquardt method is that the network's matrix of weights is updated based on the Jacobian matrix, J, collecting the partial derivatives of the network error e with respect to the weights, similar to Chapter 4 and Chapter 8. In other words, the matrix ΔW collecting the corrections of the weights in matrix W is computed according to

$$\Delta W = (J^T J + \mu I)^{-1} J^T e \tag{14.3}$$

where μ is a constant. If μ is sufficiently large, the previous algorithm is similar to the gradient descent algorithm; if μ is equal to zero, the previous rule will be a Gauss-Newton algorithm. Thus, Levenberg-Marquardt algorithm is a very flexible method.

The advantage of the Levenberg-Marquardt algorithm lies in that Levenberg-Marquardt algorithm can accelerate the network's training speed, save training time and achieve better training results, at the same time through escaping from local minima.

Sample validation. This step is also important. Once the training phase has been completed, the performance of the trained network has to be validated on an independent data set using the criteria chosen. If the difference in the error obtained using the validation set is markedly different than that obtained using the training data, it is likely that the two data sets are not representative of the same population or that the model has been over-fitted (Yu et al., 2003). Therefore, it is vital that the validation data should not have been used as part of the training process in any capacity. In our study, we use cross-validation method with "leave one out" procedure to avoid over-fitting problem.

14.3.1.4 Phase IV: Simulation and prediction

This is the last step. After completing the several previous steps, we can extrapolate the training data by simulation and obtain forecasting results. In our study, the prediction results can be obtained as long as the forecasting horizon is set in advance.

To summarize, our procedure for developing the BPNN-based forecasting model are as follows (Yu et al., 2005h, 2006f).

(a) Data collection and input sample preprocessing including sample size determination, input variable selection, data division and data normalization;

(b) Determine the parameters and architecture of network: i.e., transfer function, epoch size, error function, learning rate, momentum, network layers, and nodes of every layer.

(c) Initialize all weights randomly;

(d) Training, where the stopping criterion is either the number of iterations reached or when the error function is lower than a pre-determined value;

(e) Validating the network model by cross-validation method;

(f) Choose a network with the best performance (e.g., minimal error);

(g) Forecast future outcome.

From the previous discussion, we can see that this is a relatively complex process that relates to many professional terms about neural networks. However, easy operating is an important measurement of software system. In order to facilitate forecasters and investors, we develop this easy-use BPNN-based forex forecasting sub-system as a component of the FRFTDSS, as showed below. Fig. 14.4 presents the main user interface and training process interface of the BPNNFRFS.

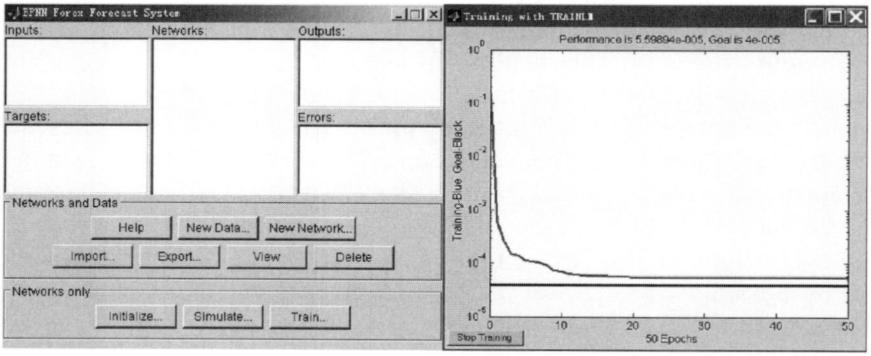

Fig. 14.4. The main user interface and training process interface of BPNNFRFS

In addition, to improve prediction accuracy and performance, rolling forecasting techniques are introduced into this sub-system. In this study, the aim of introducing rolling forecasting into the BPNN-based forex forecasting sub-system is to improve forecast accuracy and achieve satisfactory forecast performance. For space consideration, some introductions are omitted here. Interested readers can refer to Yu et al. (2003) and Lawrence and O'Connor (2000) for more details.

It is worth noting that the BPNN-based forex forecasting sub-system (BPNNFRFS) is built using the Matlab software package, which is produced by Mathworks Laboratory Corporation. For further explanation, the interface of forecasting results is shown in Fig. 14.5. The graphical results in Fig. 14.5 show that the directional forecasts of BPNNFRFS worked well, indicating that BPNNFRFS had a good potential to accurately forecast forex price in volatile foreign exchange markets. Especially, this sub-system can be used independently as a forecasting system

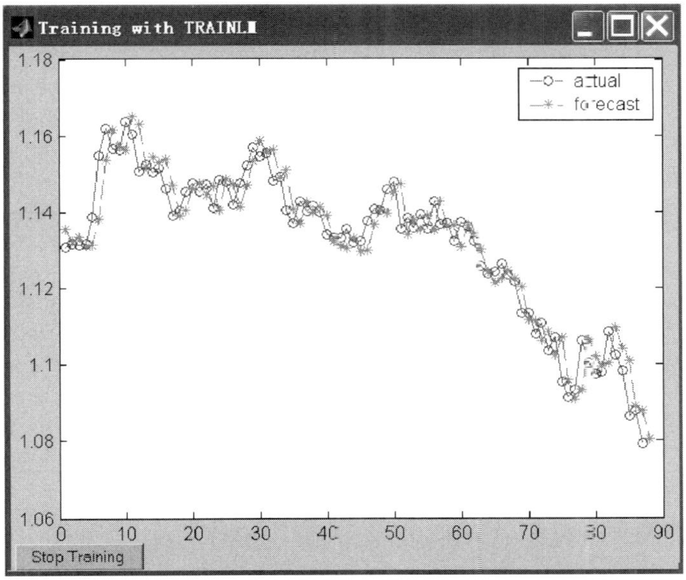

Fig. 14.5. The forecasting results of BPNNFRFS (EUR/USD)

14.3.2 Web-Based Forex Trading Decision Support System

WFTDSS is developed to support decision making in forex trading. It is designed to convert forex price information that is generated by BPNNFRFS into forex movement direction information, thus directly providing sell/buy/hold suggestions for individual investors. In this sub-system, four trading rules are presented first, and then the basic structure of WFTDSS is described (Lai et al., 2005; Yu et al., 2005h, 2006f). In following rules, y_t and $\hat{y}_t$ denote the forex actual and forecasting values at time t, respectively.

14.3.2.1 Trading rules I: Price comparison judgment criterion

The **first** rule is a simple price comparison judgment criterion. In general, this criterion is the simplest of all, and includes the following three rules.

Rule I: If $(\hat{y}_{t+1} - y_t) > 0$, then forex price trend is "*upward*" and the current trading strategy is "*buy*".

Rule II: If $(\hat{y}_{t+1} - y_t) < 0$, then forex price trend is "*downward*" and current trading strategy is "*sell*".

Rule III: If $(\hat{y}_{t+1} - y_t) = 0$, then forex price is "*unchangeable*" and the current trading strategy is "*hold and deposit*".

It is worth noting that use of this rule has no man's participation (e.g. parameter specification), but the forex trading strategies can still be generated by the sub-system automatically.

14.3.2.2 Trading rules II: Cost-adjusted filter judgment criterion

The **second** rule is a cost-adjusted filter judgment criterion. This rule introduces cost as adjusted bands. The objective of the band is to reduce the number of buy (sell) signals by eliminating weak signals when the predicted value $\hat{y}_{t+1}$ and the actual value y_t are very close. After considering funding costs and transaction costs, some trading is not worth executing. Thus, this rule can help a trading system eliminate unnecessary trading and gain more profits. Mathematically the trading rules in their simplest form can be expressed as follows.

Rule I: If $(\hat{y}_{t+1} - y_t) > c$, then the current trading strategy is "*buy*".

Rule II: If $(\hat{y}_{t+1} - y_t) < c$, then the current trading strategy is "*sell*".

Rule III: If $(\hat{y}_{t+1} - y_t) = c$, then the current trading strategy is "*hold and deposit*".

where c denotes funding cost. Here transaction costs are not included because it is noninstallment and the quantity is much larger than the difference between y_t and $\hat{y}_t$.

In this criterion, decision-makers can arbitrarily input the amount of funding costs according to the regulations and personal preferences.

14.3.2.3 Trading rules III: Probability threshold judgment criterion

The **third** rule is a probability-based threshold judgment criterion. The trading probability based on the predictable forex price return is the basis of this trading judgment criterion. Its procedures include three steps.

First, let price of a forex trade at day t be denoted by y_t, then the forex price return R_t is given by

$$R_t = (y_t - y_{t-1}) / y_{t-1} \tag{14.4}$$

If the forex price return R_{t+1} depends on the forex price returns of the previous M days, the forex price return $\hat{R}_{t+1}$ is predicted by

$$\hat{R}_{t+1} = \sum_{i=0}^{M-1} a_i^j R_{t-i} \tag{14.5}$$

where a_i^j denotes parameters and R_{t-i} the forex return on the $(t - i)^{th}$ day.

Based on the predicted value $\hat{y}_{t+1}$ that is produced by BPNNFRFS, the forex price return can also be calculated by

$$\hat{R}_{t+1} = (\hat{y}_{t+1} - y_t) / y_t \tag{14.6}$$

The difference between the two is insignificant from practical experimentation. For convenience, the latter is always chosen in the sub-system.

Second, let the "buy" and "sell" probability of the forex price return be denoted by $B_{t+1(j)}$, $S_{t+1(j)}$, respectively. As the forex price is a stochastic process, the probability $B_{t+1(j)}$ and $S_{t-1(j)}$ for the next day are calculated, respectively, by

$$B_{t+1(j)} = P\{R_{t+1(j)} > 0\} \quad (j = 1, 2, \cdots, N) \tag{14.7}$$

$$S_{t+1(j)} = P\{R_{t+1(j)} < 0\} \quad (j = 1, 2, \cdots, N) \tag{14.8}$$

where j denotes the number of forex candidates.

In the "buy" case, the basic rule is that the predicted forex price of the next day is higher than the current price, i.e. the predicted forex price return $\hat{R}_{t+1}$ is larger than zero ($\hat{R}_{t+1} > 0$). In the "sell" case, the basic criterion is that the predicted forex price of the next day is lower than the current price, i.e. the predicted forex price return should be smaller than zero ($\hat{R}_{t+1} < 0$). It is worth noting that in the "buy" case, the forex with the largest trading probability $B_{t+1(\max)}$ is chosen from the trading probability $B_{t+1(j)}$ of all N Forex candidates by

$$B_{t+1(\max)} = \max\{B_{t+1(1)}, B_{t+!(2)}, \cdots, B_{T+1(N)}\} \tag{14.9}$$

Third, the thresholds for buying and selling, θ_B and θ_S, are set to some certain values in advance.

Up until now, the corresponding trading judgment rules are given by:

Rule I: If $B_{t+1(\max)} \geq \theta_B$, then the trading strategy is "*buy*".

That is to say, the investor should buy if the trading probability $B_{t+1(\max)}$ exceeds a buying threshold θ_B.

Rule II: If $S_{t+1(j)} \geq \theta_S$, then the trading strategy is "*sell*".

In the same way, if the trading probability $S_{t+1(j)}$ of holding forex j is larger than a selling threshold θ_S, the investor should sell.

In this criterion, once the users or decision-makers specify a certain threshold, the optimal trading strategies will be presented explicitly.

14.3.2.4 Trading rules IV: Risk-adjusted trading judgment criterion

The **fourth** or **last** criterion is risk-adjusted forex trading judgment criterion, which originated from the work of Chen and Leung (2004). Their work assumed that the forex market investment returns are determined by three features: forex appreciation, forex depreciation, and interests received from the money market. Let r and r^* be the daily domestic and foreign money market rates, and y_t and $\hat{y}_{t+1}$ be the observed exchange rates at day t and the forecast of exchange rate at day (t+1) respectively. Idealistically, an optimal investment allocation can be made if the domestic and foreign money market rates as well as the observed exchange rates on the current day and those forecast for the next day are known. Nevertheless, the forex forecast is only an estimate and involves uncertainty. Given the notion that an investor is risk averse and can earn risk free interest from the domestic money market, we need to discount the expected rate of return based on a long or a short position by a risk aversion factor, γ. The risk aversion factor γ can take on any real value greater than negative one ($\gamma > -1.0$). For the risk neutral case, γ is zero ($\gamma = 0.0$). The value of γ is greater than zero ($\gamma > 0.0$) if the investor is risk averse. On the other hand, the value of γ is between zero and negative one ($-1.0 < \gamma < 0.0$) if the investor is a risk lover.

Based on this logical framework, we develop a risk-adjusted forex trading rule. The mathematical expressions are as follows (for derivation of these conditions, refer to Chen and Leung (2004)).

Rule I: If $\ln\left(\frac{\hat{y}_{t+1}}{y_t}\right) < (r^* - r)(1+\gamma)$ and $(r^* - r)(1+\gamma) < 0$, then the trading strategy is "buy".

Rule II: If $\ln\left(\frac{\hat{y}_{t+1}}{y_t}\right) \leq \frac{(r^* - r)(1+\gamma)}{2}$ and $(r^* - r)(1+\gamma) \geq 0$, then the trading strategy is "buy".

Rule III: If $\ln\left(\frac{\hat{y}_{t+1}}{y_t}\right) > \frac{(r^* - r)(1+\gamma)}{2}$ and $\ln\left(\frac{\hat{y}_{t+1}}{y_t}\right) > 0$, then the trading strategy is "sell".

Rule IV: If $\ln\left(\frac{\hat{y}_{t+1}}{y_t}\right) \geq (r^* - r)(1+\gamma)$ and $\ln\left(\frac{\hat{y}_{t+1}}{y_t}\right) \leq 0$, then the trading strategy is "hold and deposit".

Compared with the previous three criteria, this criterion seems to be relative complex. However, once the values of r^*, r and γ are specified by users or decision-makers, the optimal trading strategies will also be generated automatically regardless of considering the meaning of conditions.

To summarize, different trading rules may result in different trading strategies. It is hard to say which one is better for the four criteria because of different perspectives. From our experiment evaluations, the simple price comparison judgment criterion is preferred by most business users because it is simple and easy to understand.

Fig. 14.6 shows the basic architecture of WFTDSS sub-system and the data flows between user interfaces, knowledge bases (KB), model bases (MB) and databases (DB). The components of the sub-system are described below.

User-friendly interface. Used by decision-makers and investors to input some data and query conditions, and to select some trading decision methods and retrieve answers from related services. It also displays to the users and allows them to interact with the sub-system.

Model bases. Used to handle models of the many different types, which may be chosen.

Knowledge bases. Used to judge the forex change tendency and determine forex trading decisions.

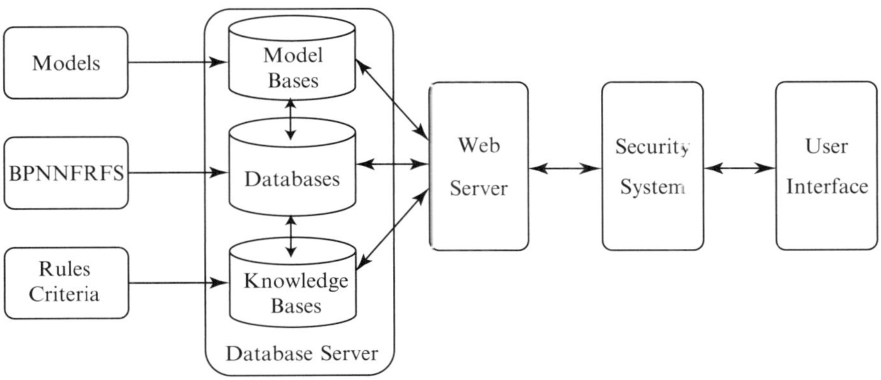

Fig. 14.6. The basic architecture and data flow of the WFTDSS

Databases. Used as a repository of historical data and forecasting data for all model programs. In the meantime, it provides data support for the KB and MB needed.

Security system. Not only used for user protection purposes, but also checks for user access levels to determine the level of flexibility and interaction that the sub-system would allow or provide.

According to the previous WFTDSS architecture and data flow, we develop the web-based forex trading decision sub-system using latest web technology and client/server (C/S) mode, which contains a three-tier structure. The basic construction of three-tier structure is shown in Fig. 14.7. One sample user interfaces of WFTDSS is shown in Fig. 14.8.

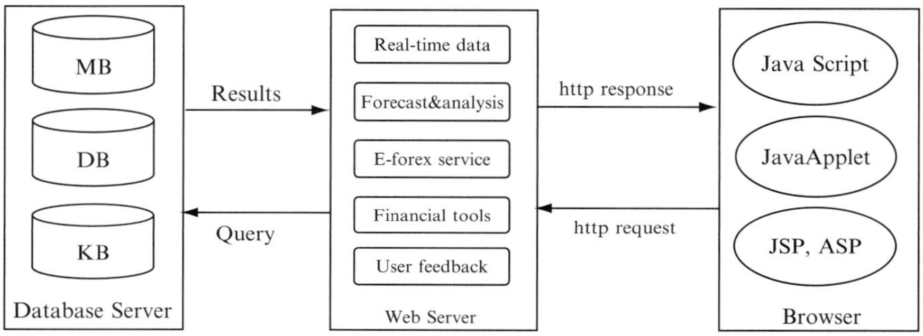

Fig. 14.7. Three-tier structure of WFTDSS

The left part of Fig. 14.8 shows the interface of the trading decision criterion selection, which presents four types of decision criteria alternatives. Of the four, only one is selected by users every time. As shown in right part of Fig. 14.8, the interface of the fourth trading criterion mentioned previously is presented. In the interface, the corresponding optimal trading strategies will be generated by the WFTDSS as long as users or decision-makers can input a value or select one of "option buttons". In the same way, WFTDSS can also be used as an individual trading support system.

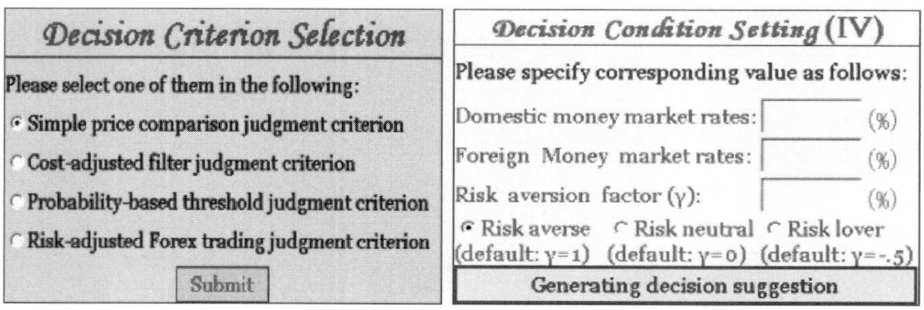

Fig. 14.8. A sample user interface of WFTDSS

14.4 Development and Implementation of FRFTDSS

14.4.1 Development of the FRFTDSS

Based on the system framework and the description of two previous sub-systems, an intelligent forex rolling forecasting and trading decision support system (FRFTDSS) that incorporates BPNNFRFS and WFTDSS is developed. BPNNFRFS provides forex forecasting data; and WFTDSS provides trading rules and trading decisions using prediction results generated by BPNNFRFS. Thus, the integration process is that BPNNFRFS is linked with WFTDSS by data coupling.

As noted earlier, the BPNNFRFS is a forex forecasting sub-system that is developed using Matlab software and a neural network toolbox. This sub-system adopts client/server (C/S) mode. The development and implementation environment of the BPNNFRFS are shown in Table 14.2.

Table 14.2. The development and implementation environment of BPNNFRFS

(1) Development environment	
Development tools:	Matlab 6.1 software package
Database server:	MS SQL Server 7.0
(2) Implementation environment	
Operation system:	Windows series, e.g. Windows9x, 2000 and XP
Client software:	Software developed by this project team

In order to facilitate using the forecasting results produced by BPNNFRFS and transforming the forecasting results into corresponding trading decision information, the WFTDSS is developed for all kinds of users based on web technology. The WFTDSS can be seen as an extension of BPNNFRFS to some extent. We develop WFTDSS with a three-tier structure adopting web technology and client/server (C/S) mode. The advantage of this mode lies in easy operation and easy maintenance because all programs are installed on the server site. Moreover, programs can be edited and revised remotely in the client site if access is authorized. At the client site, the web browser and Java applet handle the user interface and the presentation logic. At the server site, the web server receives all http requests from the web user and propagates the requests to the application server. Communication between the web server and the application logic can be done through CGI, ASP and JSP or other gateway tools. At the database server site, model bases, knowledge bases and databases, which are stored in the database server, are the basis of the WFTDSS. This system mainly provides forex forecasting results, forex market trend analysis and forex trade strategies for business practitioners. At the same time, a series of financial tools are also provided for investors. In addition, the system provides user feedback as an accessory function for users to ask questions, or contact web administrators. In general, the web application server sends the query to the database server and gets the result sets. It prepares the response after a series of processes and returns to the web server for the response to be sent back to the client site. Since the system's kernel is placed on the web server, many investors therefore can use a web browser (e.g. Internet Explorer) to access the web server through HTML language and HTTP protocol without needing to pay attention to any technical details.

With respect to the development of WFTDSS, ASP, JSP and Java technologies are used extensively (Chou, 1998; Fan et al., 1999; Li, 1998; Ralph and George, 2000; Fan et al, 2000; Saatcioglu et al., 2001). The integrated web application development tools (Visual InterDev 6.0), network operation system (Windows 2000 Server edition) and Database server (MS SQL Server 7.0) were also used. The development and implementation environment of WFTDSS is presented in Table 14.3 below.

Table 14.3. The development environment and tools of WFTDSS

(1) Software in server	
Operation system:	Windows 2000 Server
Database server:	MS SQL Server 7.0
Web server:	MS IIS 4.0
(2) Software in client	
Operation system:	Windows series edition, e.g. Windows9x, 2000 and XP.
Browser:	Internet Explorer 4.01 (or higher edition)
(3) Development tools:	Visual InterDev 6.0, FrontPage XP, JSP, ASP (JavaScript and VBScript), Java

14.4.2 Implementation of the FRFTDSS

BPNNFRFS is incorporated into WFTDSS by data coupling in the integrated system. That is, WFTDSS receives the forex data from related website and provide data support for BPNNFRFS training; while the results of BPNNFRFS are sent to database server for WFTDSS. The operation of FRFTDSS is described below.

First of all, we enter the main interface of the FRFTDSS through Internet access, as shown in Fig. 14.9.

Then, we can enter "Data Management" interface by click the "Data Management" button on the client site after being verified by password, and the detailed interface is shown in Fig. 14.10.

Subsequently, through inputting the corresponding computational parameters (e.g., sample size, forecasting horizon, sampling interval, hidden nodes, epoch size etc.) after importing the forex data; then click the link "Server training program activation", the software of BPNNFRFS in the server site is activated and the BPNNFRFS is implemented in the server site according to the corresponding instruction and parameters, as demonstrated in Fig. 14.4.

After training by neural networks, the forecasting results are obtained, as shown in Fig. 14.5. At the same time, the results generated are sent into the corresponding database of the database server.

The forecasting results are utilized by WFTDSS and the corresponding trading strategies are presented by the WFTDSS. By accessing the Internet, these trading decision strategies can be received on-line for organizational investors and individual investors around the world. Fig. 14.11 illustrates two typical FRFTDSS interfaces when the first trading judgment criterion is selected.

To summarize, the intelligent FRFTDSS system has the following advantages and implications. First, this intelligent system can not only predict forex rates and forex movement direction, but also make forex trading judgment to provide suggestions for investors, which is more accurate than some other similar systems. Second, this system can not only operate in an Intranet environment, but also in the Internet environment, which is more flexible than some other existing systems. Third, the intelligent system is easy to maintain because it can operate in the Internet and be maintained from different locations. Fourth, both BPNNFRFS and WFTDSS subsystems can be used separately for different purposes. Thus, the FRFTDSS is modularized. This unique feature is also good for system maintenance. Finally, the intelligent system provides an important e-tool for forex traders, which advances the development of e-business intelligence.

Fig. 14.9. The main interface of the FRFTDSS

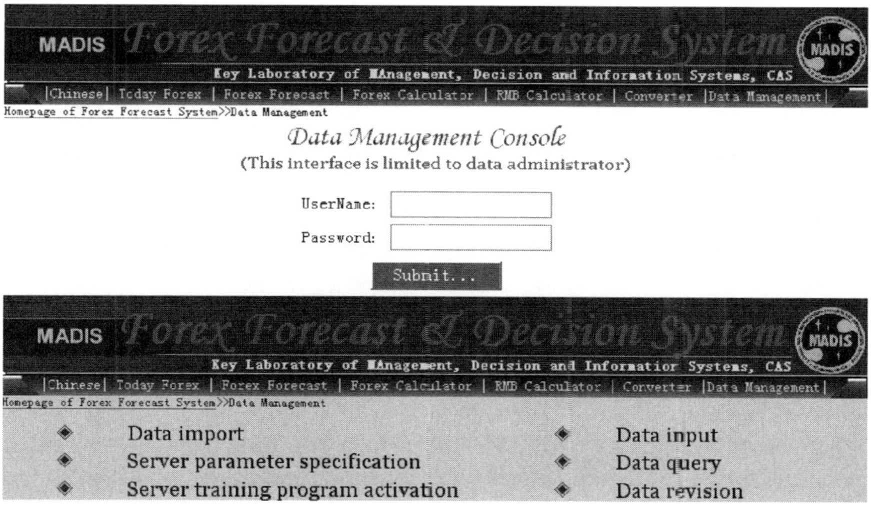

Fig. 14.10. Data management interface

Five days ahead forecast (JPY/USD)

Date	Actual	Forecast	Error	Direction	Strategy
2004-04-10	106.52	105.67	0.008		
2004-04-11	106.21	105.81	0.0038		
2004-04-12	106.25	105.96	0.0027		
2004-04-13	106.96	106.16	0.0075		
2004-04-14	106.90	106.54	0.0034		
2004-04-15		106.69		Down	Sell
2004-04-16		106.51		Down	Sell
2004-04-17		106.73		Down	Sell
2004-04-18		106.82		Down	Sell
2004-04-19		106.93		Up	Buy

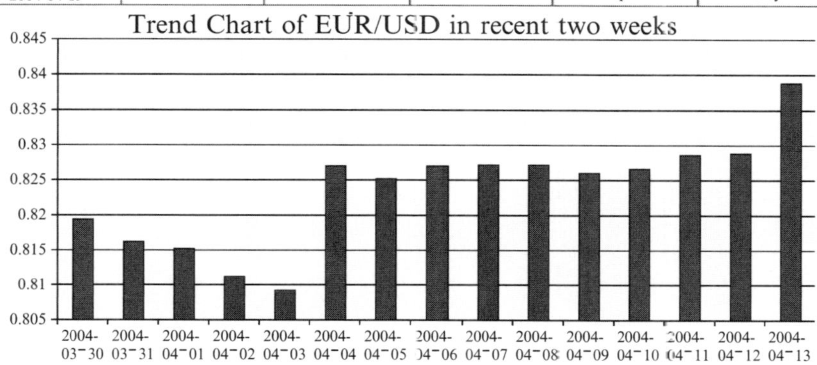

Fig. 14.11. Two user interfaces of the FRFTDSS

14.5 Conclusions

This chapter presents an integrated framework for forex rolling forecasting and trading decision support, and develops an intelligent forex rolling forecasting and trading decision support system (FRFTDSS), integrating a BPNN-based forex rolling forecasting sub-system (BPNNFRFS) and a web-based forex trading decision support sub-system (WFTDSS). The FRFTDSS system has been put into operation for more than one year so far, and has helped improve the overall performance of forecasting and decision process in practice. A comprehensive assessment on the actual forecasting performance of the developed FRFTDSS, as well as the comparison with other similar forex forecasting systems will be presented in the next chapter.

It is worth noting, however, that the system could be further improved in the future, such as in areas of on-line real time forecasting with BPNN and the precision of long-term forecasting. Future studies will look into these important issues.

15 Developing an Intelligent Forex Rolling Forecasting and Trading Decision Support System II: An Empirical and Comprehensive Assessment

15.1 Introduction

In the previous chapter, the system framework of the intelligent forex rolling forecasting and trading decision support system (FRFTDSS) was presented, and the FRFTDSS integrating a BPNN-based forex rolling forecasting system (BPNNFRFS) and a web-based forex trading decision support system (WFTDSS) was developed and implemented (Yu et al., 2006f). The main goals of this intelligent integrated system are to present short-term forex forecasting and make corresponding e-trading decisions for forex traders and various investors. Specifically, the general framework architecture of the FRFTDSS is shown in Fig. 15.1. Interested readers can visit the website (http://forex.iss.ac.cn) for more details about FRFTDSS. FRFTDSS has now been put into operation for more than two year. This part mainly provides a comprehensive assessment on the performance and operation of FRFTDSS, as compared to some typical forecasting model and other existing forex forecasting decision support systems.

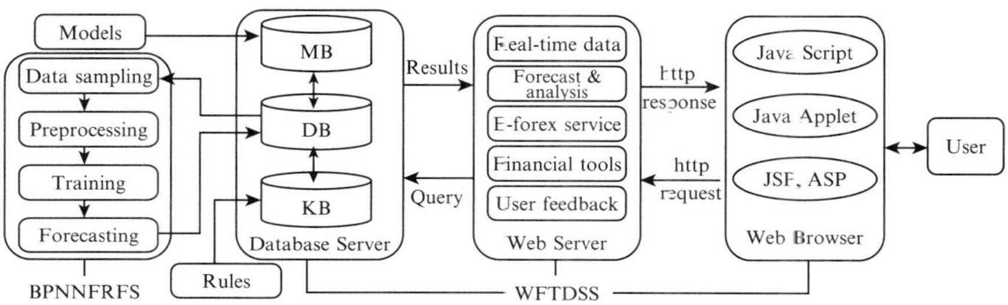

Fig. 15.1. The general framework architecture of the FRFTDSS

The systematic and comprehensive assessment is conducted from two different perspectives. The first one is to assess the performance of operation of FRFTDSS itself, and the other one is to evaluate the performance of FRFTDSS through a comparison with some classical forecasting models and other existing forex forecasting and trading decision support systems.

The reminder of this chapter is organized as follows. Section 15.2 describes assessment methods and presents assessment results. Sections 15.3 and 15.4 compare the performance of FRFTDSS with that of some classical forecasting models and other existing forex forecasting and trading decision support systems, respectively. Section 15.5 concludes the chapter.

15.2 Empirical Assessment on Performance of FRFTDSS

A parametric and a nonparametric evaluation method were used to assess the performance of FRFTDSS in this section. In this paper, we take daily data from 2 January 1990 till 30 April 2003 as in-sample data sets with a total of 3190 observations (the data from 1 January 2001 to 30 April 2003 as cross-validation data sets with 586 observations), and meantime we take the data from 1 May 2003 to 30 April 2004 as evaluation testing sets with 252 observations, which is used to evaluate the good or bad performance of prediction based on some evaluation measurement and certain trade strategies. In order to save space, the original data are not listed in the paper, and detailed data can be obtained from authors or the following website (http://fx.sauder.ubc.ca/data.html).

15.2.1 Parametric Evaluation Methods

The parametric evaluation method was based on assumptions of the probability distribution of estimators and it contains a number of classical statistics, including mean absolute error (MAE), mean square error (MSE), mean absolute percentage error (MAPE), root mean square error (RMSE), and normalized mean square error (NMSE). Table 15.1 presents the statistics and computational formulae.

Table 15.1. Some typical parametric evaluation criteria and their formulae

Statistics	Formula	Symbol Remarks
Mean absolute error (MAE)	$MAE = \frac{1}{n}\sum_{t=1}^{n}\lvert x_t - \hat{x}_t\rvert$	x_t : Actual value at time t; $\hat{x}_t$: Forecasting value at time t; n : Forecasting period.

Mean square error (MSE)	$MSE = \frac{1}{n}\sum_{t=1}^{n}(x_t - \hat{x}_t)^2$	x_t : Actual value at time t; $\hat{x}_t$: Forecasting value at time t; n : Forecasting period.
Mean absolute percentage error (MAPE)	$MAPE = \frac{1}{n}\sum_{t=1}^{n}\left\lvert\frac{x_t - \hat{x}_t}{x_t}\right\rvert \times 100$	x_t : Actual value at time t; $\hat{x}_t$: Forecasting value at time t; n : Forecasting period.
Root mean square error (RMSE)	$RMSE = \sqrt{\frac{1}{n}\sum_{t=1}^{n}(x_t - \hat{x}_t)^2}$	x_t : Actual value at time t; $\hat{x}_t$: Forecasting value at time t; n : Forecasting period.
Normalized mean square error (NMSE)	$NMSE = \frac{\sum_{t=1}^{n}(x_t - \hat{x}_t)^2}{\sum_{t=1}^{n}(x_t - \bar{x}_t)^2}$	x_t : Actual value at time t; $\hat{x}_t$: Forecasting value at time t; $\bar{x}_t$: Mean of x_t; n : Forecasting period.

According to the design of FRFTDSS, three levels of forecasting horizons, i.e., one-day-ahead, three-days-ahead and five-days-ahead, and five currencies, i.e., euros (EUR), British pounds (GBP), Japanese yen (JPY), Canadian dollars (CAD) and Australian dollars (AUD), were used. The detailed assessment results are presented in Table 15.2.

Table 15.2. Performance summary of the FRFTDSS in levels*

Currency	Ahead	MAE ($\times10^{-3}$)	MSE ($\times10^{-5}$)	MAPE($\times10^{-3}$)	RMSE($\times10^{-3}$)	NMSE($\times10^{-2}$)
EUR	1 day	3.5851	2.3147	4.2419	4.8111	1.7793
	3 days	4.2447	2.9172	5.0236	5.4011	2.2425
	5 days	4.6720	2.5318	5.5348	5.9429	2.7149
GBP	1 day	2.0488	0.8169	3.5361	2.8582	0.8033
	3 days	2.5128	1.2352	4.3550	3.5145	1.2145
	5 days	2.9251	1.4117	5.0485	3.7572	1.3881
JPY	1 day	34.954	2465.6	3.1358	49.655	0.9550
	3 days	46.362	3647.7	4.1356	60.396	1.4128
	5 days	52.959	4898.8	4.7476	69.991	1.8974
CAD	1 day	4.5071	3.8423	3.3707	6.1986	3.4245
	3 days	5.0948	4.7014	3.8125	6.8567	4.1902
	5 days	5.3262	5.0606	3.9843	7.1138	4.5104
AUD	1 day	5.8271	6.8196	4.1250	8.2581	0.7688
	3 days	6.3530	7.3993	4.5038	8.6019	0.8342
	5 days	6.7964	8.3731	4.8193	9.1505	0.9440

*EUR: euro; GBP: British pound; JPY: Japanese yen; CAD: Canadian dollar; AUD: Australian dollar; MAE: mean absolute error; MSE: mean square error; MAPE: mean absolute percentage error; RMSE: root mean square error; NMSE: normalized mean square error.

From Table 15.2, we can see that (1) with the increase of the forecasting horizon, the errors increase in currencies and error measurements; and (2) of the five currencies, GBP's forecasting results are the best in all evaluation categories, while JPY the worst. Possible reasons for this could be that (a) uncertainty increases with the increase of forecasting horizon in forex market; (b) the volatility of JPY is generally higher than that of GBP and other currencies during the time period; (c) from the point of input variables, appropriate exogenous variables often increase the forecasting accuracy. In the Part I of the paper, we select five external factors (GDP, exports and imports, CPI and interest rates) as input variables for prediction. Actually, due to the daily data availability of these macroeconomic variables, these explanatory variables are used as dumb variables to adjust the neural network forecasting model.

15.2.2 Nonparametric Evaluation Methods

As mentioned before, in forex market, the forecasting of forex trend change and turning point change are sometimes more important than the precision of goodness-of-fit forecasting. A nonparametric evaluation is used for this purpose.

Nonparametric evaluation measurements are distribution-free tests and refer to various possibilities for obtaining types of information concerning the performance of a model. Typical measurement rules and techniques include directional change statistics (D_{stat}) analysis (Yao and Tan, 2000; Yu et al., 2005c) and turning point analysis (TPA) (Yu et al., 2006f).

Directional change statistics has been introduced in previous chapter. Here our focus is turning point analysis (TPA). More specifically, a 4×4 contingency table constructed by Naik and Leuthold (1986) is used to verify the ability of the estimated model to explain turning points of fluctuations. The 4×4 contingency table describes four different directional changes for both actual values and forecasts; i.e., peak turning point (PTP), trough turning point (TTP), upward no turning point (UNTP), and downward no turning point (DNTP). This leads to 16 different pairs of turning points, which are denoted as F11, F12, …, F44 as shown in Table 15.3. Assuming that x_{i-1}, x_i, x_{i+1} are any three numbers of a data series; then these four directional changes can be judged by following criteria:

Criterion I: If $x_i - x_{i-1} > 0$ and $x_{i+1} - x_i < 0$, then x_i is PTP.
Criterion II: If $x_i - x_{i-1} < 0$ and $x_{i+1} - x_i > 0$, then x_i is TTP.
Criterion III: If $x_i - x_{i-1} > 0$ and $x_{i+1} - x_i > 0$, then x_i is UNTP.
Criterion IV: If $x_i - x_{i-1} < 0$ and $x_{i+1} - x_i < 0$, then x_i is DNTP.

Table 15.3. A 4×4 Contingency table for the performance evaluation*

Actual value	Forecasting value			
	PTP	TTP	UNTP	DNTP
PTP	F11	F12	F13	F14
TTP	F21	F22	F23	F24
UNTP	F31	F32	F33	F34
DNTP	F41	F42	F43	F44

* PTP is peak turning point; TTP is trough turning point; UNTP is upward no turning point; DNTP: downward no turning point

Accurate and worst forecast ratios are the two ratios being of interest here. The accurate forecast (AF) is represented by elements F11, F22, F33, and F44; and the worst forecast (WF) by elements F12, F21, F34, and F43. The sum of the four elements on each of the above sets divided by the sum of the 16 elements of the table gives two key indicators: the accurate forecast ratio (AFR) and the worst forecast ratio (WFR). That is,

$$AFR = \frac{\sum_{i=1}^{4} F_{ii}}{\sum_{i=1}^{4}\sum_{j=1}^{4} F_{ij}} \times 100\% \tag{15.1}$$

$$WFR = \frac{F_{12} + F_{21} + F_{34} + F_{43}}{\sum_{i=1}^{4}\sum_{j=1}^{4} F_{ij}} \times 100\% \tag{15.2}$$

Accordingly, the assessment results for D_{stat} and the indicators of TPA are shown in Table 15.4 below.

From Table 15.4, we see that the shorter the forecasting horizon, the higher the directional forecasting accuracy for all currencies in terms of D_{stat}. In turning point analysis, although the AFR is not very satisfactory but the WFR is quite satisfactory. For short-term forecasting, the WFR is about 10 percent. Therefore, FRFTDSS can be quite useful in practice.

In the following two sections, more detailed comparison between FRFTDSS and some similar models and systems is presented.

Table 15.4. Summary results for D_{stat} and TPA indicators for FRFTDSS*

Currency	Days ahead	D_{stat} (%)	AFR (%)	WFR (%)
EUR	1-day-ahead	70.7650	45.63	9.84
	3-days-ahead	63.6612	42.62	12.02
	5-days-ahead	60.9290	42.35	14.48
GBP	1-day-ahead	72.6776	48.63	11.48
	3-days-ahead	70.4918	53.28	9.56
	5-days-ahead	62.2951	46.45	10.93

JPY	1-day-ahead	71.0383	38.80	12.84
	3-days-ahead	62.5683	36.61	17.76
	5-days-ahead	58.7432	37.70	18.03
CAD	1-day-ahead	71.8579	46.45	11.20
	3-days-ahead	67.7596	45.63	14.48
	5-days-ahead	63.9344	42.90	12.84
AUD	1-day-ahead	65.8470	50.27	6.28
	3-days-ahead	59.8361	51.09	5.74
	5-days-ahead	58.4670	50.82	7.65

*EUR: euro; GBP: British pound; JPY: Japanese yen, CAD: Canadian dollar; AUD: Australian dollar; D_{stat}: directional change statistics; AFR: accurate forecast ratio; WFR: worst forecast ratio.

15.3 Performance Comparisons with Classical Models

15.3.1 Selection for Comparable Classical Models

For providing more convincing evidence that the proposed forecasting method outperforms the classical forecasting methods, some classical models, such as ARIMA (Box and Jenkins, 1976), Holt-Winter (Chatfield and Yar, 1988), polynomial regression (Chang and Hsu, 2005) and naïve random walk model (Franch and Opong, 2005), are selected to the following comparison analysis. The descriptions of the above four classical methods can refer to Box and Jenkins (1976), Chatfield and Yar (1988), Chang and Hsu (2005), Franch and Opong (2005), Harvey (1989) and Makridakis et al. (1984) for more details.

15.3.2 Performance Comparison Results with Classical Models

For convenience of comparisons, several typical performance evaluation indicators, *RMSE*, D_{stat}, *AFR* and *WFR*, are chosen. Table 15.5 shows the comparison results for the one-day-ahead forecasts.

From Table 15.5, several conclusions can be made in the following: (a) Focusing on the RMSE indicator, the proposed FRFTDSS performs the best, followed by Holt-Winter model, ARIMA, naïve random walk and polynomial regression model from a rough view; (b) From the D_{stat} point of view, the FRFTDSS model really outperforms the comparable classical models; (c) According to the AFR and WFR, the performances of the proposed system are better than those of the classical methods. The previous three propositions will lead to the final conclusion: (d) In general, the proposed FRFTDSS model clearly performs better than the four classical methods in terms of both parametric and nonparametric indicators.

Table 15.5. Comparison results between FRFTDSS and some classical models

Models	Objects	RMSE	D_{stat}(%)	AFR(%)	WFR(%)
FRFTDSS	EUR	0.0048	70.76	45.63	9.84
	GBP	0.0029	72.68	48.63	11.48
	JPY	0.0497	71.04	38.80	12.84
	CAD	0.0062	71.86	46.45	11.20
	AUD	0.0083	65.85	50.27	6.28
Holt-Winter	EUR	0.0097	61.25	33.58	14.58
	GBP	0.0056	62.53	34.87	19.63
	JPY	0.0125	51.24	25.65	21.25
	CAD	0.0114	53.26	29.42	19.63
	AUD	0.0219	50.18	27.69	20.54
ARIMA	EUR	0.0108	60.58	36.96	11.54
	GBP	0.0096	60.97	33.24	10.89
	JPY	0.0563	49.68	28.36	18.35
	CAD	0.0227	55.35	30.54	16.44
	AUD	0.0315	52.34	26.96	19.67
Polynomial regression	EUR	0.0225	55.63	31.25	25.36
	GBP	0.0132	56.72	33.73	20.11
	JPY	0.0631	52.54	27.39	22.57
	CAD	0.0544	51.36	31.24	23.35
	AUD	0.0426	48.14	30.05	24.66
Naïve random walk	EUR	0.0208	51.35	36.85	21.36
	GBP	0.0117	54.63	41.68	18.35
	JPY	0.0358	47.35	31.24	19.57
	CAD	0.0415	48.91	33.59	17.85
	AUD	0.0226	48.06	30.73	19.91

15.4 Performance Comparisons with Other Systems

15.4.1 Searching for Existing Forex Forecasting Systems

Because we are interested in studies of ANNs on forex forecasting and trading decisions, the literature search was involved with two processes. First, ten computer and/or information systems (IS) databases (Science Citation Index, Social Science Citation Index, ScienceDirect, Wiley InterScience, IEEE Xplore, JSTOR, Kluwer online, ProQuest Database, Springerlink, and

Academic Search Premier) were searched using the keywords "neural network(s) (in) exchange rates forecasting (prediction) and trading (decision)" for 33 years' time period covering 1971–2004. These ten computer databases included more than 5000 different computer- and/or IS-related international journals. The comprehensive literature search resulted in about 300 relevant journal papers.

Second, a reference search of textbooks on neural networks and exchange rate forecasting and trading decision was conducted, which retrieved nine textbooks (Beltratli et al., 1996; Gately, 1996; Kacapyr, 1996; Trippi and Turban, 1996; Kindon, 1997; Kovalerchuk and Vityaev, 2000; Graf, 2002; Smith and Gupta, 2002; Soofi and Cao, 2002). However, most books, except for Smith and Gupta (2002), were related to forecasting of stock prices, interest rate and bank failure. Furthermore, some related studies in Smith and Gupta's book were the same as in their published journal articles. Therefore, these books were excluded from our investigations.

Not all the retrieved 300 articles were included in our comparison analysis. An article was chosen for the comparison analysis that must meet following three conditions: (1) it presents the applications of neural networks in exchange rate forecasting; (2) it has detailed discussions on the development process of neural networks for exchange rate forecasting; and (3) it has performance results of proposed forex forecasting decision support systems. As a result, we identified a total of 32 studies (Refenes et al., 1993; Kuan and Liu, 1995; Poddig and Rehkugler, 1996; Tenti, 1996; Gencay, 1999; El Shazly and El Shazly, 1997, 1999; Leung et al., 2000; Chen and Leung, 2004; Yao and Tan, 2000; Zhang and Berardi, 2001; Rivas et al., 2003; Li et al., 1999; Zhang, 2003; Yu et al., 2005c, Chun and Kim, 2003; Dunis and Huang, 2002; Franses and van Homelen, 1998; Hong and Lee, 2003; Hu and Tsoukalas, 1999; Hu et al., 1999; Huang et al., 2003; Jasic and Wood, 2003; Kodogiannis and Lolis, 2002; Lisi and Schiavo, 1999; Nag and Mitra, 2002; Parhizgari and De Boyrie, 1997; Qi and Wu, 2003; Shin and Han, 2000; Walczak, 2001; Wu, 1995; Yao and Tan, 1997). For the comparison purpose of this study, we limited our selection of studies to the ones that included three to six currencies. Thus, only sixteen studies (Kuan and Liu, 1995; Poddig and Rehkugler, 1996; Gencay, 1999; El Shazly and El Shazly, 1997, 1999; Leung et al., 2000; Chen and Leung, 2004; Yao and Tan, 2000; Yu et al., 2005c; Franses and van Homelen, 1998; Hong and Lee, 2003; Lisi and Schiavo, 1999; Nag and Mitra, 2002; Parhizgari and De Boyrie, 1997; Qi and Wu, 2003; Walczak, 2001) were used in the following comparison analysis.

15.4.2 Performance Comparisons with Other Existing Systems

Different systems may have different research objectives. We only present the comparison with the same system objectives in forex forecasting. 16 references (Kuan and Liu, 1995; Poddig and Rehkugler, 1996; Gencay, 1999; El Shazly and El Shazly, 1997, 1999; Leung et al., 2000; Chen and Leung, 2004; Yao and Tan, 2000; Yu et al., 2005c; Franses and van Homelen, 1998; Hong and Lee, 2003; Lisi and Schiavo, 1999; Nag and Mitra, 2002; Parhizgari and De Boyrie, 1997; Qi and Wu, 2003; Walczak, 2001) were used for the comparison analysis. Table 15.6 presents the results of comparison analysis for the one-day-ahead forecasts.

Table 15.6. Comparison results between FRFTDSS and other existing systems*

Ref.	Objects	Data	MAE	MAPE	RMSE	NMSE	D_{stat}(%)	AFR(%)	WFR(%)
FRFT-	EUR		0.0036	0.0042	0.0048	0.0178	70.76	45.63	9.84
DSS	GBP		0.0021	0.0035	0.0029	0.0080	72.68	48.63	11.48
	JPY	Daily	0.0350	0.0031	0.0497	0.0095	71.04	38.80	12.84
	CAD		0.0045	0.0034	0.0062	0.0342	71.86	46.45	11.20
	AUD		0.0058	0.0041	0.0083	0.0077	65.85	50.27	6.28
Kuan &	GBP		-	-	0.5972	-	64.00	-	-
Liu	CAD		-	-	0.1372	-	56.00	-	-
(1995)	DEM	Daily	-	-	0.4353	-	62.00	-	-
	JPY		-	-	0.3417	-	66.00	-	-
	CHF		-	-	0.4513	-	66.00	-	-
Poddig &	USD		-	-	0.0751	-	66.67	-	-
Rehkugler	DEM	Monthly	-	-	0.0789	-	87.50	-	-
(1996)	JPY		-	-	0.0603	-	70.83	-	-
Gencay	GBP		-	-	0.9576	-	55.00	-	-
(1999)	DEM		-	-	0.9571	-	56.00	-	-
	FRF	Daily	-	-	0.9607	-	54.00	-	-
	JPY		-	-	0.9534	-	56.00	-	-
	CHF		-	-	0.9690	-	55.00	-	-
El Shazly	GBP		-	0.0116	-	-	62.50	-	-
& Shazly	DEM	Weekly	-	0.0114	-	-	62.50	-	-
(1997)	JPY		-	0.1151	-	-	53.13	-	-
El Shazly	GBP		-	0.0420	-	-	66.25	-	-
& Shazly	DEM	Quarterly	-	0.0471	-	-	71.25	-	-
(1999)	JPY		-	0.0480	-	-	61.00	-	-
	CHF		-	0.0568	-	-	63.25	-	-
Leung	CAD		0.0087	-	0.0106	-	-	-	-
et al.	JPY	Monthly	0.0209	-	0.2687	-	-	-	-
(2000)	GBP		0.0216	-	0.0286	-	-	-	-

Chen &	CAD	Monthly	-	-	0.0291	-	-	-	-
Leung	JPY		-	-	5.0645	-	-	-	-
(2004)	GBP		-	-	0.0338	-	-	-	-
Yao &	AUD		-	-	-	0.0543	55.00	-	-
Tan(2000)	CHF		-	-	-	0.1100	56.00	-	-
	DEM	Weekly	-	-	-	0.3153	51.00	-	-
	GBP		-	-	-	0.1555	54.74	-	-
	JPY		-	-	-	0.1146	53.40	-	-
Yu et al.	DEM		-	-	-	0.0156	83.33	-	-
(2005c)	GBP	Monthly	-	-	-	0.0357	86.11	-	-
	JPY		-	-	-	0.1391	75.00	-	-
Franses &	USD		-	-	-	-	50.84	-	-
van	CAD	Daily	-	-	-	-	52.28	-	-
Homelen	GBP		-	-	-	-	52.41	-	-
(1998)	JPY		-	-	-	-	53.01	-	-
Hong &	CAD		-	0.467	0.6041	-	51.00	-	-
Lee(2003)	DEM		-	1.137	1.4896	-	49.70	-	-
	GBP	Weekly	-	1.046	1.4328	-	47.90	-	-
	JPY		-	1.146	1.5707	-	53.90	-	-
	FRF		-	1.097	1.4429	-	50.00	-	-
Lisi &	FRF		-	-	-	0.864	-	-	-
Schiavo	DEM	Monthly	-	-	-	0.808	-	-	-
(1999)	ITL		-	-	-	0.891	-	-	-
	GBP		-	-	-	0.658	-	-	-
Nag &	DEM		0.0061	0.3364	0.0087	-	-	-	-
Mitra	GBP	Daily	0.0051	0.3097	0.0069	-	-	-	-
(2002)	JPY		0.6307	0.4969	0.8497	-	-	-	-
Parhizgari	GBP		-	-	1.3531	-	-	-	-
& De	CAD		-	-	0.9887	-	-	-	-
Boyrie	DEM	Daily	-	-	1.0650	-	-	-	-
(1997)	JPY		-	-	0.9420	-	-	-	-
	CHF		-	-	1.0043	-	-	-	-
Qi & Wu	JPY		-	-	0.0347	-	50.55	-	-
(2003)	DEM	Monthly	-	-	0.0343	-	60.44	-	-
	GBP		-	-	0.0329	-	50.55	-	-
	CAD		-	-	0.0117	-	59.34	-	-
Walczak	GBP		-	-	-	-	62.40	-	-
(2001)	DEM		-	-	-	-	61.60	-	-
	JPY	Daily	-	-	-	-	59.20	-	-
	FRF		-	-	-	-	59.20	-	-
	CHF		-	-	-	-	57.60	-	-
	ITL		-	-	-	-	57.60	-	-

*EUR: euro; GBP: British pound; JPY: Japanese yen, CAD: Canadian dollar; AUD: Australian dollar; CHF: Swiss franc; DEM: Deutsche Mark; FRF: French franc; ITL: Italian lira; USD: U.S. dollar; MAE: mean absolute error; MSE: mean square error; MAPE: mean absolute percentage error; RMSE: root mean square error; NMSE: normalized mean square error; D_{stat}: directional change statistics; AFR: accurate forecast ratio; WFR: worst forecast ratio; -.: Not available.

From Table 15.6, we can see that FRFTDSS is one of the best decision support systems in short-term forex forecasting from a rough perspective, as compared to other existing decision support systems.

15.4.3 A Comprehensive Comparison Analysis

15.4.3.1 The comprehensive comparison analysis model

Table 15.6 shows that using the parametric and nonparametric analysis methods, FRFTDSS outperformed the other DSS systems in some performance measurements while the other DSS systems were better than FRFTDSS in some other performance measurements. It was therefore necessary to conduct a comprehensive comparison analysis. Here we used a principal component analysis (PCA) technique for this purpose, which is illustrated below (Yu et al., 2006f).

The PCA technique is a multivariate statistical method to reduce the dimension through transforming multiple indices into fewer comprehensive indices (see Jolliffe (1986) for more details). The basic idea in PCA is to find principal components (s_1, s_2, $\cdots$, s_p) that can explain the maximum amount of variance possible by p linearly transformed components from a data vector with q dimensions ($p < q$). However, the aim of PCA used in this study is not to emphasize on decreasing indice, but to concentrate on the comprehensive evaluation of different models and systems. The detailed computational process can be divided into following five steps.

(1) Sample selection

The tasks of this step are to select comparative indices and collect corresponding data.

(2) Data preprocessing

In original data, there may be some positive indices (the larger the index value, the better) and some inverted indices (the smaller the index value, the better). Furthermore, original data may have different dimensions and characteristics. It is, therefore, necessary to process these original data. In general, as for inverted indices, we use reciprocals of the indices as evaluation indices. As for different dimensions and characteristics of data, standardization is an effective way to eliminate differences:

$$x_i^* = \frac{x_i - \bar{x}_i}{\sigma_i} \tag{15.3}$$

where x_i^* is the standardized data, x_i is the original data, $\bar{x}_i$ is the mean of the ith index, σ_i is standard deviation of the ith index.

(3) PCA computation

After preprocessing the data, the principal component analysis can be performed and corresponding principal component identified. Assuming that there are n individual models and systems and that every model and system contains m evaluation indices, the transformed evaluation matrix (X) can be represented as

$$\mathrm{X} = \begin{bmatrix} x_{11} & x_{12} & \cdots & x_{1m} \\ x_{21} & x_{22} & \cdots & x_{2m} \\ \vdots & \vdots & \ddots & \vdots \\ x_{n1} & x_{n2} & \cdots & x_{nm} \end{bmatrix} \tag{15.4}$$

where x_{ij} is the jth index value with the ith model or system.

Next, we deal with the evaluation matrix using the PCA technique. First, eigenvalues ($\lambda_1, \lambda_2, \cdots, \lambda_n$) and corresponding eigenvectors $C = (\boldsymbol{c_1}, \boldsymbol{c_2}, \cdots, \boldsymbol{c_n})$ can be generated from the evaluation matrix. The individual contribution rate and cumulative contribution can then be obtained:

$$a_i = \frac{\lambda_i}{\sum_{i=1}^{n} \lambda_i} \tag{15.5}$$

$$a_{mn} = \frac{\sum_{i=1}^{m} \lambda_i}{\sum_{i=1}^{n} \lambda_i} \tag{15.6}$$

where a_i is the ith contribution rate, a_{mn} is the cumulative variance contributions of the principal factors x_1 to x_m. When a_{mn} is sufficiently large or a_{mn} satisfies a certain cutoff, we can choose several principal factors to construct a comprehensive evaluation function. In general, a_{mn} is larger than 0.90 in practice.

(4) Constructing a comprehensive evaluation function

According to the previous computation and analysis, the comprehensive evaluation function of the principal factors can be expressed as

$$Y_i = \sum CX = \sum_{j=1}^{k} c_j x_i \quad (i = 1, 2, \ldots n) \tag{15.7}$$

where k represents the k-first principal axes related to the cutoff obtained from the previous step.

It is worth noting that the transformation of equation (15.7) is necessary if the indices are standardized previously.

(5) Calculating all values of comprehensive evaluation function for every comparative model and system and ranking corresponding assessments.

15.4.3.2 A comprehensive comparison analysis results

A comprehensive assessment was done according to the above-mentioned five steps.

First of all, we selected five evaluation measurements (i.e., MAE, MAPE, RMSE, NMSE and D_{stat}) as comparative indices. Because AFR and WFR are seldom used in other comparative forecasting systems, they were excluded in this analysis.

Next, related data were collected. As the original data of some performance measurements were not available in the existing forecasting systems, the average value of related measurements was used as a surrogate. That is, if a certain data are not available, an average value is used as a surrogate data for comparison purpose. At the same time, the D_{stat} was transformed into an inverted index. In this sense, the smaller the comprehensive evaluation value, the better the evaluation objects.

Subsequently, principal component analysis was performed as mentioned above, the eigenvalues $(\lambda_1, \lambda_2, \cdots, \lambda_5)$ = (0.0000, 0.0015, 0.0245, 0.0347, 0.2228) and corresponding eigenvectors C were generated. The individual contribution rate of five measurements was (0.0000, 0.0054, 0.0863, 0.1224, 0.7859), and the cumulative contribution rate of principal factor was (0.7859, 0.9083, 0.9946, 1.0000, 1.0000). Assuming that the cutoff value is 0.90, the three error measurements (MAE, MAPE, RMSE) should be excluded. If the cutoff is 0.95, the MAE and MAPE should be excluded. Here the cutoff was set at 0.95 for further interpretation. That is, the principal components are RMSE, NMSE and D_{stat} in this study. Of course, other cutoff can also be set, but the sufficient large cumulative contribution rate must be guaranteed, as earlier noted.

Finally, using the eigenvectors C of every system generated previous step and its corresponding index values, the comprehensive evaluation value can be calculated with Equation (8). For example, the eigenvector of FRFTDSS is C = (0.1329, 0.8905, 1.3412), the average of several main principal components (as a surrogate) is x = (0.01438, 0.01544, 0.014197), then comprehensive evaluation value $Y_1 = 0.1329\times 0.01438+0.8905\times 0.01544+1.3412\times 0.014197=0.0347$. Accordingly, other comprehensive assessment results on the seventeen DSS systems are reported in Table 15.7.

Table 15.7. The results of the comprehensive assessment

Ref.	FRFTDSS	Kuan	Poddig	Gencay	Shazly	Shazly	Leung	Chen
Value	0.0347	0.8966	0.6519	1.3288	0.8680	0.8655	0.6703	1.9675
Rank	1	8	3	13	7	6	4	16
Yao	Yu	Franses	Hong	Lisi	Nag	Parhizgari	Qi	Walczak
0.9137	0.8158	1.0506	2.218	1.7395	0.9101	1.4134	0.6192	1.0506
10	5	11	17	15	9	14	2	11

In terms of the comprehensive evaluation performance, Table 15.7 shows that FRFTDSS is the best, followed by Qi and Wu (2003) and (Poddig and Rehkugler, 1996); the worst is Hong and Lee (2003).

15.5 Discussions and Conclusions

As compared to other existing DSS systems, FRFTDSS has following different features.

- FRFTDSS integrates both forex forecasting (the BPNNFRFS subsystem) and forex trading decisions (the WFTDSS subsystem).
- FRFTDSS cannot only provide forex forecasting data and forex real-time price information, but also present forex movement direction information and forex trading strategies, which can meet the needs of different users.
- The FRFTDSS can be run in both Internet and Intranet environments, which is more flexible than some other existing decision support systems.
- The FRFTDSS is modularized. BPNNFRFS and WFTDSS in the integrated system can be used separately to meet the needs of forecasting and decision-making. Meanwhile, the two subsystems can also be combined to provide integrated services for investors.

Based on the above systematic and comprehensive assessment, FRFTDSS demonstrates to be one of the best forex forecasting decision support systems in terms of short-term forex forecasting. In fact, it is the best DSS system in terms of comprehensive assessment result done above, as compared to other existing DSS systems. Further, Tables 15.4 and 15.5 show that the forecasting accuracy of decision strategies of FRFTDSS reaches to 70%, which suggests that FRFTDSS can be a useful decision tool for investors in forex market.

It is worth noting, however, that several possible improvements of the system, such as forecast horizons, prediction accuracy, and online real time forex trading decision, can be done in the future. In addition, the system could also be further extended into some related fields, such as crude oil markets, real option markets and some emerging markets, which can be used as a future development direction.

References

Abiteboul, S., Cluet, S., Milo, T., Mogilevsky, P., Simeon, J., Zohar, S. (1999). Tools for translation and integration. IEEE Data Engineering Bulletin 22, 3–8.

Abraham, A. (2004). Meta learning evolutionary artificial neural networks. Neurocomputing 56, 1–38.

Achelis, S.B. (1995). Technical Analysis from A to Z. Probus Publishing, Chicago.

Akaike, H. (1973). Information theory and an extension of the maximum likelihood principle. In: Petrov, B.N., Csaki, F. (eds). The Proceedings of 2nd International Symposium of Information Theory, Akademia Kiado, Budapest, pp. 267–281.

Allred, L.G., Kelly, G.E. (1990). Supervised learning techniques for backpropagation networks. Proceedings of International Joint Conference on Neural Networks 1, 702–709.

Alon, I. (1997). Forecasting aggregate retail sales: the Winters' model revisited. In: Goodale, J.C. (ed.): The 1997 Annual Proceedings. Midwest Decision Science Institute, pp. 234–236.

Armstrong, J.S. (1983). Strategic planning and forecasting fundamentals. In: Albert, K. (ed.). Strategic Management Handbook. McGraw-Hill, New York, pp. 2–1 - 2–32.

Ashley, D.W., Allegrucci, A. (1999). A spreadsheet method for interactive stepwise multiple regression. Proceedings, Western Decision Sciences Institute, pp. 594–596.

Azimi-Sadjadi, M.R., Stricker, S.A. (1994). Detection and classification of buried dielectric anomalies using neural networks – further results. IEEE Transaction on Instrumentations and Measurement 43, 34–39.

Azoff, E.M. (1994). Neural network time series: forecasting of financial markets. John Wiley & Sons.

Baillie, R., Bollerslev, T. (1989). Common stochastic trends in a system of exchange rates. Journal of Finance 44, 167–181.

Batchelor, R., Dua, P. (1995). Forecaster diversity and the benefits of combining forecasts. Management Science 41, 68–75.

Bates, J.M., Granger, C.W.J. (1969). The combination of forecasts. Operations Research Quarterly 20, 451–468.

Batista, G.E., Monard, M.C. (2003a). Experimental comparison of K-nearest neighbor and mean or mode imputation methods with the internal strategies used by C4.5 and CN2 to treat missing data", Technical Report 186, ICMC USP.

Batista, G.E., Monard, M.C. (2003b). An analysis of four missing data treatment methods for supervised learning. Applied Artificial Intelligence 17, 519–533.

Baumgarten, A. (1999). Probabilistic solution to the selection and fusion problem in distributed information retrieval. Proceedings of SIGIR'99, pp. 246–253.

Bekaert, G., Hodrick, R.J. (1992). Characterizing predictable components in excess returns on equity and foreign exchange markets. Journal of Finance 47, 467–509.

Beltratli, A., Margarita, S., Terna, P. (1996). Neural networks for economic and financial modeling. International Thomson Publishing Inc. London.

Benediktsson, J.A., Sveinsson, J.R., Ersoy, O.K., Swain, P.H. (1997). Parallel consensual neural networks. IEEE Transactions on Neural Networks 8, 54–64.

Bhattacharyya, S., Pendharkar, P.C. (1998). Inductive, evolutionary and neural techniques for discrimination: a comparative study. Decision Sciences 29 (4), 900–971.

Bishop, C.M. (1995). Neural Networks for Pattern Recognition. Oxford University Press.

Bolland, P.T., Connor, J.T. (1997). A constrained neural network kalman filter for price estimation in high frequency financial data, 8(4), 399–415.

Box, G.E.P., Jenkins, G. (1970). Time Series Analysis: Forecasting and Control. Holden-Day, San Francisco.

Breiman L. (1994). Bias, variance, and arcing classifiers. Technical Report TR460, Department of Statistics, University of California.

Breiman, L. (1996a). Bagging predictors. Machine Learning 24, 123–140.

Breiman, L. (1996b). Stacked regressions. Machine Learning 24, 49–64.

Breiman, L. (1999). Combining predictors. In: Sharkey, A.J.C. (ed.): Combining Artificial Neural Nets – Ensemble and Modular Multi-net Systems. Springer, Berlin, pp. 31–50.

Brent, R.P. (1991). Fast training algorithms for multilayer neural nets. IEEE Transactions on Neural Networks 2, 346–35.

Brooks, C. (1997). Linear and nonlinear (non-) forecastability of high frequency exchange rates. Journal of Forecasting 16, 125–145.

Broomhead, D.S., Lowe, D. (1988). Multivariable functional interpolation and adaptive networks. Complex Systems 2, 321–355.

Brown, R.G. (1959). Statistical Forecasting for Inventory Control. McGraw-Hill, New York.

Buckland, S.T., Burnham, K.P., Augustin, N.H. (1995). Model selection: An integral part of inference. Biometrics 53, 603–618.

Burden, F.R., Brereton, R.G., Walsh, P.T. (1997). Cross-validatory selection of test and validation sets in multivariate calibration and neural networks as applied to spectroscopy. Analyst 122(10), 1015–1022.

Burges, C.J.C. (1998). A tutorial on support vector machines for pattern recognition. Data Mining and Knowledge Discovery 2(2), 955–974.

Campbell, J.Y., Grossman, S.J., Wang, J. (1993). Trading volume and serial correlation in stock returns. Quarterly Journal of Economics 107, 905–939.

Carney, J., Cunningham, P. (2000). Tuning diversity in bagged ensembles. International Journal of Neural Systems 10, 267–280.

Chan, P., Stolfo, S. (1993). Meta-learning for multistrategy and parallel learning. Proceedings of the Second International Workshop on Multistrategy Learning, pp. 150–165.

Chang, A.C., Hsu, J.P. (2005). A polynomial regression model for the response of various accelerating techniques on maize wine maturation. Food Chemistry, in press.

Chang, J., Jung, Y., Yeon, K., Jun, J., Shin, D., Kim, H. (1996). Technical Indicators and Analysis Methods, Jinritamgu Publishing, Seoul.

Chase, R.B., Quilano, A.N.J., Jacobs, R.F. (1998). Production and Operations Management: Manufacturing and Services, McGraw-Hill, 1998.

Chatfield, C., Yar, M. (1988). Holt–Winters forecasting: some practical issues. The Statistician 37, 129–140.

Chaudhuri, S., Dayal, U. (1997). A overview of data warehousing and OLAP technology. SIGMOD Record 26, 65–74.

Chen, A.S., Leung, M.T. (2004). Regression neural network for error correction in foreign exchange forecasting and trading. Computers & Operations Research 31, 1049–1068.

Chen, G., Ogmen, H. (1993). Modified extended Kalman filtering for supervised learning. International Journal of System Science 24, 1207–1214.

Choi, J. (1995). Technical Indicators. Jinritamgu, Seoul.

Chou, S.C.T. (1998). Migrating to the web: a web financial information system server. Decision Support Systems 23, 29–40.

Chun, S.H., Kim, S.H. (2003). Impact of momentum bias on forecasting through knowledge discovery techniques in the foreign exchange market. Expert Systems with Applications 24, 115–122.

Clemen, R. (1989). Combining forecasts: a review and annotated bibliography with discussion. International Journal of Forecasting 5, 559–608.

Davis, J.T., Episcopos, A., Wettimuny, S. (2001). Predicting direction shifts on Canada-US exchange rates with artificial neural networks. International Journal of Intelligent Systems in Accounting, Finance & Management 10, 83–96.

Dawson, C.W., Wilby, R. (1998). An artificial neural network approach to rainfall-runoff modeling. Hydrological Sciences Journal 43(1), 47–66.

De Matos, G. (1994). Neural networks for forecasting exchange rate. MSc. Thesis, University of Manitoba, Canada.

De Noord, O.E. (1994). The influence of data preprocessing on the robustness and parsimony of multivariate calibration models. Chemometrics and Intelligent Laboratory Systems 23, 65–70.

Dempster, M.A.H., Payne, T.W., Romahi, Y., Thompson, G.W.P. (2001). Computational learning techniques for intraday FX trading using popular technical indicators. IEEE Transactions on Neural Networks 12(4), 744–754.

Denton J.W. (1995). How good are neural networks for causal forecasting? Journal of Business Forecasting 14, 17–20.

DeWitt, J. (1994). Adaptive filtering network for associative memory data preprocessing. World Congress on Neural Networks, San Diego, CA, vol. IV, pp. 34–38.

Diamantaras, K.I., Kung, S.Y. (1996). Principal Component Neural Networks: Theory and Applications. John Wiley and Sons, Inc, New York.

Dickey, D.A., Fuller, W.A. (1979). Distribution of the estimators for autoregressive time series with a unit root. Journal of the American Statistical Association 74, 427–431.

Diebold, F.X., Nason, J.A. (1990). Nonparametric exchange rate prediction. Journal of International Economics 28, 315–332.

Draper, D. (1995). Assessment and propagation of model uncertainty. Journal of the Royal Statistical Society: Series B 57, 45–97.

Dunis, C.L., Huang, X.H. (2002). Forecasting and trading currency volatility: an application of recurrent neural regression and model combination. Journal of Forecasting 21, 317–354.

El Shazly, M.R., El Shazly, H.E. (1997). Comparing the forecasting performance of neural networks and forward exchange rates. Journal of Multinational Financial Management 7, 345–356.

El Shazly, M.R., El Shazly, H.E. (1999). Forecasting currency prices using a genetically evolved neural network architecture. International Review of Financial Analysis 8(1), 67–82.

Ergezinger, S., Thomsen, E. (1995). An accelerated learning algorithm for multilayer perceptrons: optimization layer by layer. IEEE Transactions on Neural Networks 6, 31–42.

Esposito, A., Marinaro, M., Oricchio, D., Scarpetta, S. (2000). Approximation of continuous and discontinuous mappings by a growing neural RBF based algorithm. Neural Networks 13, 651–665.

Fadlalla, A., Lin, C.H. (2001). An analysis of the application of neural networks in finance. Interfaces 31, 112–122.

Fahlman, S.E. (1988). An empirical study of learning speed in back-propagation networks. Technical Report CMU-CS-88-162, Computer Science Department, Carnegie-Mellon University.

Famili, A., Shen, W., Weber, R., Simoudis, E. (1997). Data preprocessing and intelligent data analysis. Intelligent Data Analysis 1, 3–23.

Fan, M., Stallaert, J., Whinston, A.B. (1999). Implementing a financial market using Java and Web-based distributed computing. IEEE Computer 32, 64–70.

Fan, M., Stallaert, J., Whinston, A.B. (2000). The internet and the future of financial markets. Communications of the ACM 43, 82–88.

Faraway, J. Chatfield, C. (1998). Time series forecasting with neural networks: a comparative study using the airline data. Applied Statistics 47(2), 231–250.

Fayyad, U., Irani, K. (1993). Multi-interval discretization of continuous-valued attributes for classification learning. Proceedings of 13th International Joint Conference on Artificial Intelligence, Chambery, France, pp. 1022–1027.

Fayyad, U.M., Piatetsky-Shapiro, G., Smyth, P., Uthurusamy, R. (1996). Advances in Knowledge Discovery and Data Mining, AAAI Press, Menlo Park, CA.

Firth, D. (1990). Generalized linear models. In: Hinkley, D,V,, Reid, N., Snell, E.J. (Eds.). Statistical Theory and Modelling. London, Chapman and Hall, London, pp. 55–82.

Fletcher, R. (1987). Practical Methods of Optimization. Wiley, New York.
Fong, W.M., Koh, S.K., Ouliaris, S. (1997). Joint variance ratio tests of the martingale hypothesis for exchange rates. Journal of Business and Economic statistics 15, 51–59.
Fong, W.M., Ouliaris, S. (1995). Spectral test of martingale hypothesis for exchange rates. Journal of Applied Econometrics 10, 255–271.
Forster, M.R. (2000). Key concepts in model selection: Performance and generalizability. Journal of Mathematical Psychology 44, 205–231.
Foster, B., Collopy, F., Ungar, L. (1992). Neural network forecasting of short, noisy time series. Computers and Chemical Engineering 16, 293–297.
Franch, J.B., Opong, K.K. (2005). Some evidence of random walk behavior of Euro exchange rates using ranks and signs. Journal of Banking & Finance 29, 1631–1643.
Franses, P.H., Van Homelen, P. (1998). On forecasting exchange rates using neural networks. Applied Financial Economics 8, 589–596.
Freeman, J.A., Skapura, D.M. (1991). Neural Networks: Algorithms, Applications and Programming Techniques. Addison-Wesley, MA.
Freund, Y., Schapire, R. (1997). A decision-theoretic generalization of online learning and an application to boosting. Journal of Computer and System Sciences 55, 119–139.
Gamberger, D., Lavrac, N., Dzeroski, S. (2000). Noise detection and elimination in data preprocessing: experiments in medical domains. Applied Artificial Intelligence 14, 205–223.
Gardner, E.S. (1985). Exponential smoothing: the state of the art. Journal of Forecasting 4, 1–28.
Gardner, M.W., Dorling, S.R. (1998). Artificial neural networks (the multilayer perceptron) – A Review of applications in the atmospheric sciences. Atmospheric Environment 32, 2627–2636.
Gately, E.J. (1996). Neural networks for financial forecasting. John Wiley & Sons, Inc. New York.
Geman, S. Bienenstock, E. and Doursat, R. (1992). Neural networks and the bias/variance dilemma. Neural Computation 4, 1–58.
Gencay, R. (1999). Linear, non-linear and essential foreign exchange rate prediction with simple technical trading rules. Journal of International Economics, 47, 91–107.
George, E.I., McCulloch, R.E. (1997). Approaches for Bayesian variable selection. Statistica Sinica 7, 339–373.
Gifford, E. (1995). Investor's Guide to Technical Analysis: Predicting Price Action in the Markets. Pitman Publishing, London.
Giles, C.L., Lawrence, S., Tsoi, A.C. (2001). Noisy time series prediction using a recurrent neural network and grammatical inference. Machine Learning 44, 161–183.
Ginzburg, I., Horn, D. (1994). Combined neural networks for time series analysis. Neural Information Processing Systems 6, 224–231.
Goldberg, D.E. (1989). Genetic Algorithm in Search, Optimization, and Machine Learning. Addison-Wesley, Reading, MA.

Graf, H.G. (2002). Economic forecasting for management: possibilities and limitations. Quorum, CT.

Granger, C.W.J. (1969). Investigating causal relations by econometric models and cross-spectral methods. Econometrica 37, 424–438.

Granger, C.W.J., Ramanathan, R. (1984). Improved methods of forecasting. Journal of Forecasting 3, 197–204.

Grudnitski, G., Osburn, L. (1993). Forecasting S&P and gold futures prices: An application of neural networks. Journal of Futures Market 13, 631–643.

Gujarati, D.N. (1995). Basic Econometrics (3rd Edition). McGraw-Hill, Inc. pp. 741–746.

Hagan, M.T., Menhaj, M. (1994). Training feedforward networks with marquardt algorithm. IEEE Transactions on Neural Networks 5, 989–993.

Han, J., Fu, Y. (1994). Dynamic generation and refinement of concept hierarchies for knowledge discovery in database. Proceedings of AAAI'94 Workshop on Knowledge Discovery in Database, Seattle, WA, pp. 157–168.

Hann, T.H., Steurer, E. (1996). Much ado about nothing? Exchange rates forecasting: neural networks vs. linear models using monthly and weekly data. Neurocomputing 10, 323–339.

Hansen, L.K., Salamon, P. (1990). Neural network ensembles. IEEE Transactions on Pattern Analysis and Machine Intelligence 12, 993–1001.

Harvey, A.C. (1989). Forecasting, structural time series models and the Kalman filter. Cambridge University Press, Cambridge.

Hashem, S. (1997). Optimal linear combination of neural networks. Neural Networks 10, 599–614.

Haykin, S. (1999). Neural Networks: A Comprehensive Foundation. Prentice-Hall Inc., Englewood Cliffs, New-Jersey.

Hertz, J., Krogh, A., Palmer, R.G. (1989). Introduction to the Theory of Neural Computation, Addison-Wesley, Reading, MA.

Hiemstra, C., Jones, J.D. (1994). Testing for linear and nonlinear Granger causality in the stock price–volume relation. Journal of Finance 49 (5), 1639–1664.

Hoeting, J.A., Madigan, D., Raftery, A.E., Volinsky, C.T. (1999). Bayesian model averaging: A tutorial. Statistical Science 14 (4), 382–417.

Holland, J. H. (1992). Genetic algorithms. Scientific American 267, 66–72.

Hong, Y.M., Lee, T.H. (2003). Inference on Predictability of Foreign Exchange Rates via Generalized Spectrum and Nonlinear Time Series Models. Review of Economics and Statistics 85(4), 1048–1062.

Horik, K., Stinchcombe, M., White, H. (1989). Multilayer feedforward networks are universal approximators. Neural Networks 2(5), 359–366.

Hsieh, D.A. (1988). The statistical properties of daily foreign exchange rates: 1974-1983. Journal of International Economics 24, 129–145.

Hsieh, D.A. (1989). Testing for nonlinear dependence in daily foreign exchange rates. Journal of Business 62, 339–368.

Hsieh, D.A. (1993). Implications of nonlinear dynamics for financial risk management. Journal of Financial and Quantitative Analysis 28, 41–64.

Hsu, W., Hsu, L.S., Tenorio, M.F. (1995). A neural network procedure for selecting predictive Indicators in currency trading. In: Refenes, A.N. (ed.). Neural Networks in the Capital Markets, Ch. 17, Wiley, New York.

Hu, M.Y., Tsoukalas, C. (1999). Combining conditional volatility forecasts using neural networks: an application to the EMS exchange rates. Journal of International Financial Markets, Institutions and Money, 9, 407–422.

Hu, M.Y., Zhang, G., Jiang, C.X., Patuwo, B.E. (1999). A cross-validation analysis of neural network out-of-sample performance in exchange rate forecasting. Decision Sciences 30(1), 197–216.

Hu, X. (2003). DB-H reduction: A data preprocessing algorithm for data mining applications. Applied Mathematics Letters 16, 889–895.

Huang, W., Lai, K.K., Nakamori, Y., Wang, S.Y. (2003). An empirical analysis of sampling interval for exchange rate forecasting with neural networks. Journal of Systems Science and Complexity, 16(2), 165–176.

Huang, W., Lai, K.K., Nakamori, Y., Wang, S.Y. (2004a). Forecasting foreign exchange rates with artificial neural networks: a review. International Journal of Information Technology and Decision Making, 3(1), 145–165.

Huang, W., Lai, K.K., Nakamori, Y., Wang, S.Y., Yu, L. (2006). Neural networks in finance and economics forecasting. International Journal of Information Technology and Decision Making 5(4), forthcoming.

Huang, W., Nakamori, Y., Wang, S.Y. (2004b). Seeking appropriate training set for neural network forecasting, The 2004 International Multi-Conference in Computer Science & Computer Engineering, Las Vegas, Nevada, USA, June 21-24, 2004.

Huang, W., Nakamori, Y., Wang, S.Y. (2005). Forecasting stock market movement direction with support vector machine. Computers & Operations Research 32(10), 2513–2522.

Huang, Y.S., Liu, K., Suen, C.Y. (1995). The combination of multiple classifiers by a neural network approach. International Journal of Pattern Recognition and Artificial Intelligence 9, 579–597.

Huang, Y.S., Suen, C.Y. (1995). A method of combining multiple experts for the recognition of unconstrained handwritten numerals. IEEE Transactions on Pattern Analysis and Machine Intelligence 17, 90–94.

Hunt, K.J., Sbarbaro, D., Bikowski, R., Gawthrop, P.J. (1992). Neural networks for control systems – A survey. Automatica 28, 1083–1112.

Hwang, H.B. (2001). Insights into neural-network forecasting time series corresponding to ARMA (p, q) structures. Omega 29, 273–289.

Iiguni, Y., Sakai, H., Tokumaru, H. (1992). A real-time learning algorithm for a multilayered neural network based on the extended Kalman filter. IEEE Transactions on Signal Processing 40, 959–966.

Ince, H., Trafalis, T.B. (2005). A hybrid model for exchange rate prediction. Decision Support Systems, in press, available online 20 October 2005.

Ingwersen, P. (1992). Information Retrieval Interaction. Taylor Graham, London.

Jacobs, R.A. (1988). Increase rates of convergence through learning rate adaptation. Neural Networks 1, 295–307.

Jain, L., Johnson, R., van Rooij, A. (1996). Neural Network Training Using Genetic Algorithms, World Scientific.

Jamal, A.M.M., Sundar, C. (1997). Modeling exchange rates with neural networks. Journal of Applied Business Research, 14(1), 1–5.

Jasic, T., Wood, D. (2003). Neural network protocols and model performance. Neurocomputing, 55, 747–753.

John, G.H. (1995). Robust decision trees: removing outliers from data. Proceedings of First International Conference on Knowledge Discovery and Data Mining, pp. 174–179.

Jolliffe, I.T. (1986). Principal Component Analysis. Springer-Verlag, New York.

Joo, D., Choi, D., Park, H. (2000). The effects of data preprocessing in the determination of coagulant dosing rate. Water Research 34, 3295–3302.

Juditsky, A., Nemirovski, A. (2000). Functional aggregation for nonparametric estimation. The Annals of Statistics 28(3), 681–712.

Kaashock, J.F., Van Dijk, H.K. (2002). Neural network pruning applied to real exchange rate analysis. Journal of Forecasting, 21, 559–577.

Kaastra, I., Boyd, M.S. (1995). Forecasting futures trading-volume using neural networks. The Journal of Futures Markets, 15(8), 953–970.

Kacapyr, E. (1996). Economic forecasting: the state of the art. Sharpe, Inc. New York.

Kalman, B.L., Kwasney, S.C. (1992). Why tanh? Choosing a sigmoidal function. Proceedings of the International Joint Conference on Neural Network, Baltimore, MD IEEE, New York.

Kanas, A. (2001). Neural network linear forecasts for stock returns. International Journal of Finance and Economics 6, 245–254.

Kanas, A., Yannopoulos, A. (2001). Comparing linear and nonlinear forecasts for stock returns. International Review of Economics and Finance 10, 383–398.

Kang B.H. (1986). Unstable weights in the combination of forecasts. Management Science 32, 683–695.

Karaboga, D., Pham, D. (2000). Intelligent Optimization Techniques: Genetic Algorithms, Tabu Search, Simulated Annealing and Neural Networks, Springer Verlag, New York.

Karhunen, J., Joutsensalo, J. (1995). Generalizations of principal component analysis, optimization problems, and neural networks. Neural Networks 8(4), 549–562.

Karunanithi, N., Grenney, W.J., Whitley, D., Bovee, K. (1994). Neural networks for river flow prediction. Journal of Computing in Civil Engineering 8(2), 201–220.

Kean, P.G.W., Scott Morton, M.S. (1978). Decision Support Systems: An Organizational Perspective, Addison-Wesley, Reading, MA.

Kim, K.J. (2003). Financial time series forecasting using support vector machines. Neurocomputing 55, 307–319.

Kim, T.Y., Oh, K.J., Sohn, I., Hwang, C. (2004). Usefulness of artificial neural networks for early warning systems of economic crisis. Expert Systems with Applications 26, 583–590.

Kim, Y., Street, W.N. (2004). An intelligent system for customer targeting: a data mining approach. Decision Support Systems 37, 215–228.

Kindon, J. (1997). Intelligent systems and financial forecasting, Springer-verlag, London.

Kittler, J., Hatef, M., Dui, R.P.W., Matas, J. (1998). On combining classifiers. IEEE Transactions on Pattern Analysis and Machine Intelligence 20, 226–239.

Klein, B.D., Rossin, D.F. (1999). Data quality in neural network models: Effect of error rate and magnitude of error on predictive accuracy. OMEGA 27, 569–582.

Kodogiannis, V., Lolis, A. (2002). Forecasting financial time series using neural network and fuzzy system-based techniques. Neural Computing & Applications, 11, 90–102.

Kovalerchuk, B., Vityaev, E. (2000). Data mining in finance: advances in rational and hybrid methods. Kluwer Academic Publishers, Massachusetts.

Krogh, A., Sollich, P. (1997). Statistical mechanics of ensemble learning. Physical Review E 55(1), 811–825.

Krogh, A., Vedelsby, J. (1995). Neural network ensembles, cross validation, and active learning. In: Tesauro, G., Touretzky, D., Lean T. (eds). Advances in Neural Information Processing Systems. MIT Press, Cambridge, Massachusetts, pp. 231–238.

Krycha, K.A., Wagner, U. (1999). Applications of artificial neural networks in management science: A survey. Journal of Retailing and Consumer Services 6, 185–203.

Kuan, C.M., Liu, T. (1995). Forecasting exchange rates using feedforward and recurrent neural networks. Journal of Applied Econometrics 10, 347–364.

Kudo, M., Sklansky, J. (2000). Comparison of algorithms that select features for pattern classifiers. Pattern Recognition 33, 25–41.

Kumar, A., Agrawal, D.P., Joshi, S.D. (2003). Study of Canada/US dollar exchange rate movements using recurrent neural network model of FX-market. Lecture Notes in Computer Science, 2810, 409–417.

Kuo, R.J., Chen, C.H., Hwang, Y.C. (2001). An intelligent stock trading decision support system through integration of genetic algorithm based fuzzy neural network and artificial neural network. Fuzzy Sets and Systems 118, 21–45.

Kupinski, A.M., Giger, M.L. (1999). Feature selection with limited datasets. Medical Physics 26, 2176–2182.

Lai, K.K., Yu, L., Wang, S.Y. (2005). A neural network & web-based decision support system for forex forecasting and trading. Lecture Notes in Artificial Intelligence 3327, 243–253.

Lai, K.K., Yu, L., Huang, W., Wang, S.Y. (2006a). Multistage neural network metalearning with application to foreign exchange rates forecasting. Lecture Notes in Artificial Intelligence, forthcoming.

Lai, K.K., Yu, L., Wang, S.Y., Huang, W. (2006b). Hybridizing exponential smoothing and neural network for financial time series prediction. Lecture Notes in Computer Science 3994, 493–500.

Lai, K.K., Yu, L., Wang, S.Y., Huang, W. (2006c). A noel nonlinear neural network ensemble model for financial time series forecasting. Lecture Notes in Computer Science 3991, 790–793.

Lai, K.K., Yu, L., Wang, S.Y., Zhou, L.G. (2006d). Credit risk analysis using a reliability-based neural network ensemble model. Lecture Notes in Computer Science 4132, 682–690.

Lapedes, A., Farber, R. (1987). Nonlinear signal processing using neural network: prediction and system modeling. Technical report LA-UR-87-2662, Los Alamos National Laboratory.

Lawrence, M., O'Connor, M. (2000).Sales forecasting updates: how good are they in practice? International Journal of Forecasting 16, 369–382.

Lawrence, S., Giles, C.L., Tsoi, A.C. (1997). Lessons in neural network training: Overfitting may be harder than expected. Proceedings of the Fourteenth National Conference on Artificial Intelligence, AAAI-97, AAAI Press, Menlo Park, California, pp. 540–545.

LeBaron, B. (1999). Technical trading rules profitability and foreign exchange intervention. Journal of International Economics, 49(1), 125–143.

LeBlanc, M., Tibshirani, R. (1996). Combining estimates in regression and classification. Journal of the American Statistical Association 91, 1641–1650.

Lee, B.W., Shen, B.J. (1991). Hardware annealing in electronic neural networks. IEEE Transactions on Circuits and Systems 38, 134–137.

Lee, R.C.T., Slagle, J.R., Mong, C.T. (1976). Application of clustering to estimate missing data and improve data integrity. Proceedings of International Conference of Software Engineering, pp. 539–544.

Lemke, F., Muller, J.A. (2003). Self-organizing data mining. Systems Analysis Modelling Simulation 43, 231–240.

Leung, M.T., Chen, A.S., Daouk, H. (2000). Forecasting exchange rates using general regression neural networks. Computers & Operations Research, 27, 1093–1110.

Leung, M.T., Daouk, H., Chen, A.S. (2000). Forecasting stock indices: a comparison of Classification and level estimation models. International Journal of Forecasting 16, 173–190.

Levich, R., Thomas, L. (1993). The significance of technical trading rule profits in the foreign exchange market: a bootstrap approach. Journal of International Money and Finance, 12, 451–474.

Li, X., Ang, C.L., Gray, R. (1999). An intelligent business forecaster for strategic business planning. Journal of Forecasting, 18, 181–204.

Li, Y., Zhang, C., Zhang, S. (2003). Cooperative strategy for web data mining and cleaning. Applied Artificial Intelligence 17, 443–460.

Li, Z.M. (1998). Internet/Intranet technology and its development. Transaction of Computer and Communication 8, 73–78.

Lilien, G.L., Kotler, P. (1983). Marketing Decision Making: A Model Building Approach. Harper and Row Publishers, New York.

Lincoln, W., Skrzypek, J. (1990). Synergy of clustering multiple back propagation networks. Advances in Neural Information Processing Systems 2, 650–657.

Lisboa, P.J.G., Edisbury, B., Vellido, A. (2000). Business applications of neural networks: the state-of-the-art of real-world applications. World Scientific Publishing Company.

Lisi, F., Schiavo, R.A. (1999). A comparison between neural networks and chaotic models for exchange rate prediction. Computational Statistics & Data Analysis, 30, 87–102.

Little, R.J., Murphy, P.M. (1987). Statistical Analysis with Missing Data. John Wiley and Sons, New York.

Liu, C.Y., He, J. (1991). A variance ratio test of random walks in foreign exchange rates. Journal of Finance, 46, 773–785.

Lou, M. (1993). Preprocessing data for neural networks. Technical Analysis of Stocks & Commodities Magazine.

Luxhoj, J.T., Riis J.O., Stensballe, B. (1996). A hybrid econometric-neural network modeling approach for sales forecasting. International Journal of Production Economics 43, 175–192.

Mackay, D.J.C. (1992). The evidence framework applied to classification problems. Natural Computation 4, 720–736.

Madigan, D., York, J. (1995). Bayesian graphical models for discrete data. International Statistical Review 63,215–232.

Mahfoud, S.M. (1995). Niching methods for genetic algorithms. PhD thesis, Department of General Engineering, University of Illinois at Urbana-Champaign, Champaign, IL.

Maier, H.R., Dandy, G.C. (2000). Neural networks for the prediction and forecasting of water resources variables: a review of modeling issues and applications. Environmental Modelling & Software 15, 101–124.

Maier, H.R., Dandy, G.C. (1996). The use of artificial neural networks for the prediction of water quality parameters. Water Resources Research, 32(4), 1013–1022.

Makridakis, S., Andersen, A., Fildes, R., Carbone, R., Hibon, M., Lewandowski, R., Newton, J., Parzen, E., Winkler, R. (1984). The Forecasting Accuracy of Major Time Series Methods. Wiley, New York.

Makridakis, S., Anderson, A., Carbone, R., Fildes, R., Hibdon, M., Lewandowski, R., Newton, J., Parzen, E., Winkler, R. (1982). The accuracy of extrapolation (time series) methods: results of a forecasting competition. Journal of Forecasting 1, 111–153.

Mani, M., Bloedorn, E. (1997). Multi-document summarization by graph search and matching. Proceedings of Fifteen National Conference on Artificial Intelligence, pp. 622–628.

Markham, I.S., Rakes, T.R. (1998). The effect of sample size and variability of data on the comparative performance of artificial neural networks and regression. Computers & Operations Research 25, 251–263.

Martens, H., Naes, T. (1989). Multivariate Calibration. John Wiley & Sons Inc., New York.

Masters, T. (1995). Neural, novel & hybrid algorithms for time series prediction. John Wiley & Sons.

McAulay A.D., Li, J. (1992). Wavelet data compression for neural network pre-processing. Signal Processing, Sensor Fusion and Target Recognition SPIE 1699, 356–365.

McCullagh, P., Nelder, J.A. (1989). Generalized linear models (2nd Edition). Chapman and Hall, London.

McCulloch, W.S., Pitts, W. (1943). A logical calculus of the ideas imminent in nervous activity. Bulletin and Mathematical Biophysics 5, 115–133.

McCurdy, T.H., Morgan, I.G. (1988). Testing the martingale hypothesis in Deutsche mark futures with models specifying the form of heteroskedasticity. Journal of Applied Econometrics 3, 187–202.

McCurdy, T.H., Morgan, I.G., (1987). Tests of the martingale hypothesis for foreign currency futures with time-varying volatility. International Journal of Forecasting 3, 131–148.

Medeiros, M.C., Veiga, A., Pedreira, C.E. (2001). Modeling exchange rates: smooth transitions, neural networks, and linear models. IEEE Transactions on Neural Networks 12(4), 755–764.

Meese, R., Rogoff, K. (1983a). Empirical exchange rate models of the seventies: do they fit out of sample? Journal of International Economics, 14, 3–24.

Meese, R., Rogoff, K. (1983b). The out of sample failure of empirical exchange rate models: sampling error or misspecification? In: Frenkel, J. (ed.) Exchange Rates and International Economics, University of Chicago Press: Chicago.

Mills, T.C. (1990). Time Series Techniques for Economists. Cambridge University Press, London.

Moody, J. (1995). Economic forecasting: challenges and neural network solution. Proceedings of International Symposium on Artificial Neural Networks.

Moody, J., Yarvin, N. (1992). Network with learned unit response functions. In: Moody, J., Hanson, S.J., Lippmann, R.P. (Eds.). Advances in Neural Information Processing Systems, 4, Morgan Kaufmann, SanMateo, CA.

Murphy, J.J. (1986). Technical Analysis of the Future Markets: A Comprehensive Guide to Trading Methods and Applications. Prentice-Hall, New York.

Naftaly, U., Intrator, N., Horn, D. (1997). Optimal ensemble averaging of neural networks. Network Computation in Neural Systems 8, 283–296.

Nag, A.K., Mitra, A. (2002). Forecasting daily foreign exchange rates using genetically optimized neural networks. Journal of Forecasting, 21, 501–511.

Nahm, U.Y. (2001). Text mining with information extraction: Mining prediction rules from unstructured text. PhD Thesis.

Naik, G., Leuthold, R.M. (1986). A note on qualitative forecast evaluation. American Journal of Agricultural Economics 68, 235–243.

Narendra, K.S., Parthasarathy, K. (1990). Identification and control of dynamic systems using neural networks. IEEE Transaction on Neural Networks 1, 4–27.

Nedeljkovic, V., Milosavljevic, M. (1992). On the influence of the training set data preprocessing on neural networks training. Proceedings of 11th IAPR International Conference on Pattern Recognition, pp. 1041–1045.

Nedovic, L., Devedzic, V. (2002). Expert systems in finance — a cross-section of the field. Expert Systems with Applications 23, 49–66.

Ng, A.Y. (1997). Preventing 'overfitting' of cross validation data. Proceedings of the Fourteenth International Conference on Machine Learning, Morgan Kaufmann, Nashville, pp. 245–253.

Nguyen, H.H., Chan, C.W. (2004). A comparison of data preprocessing strategies for neural network modeling of oil production prediction. Proceedings of the Third IEEE International Conference on Cognitive Informatics, Victoria, Canada.

Ooyen, A.V. (1992). Improving the convergence of the back–propagation algorithm. Neural Networks 5, 465–571.

Parhizgari, A.M., De Boyrie, M.E. (1997). Predicting spot exchange rates in a nonlinear estimation framework using future prices. The Journal of Future Markets, 17(8), 935–956.

Parisi, R., Di Claudio, E.D., Orlandi, G., Rao, B.D. (1996). A generalized learning paradigm exploiting the structure of feedforward neural networks. IEEE Transactions on Neural Networks 7, 1450–1459.

Park, D.C., El-Sharkawi, M.A., Marks II, R.J. (1991). An adaptive training neural net-work. IEEE Transactions on Neural Networks 2, 334–345.

Partridge, D. (1996). Network generalization differences quantified. Neural Networks 9, 263–271.

Partridge, D., Yates, W.B. (1996). Engineering multiversion neural-net systems. Neural Computation 8, 869–893.

Pelikan, E., De Groot, C., Wurtz, D. (1992) Power consumption in West-Bohemia: improved forecasts with decorrelating connectionist networks. Neural Network World 2, 701–712.

Pendharkar, P.C. (2001). An empirical study of design and testing of hybrid evolutionary-neural approach for classification. Omega 29(4), 361–374.

Pendharkar, P.C., Rodger, J.A. (2003). Technical efficiency-based selection of learning cases to improve forecasting accuracy of neural networks under monotonicity assumption. Decision Support System 36, 117–136.

Perrone, M.P., Cooper, L.N. (1993). When networks disagree: ensemble methods for hybrid neural networks. In Mammone, R.J. (ed.). Neural Networks for Speech and Image Processing, Chapman-Hall, pp. 126–142.

Pickett, J. (2000). The American heritage dictionary. 4th edition, Houghton Mifflin, Boston.

Poddig, T., Rehkugler, H. (1996). A 'world' model of integrated financial markets using artificial neural networks. Neurocomputing, 10, 251-273.

Pyle, D. (1999). Data Preparation for Data Mining. Morgan Kaufmann.

Qi, M., Wu, Y. (2003). Nonlinear prediction of exchange rates with monetary fundamentals. Journal of Empirical Financ, 10, 623–640.

Qi, M., Zhang, G.P. (2001). An investigation of model selection criteria for neural network time series forecasting. European Journal of Operational Research 132, 666–680.

Quah, T.S., Tan, C.L., Raman, K.S., Srinivasan, B. (1996). Towards integrating rule-based expert systems and neural networks. Decision Support Systems 17, 99–118.

Rafiq, M.Y., Bugmann, G., Easterbrook, D.J. (2001). Neural network design for engineering applications. Computers & Structures 79, 1541–1552.

Raftery, A.E. (1995). Bayesian model selection in social research (with discussion). In: Marsden, P. V. (Ed.) Sociological Methodology. Blackwell, Cambridge, Massachusetts, pp.111–196.

Ragel, A., Cremilleux, B. (1998). Treatment of missing values for association rules. Proceedings of 2nd Pacific-Asia Conference on Knowledge Discovery and Data Mining, pp. 258–270.

Ralph, M.S., George, W.R. (2000). Principles of Information System: A Managerial Approach. China Machine Press.

Raviv, Y., Intrator, N. (1996). Bootstrapping with noise: An effective regularization technique. Connection Science 8, 355–372.

Redman, T.C. (1992). Data Quality: Management and Technology, Bantam Books, New York.

Redman, T.C. (1996). Data Quality for the Information Age. Artech House, Inc, Norwood, MA.

Refenes, A., Burgess, A.N., Bents, Y. (1997). Neural networks in financial engineering: a study in methodology. IEEE Transactions on Neural Networks 8(4), 1223–1267.

Refenes, A.N., Abu-Mostafa, Y., Moody, J., Weigend, A. (1996). Neural Networks in Financial Engineering. World Scientific Publishing Company.

Refenes, A.N., Azema-Barac, M., Chen, L., Karoussos, S.A. (1993). Currency exchange rate prediction and neural network design strategies, Neural Computing & Applications 1, 46–58.

Reid, D.J. (1968). Combining three estimates of gross domestic product. Economica 35, 431–444.

Reid, D.J. (1969). A comparative study of time series prediction techniques on economic data. Ph. D. thesis, University of Nottingham, Nottingham.

Riedmiller, M., Braun, H. (1993). A direct adaptive method for faster backpropagation learning: The RPROP algorithm. Proceedings of the IEEE International Conference on Neural Networks, pp. 586–591.

Rivas, V.M., Merelo, J.J., Castillo, P.A., Arenas, M.G., Castellano, J.G. (2003). Evolving RBF neural networks for time-series forecasting with EvRBF. Information Sciences, forthcoming.

Rojas, R. (1996). Neural Networks: A Systematic Introduction. Springer-Verlag, Berlin.

Rosen, B.E. (1996). Ensemble learning using decorrelated neural networks. Connection Science 8, 373–384.

Ruck, D.W., Rogers, S.K., Kabrisky, M., Maybeck, P.S., Oxley, M.E. (1992). Comparative analysis of backpropagation and the extended Kalman filter for training multilayer perceptrons. IEEE Transactions on Pattern Analysis and Machine Intelligence 14, 686–691.

Rumelhart, D., Hinton, G., Williams, R. (1986). Learning internal representations by error propagation. In Rumelhart, D., McClelland, J. (eds.). Parallel Distributed Processing: Explorations in the Microstructure of Cognition I, MIT Press, Cambridge, MA, pp. 318–363.

Rumelhart, D.E. (1994). The basic ideas in neural networks. Communication of the ACM 37, 87–92.

Saatcioglu, K., Stallaert, J., Whinston, A.B. (2001). Design of a financial portal. Communications of the ACM 44, 33–38.

Saravanan, M., Reghu Raj, P.C., Raman, S. (2003). Summarization and categorization of text data in high-level data cleaning for information retrieval. Applied Artificial Intelligence 17, 461–474.

Sattler, K.U., Schallehn, E. (2001). A Data preparation framework based on a multidatabase language. Proceedings of the International Symposium on Database Engineering & Applications, pp. 219–228.

Schapire, R.E. (1990). The strength of weak learnability. Machine Learning 5, 197–227.

Scheaffer, R.L., McClave, J.T. (1990). Probability and Statistics for Engineers. PWS-KENT, Boston.

Schwartz, G. (1978). Estimating the dimension of a model. The Annals of Statistics 6, 461–464.

Senol, Y., Gouch, M.P. (1992). The application of transputers to a sounding rocket instrumentation: On-board autocorrelators with neural network data analysis. Parallel Computing and Transputer Applications, 798–806.

Sha, D., Bajic, V.B. (2002). An online hybrid learning algorithm for multilayer perceptron in identification problems. Computers and Electrical Engineering 28, 587–598.

Shadbolt, J., Taylor, J.G. (2002). Neural networks and the financial markets: predicting, combining, and portfolio optimization (perspectives in neural computing). Springer Verlag.

Shah, S., Palmieri, F., Datum, M. (1992). Optimal filtering algorithms for fast learning in feedforward neural networks. Neural Networks 5, 779–787.

Sharkey, A.J.C. (1996). On combining artificial neural nets. Connection Science 8, 299–314.

Sharkey, A.J.C., Sharkey, N.E.(1997). Combining diverse neural nets. The Knowledge Engineering Review 12, 231–247.

Shen, L., Loh, H.T. (2004). Applying rough sets to market timing decisions. Decision Support Systems 37, 583–597.

Shephard, N. (1995). Generalized linear autoregressions. Economics working paper 8, Nuffield College, Oxford.

Shevade, S.K., Keerthi, S.S., Bhattacharyya, C., Murthy, K.R.K. (2000). Improvements to the SMO Algorithm for SVM Regression. IEEE Transactions on Neural Networks 11(5), 1188–1193.

Shewhart, W.A. (1931). Economic Control of Quality of Manufactured Product, New York.

Shi, S.M., Xu, L,D., Liu, B. (1999). Improving the accuracy of nonlinear combined forecasting using neural networks. Expert Systems with Applications 16, 49–54.

Shin, T., Han, I. (2000). Optimal signal multi-resolution by genetic algorithms to support artificial neural networks for exchange rate forecasting. Expert Systems with Applications, 18, 257–269.

Silva, F., Almeida, L. (1990). Speeding up backpropagation. In: Eckmiller, R. (ed). Advanced Neural Computers, pp. 151–158.

Sjoberg, J. (1992). Regularization as a substitute for preprocessing of data in neural network training. Artificial Intelligence in Real-Time Control, pp. 31–35.

Smith K.A., Gupta, J.N.D. (2002). Neural Networks in Business: Techniques and Applications. Idea Group Publishing, Hershey, Pennsylvania.

Smith, K.A., Gupta, J.N.D. (2002). Neural network in finance and investing: using artificial intelligence to improve real-world performance. Idea Group Publishing, Hershey, Pennsylvania.

Smola, A.J. (1998). Learning with Kernels, Ph.D. dissertation, GMD, Birlinghoven, Germany.

Solla, S.A. Levin, E. Fleisher, M. (1988). Accelerated learning in layered neural networks. Complex Systems 2, 625–639.

Soofi, A.S., Cao, L.Y. (2002). Modeling and forecasting financial data — Techniques of Nonlinear Dynamics. Kluwer Academic Publishers, Boston.

Soofi, A.S., Wang, S.Y., Zhang, Y.Q. (2006). Testing for long memory in the Asian foreign exchange rates, Journal of Systems Science and Complexity 19(2), 182–190.

Srinivasan, A., Muggleton, S., Bain, M. (1992). Distinguishing exceptions from noise in non-monotonic learning. Proceedings of Second International Workshop on Inductive Logic Programming, Tokyo, Japan.

Stein, R. (1993a). Selecting data for neural networks. AI Expert 8(2), 42–47.

Stein, R. (1993b). Preprocessing data for neural networks. AI Expert 8(3), 32–37.

Suen, C.Y., Lam, L. (2000). Multiple Classifier Combination Methodologies for Different Output Levels. Lecture Notes in Computer Science 1857, 52–66.

Sweeney, R.J. (1986). Beating the foreign exchange market. Journal of Finance 41, 163–182.

Tam, K.Y., Kiang M.Y. (1992). Managerial applications of neural networks: the case of bank failure predictions. Management Science 38, 926–947.

Tang, Z., Almeida, C., Fishwick, P.A. (1991). Time series forecasting using neural networks vs. Box-Jenkins methodology. Simulation 57, 303–310.

Tang, Z., Fishwick, P.A. (1993). Feedforward neural nets as models for time series forecasting. ORSA Journal on Computing 5, 374–385.

Taylor, J.W. (2000). A quantile regression neural network approach to estimating the conditional density of multiperiod returns. Journal of Forecasting 19, 299–311.

Taylor, M.P. (1995). The economics of exchange rates. Journal of Economic Literature 33, 13–47.

Tenti, P. (1996). Forecasting foreign exchange rates using recurrent neural network. Applied Artificial Intelligence 10, 567–581.

Tollenaere, T. (1990). SuperSAB: Fast adaptive back propagation with good scaling properties. Neural Networks 3, 561–573.

Tresp, V., Taniguchi, M. (1995). Combining estimators using non-constant weighting functions. In: Tesauro, G., Touretzky, D., Lean, T. (eds.). Advances in Neural Information Processing Systems. MIT Press, Cambridge, Massachusetts, pp. 419–426.

Trippi, R.R., Turban, E. (1996). Neural networks in business: techniques and applications. IRWIN Professional publishing, Chicago.

Tsaih, R., Hsu, Y., Lai, C.C. (1998). Forecasting S&P 500 stock index futures with a hybrid AI system. Decision Support Systems 23, 161–174.

Tseng, F.M., Yu, H.C., Tzeng, G.H. (2002). Combining neural network model with seasonal time series ARIMA model. Technological, Forecasting and Social Change 69, 71–87.

Tseng, S.M., Wang, K.H., Lee, C.I. (2003). A preprocessing method to deal with missing values by integrating clustering and regression techniques. Applied Artificial Intelligence 17, 535–544.

Tsoi, A.C., Tan, C.N.W., Lawrence, S. (1993). Financial time series forecasting: Application of artificial neural network techniques. Working paper.

Tumer, K., Ghosh, J. (1995). Order statistics combiners of neural network classifiers. Proceedings of World Congress on Neural Networks, INNS Press, Washington DC, pp. 31–34.

Tumer, K., Ghosh, J. (1996a). Error correlations and error reduction in ensemble classifiers. Connection Science 8, 385–404.

Tumer, K., Ghosh, J. (1996b). Analysis of decision boundaries in linearly combined neural classifiers. Pattern Recognition 29(2), 341–348.

Tuv, E., Runger, G. (2003). Preprocessing of high-dimensional categorical predictors in classification setting. Applied Artificial Intelligence 17, 419–429.

Ueda, N. (2000). Optimal linear combination of neural networks for improving classification performance. IEEE Transactions on Pattern Analysis and Machine Intelligence 22(2), 207–215.

Vapnik, V.N. (1995). The Nature of Statistical Learning Theory. Springer, New York.

Vojinovic, Z., Kecman, V., Seidel, R. (2001). A data mining approach to financial time series modeling and forecasting. International Journal of Intelligent Systems in Accounting, Finance & Management, 10, 225–239.

Voort, M.V.D., Dougherty, M., Watson, S. (1996). Combining Kohonen maps with ARIMA time series models to forecast traffic flow. Transportation Research Part C: Emerging Technologies 4, 307–318.

Walczak, S. (2001). An empirical analysis of data requirements for financial forecasting with neural networks. Journal of Management Information Systems 17(4), 203–222.

Walczak, S., Cerpa, N. (1999). Heuristic principles for the design of artificial neural networks. Information and Software Technology 41(2), 109–119.

Wang, G.J., Chen, C.C. (1996). A fast multilayer neural networks training algorithm based on the layer-by-layer optimizing procedures. IEEE Transactions on Neural Networks 7, 768–775.

Wang, J., Zhang, C., Wu, X., Qi, H., Wang, J. (2003). SVM-OD: A new SVM algorithm for outlier detection. Proceedings of ICDM'03 Workshop on Foundations and New Directions of Data Mining, Melbourne, Florida, USA, pp. 203–209.

Warner, B., Misra, M. (1996). Understanding neural networks as statistical tools. American Statistician 50 (4), 284–293.

Wedding II, D.K., Cios, K.J. (1996). Time series forecasting by combining RBF networks, certainty factors, and the Box-Jenkins model. Neurocomputing 10, 149–168.

Weigend, A.S., Gershenfeld, N.A. (1994). Time Series Prediction: Forecasting the Future and Understanding the Past. Addison-Wesley.

White, H. (1990). Connectionist nonparametric regression: multiplayer feedforward network can learn arbitrary mappings. Neural Netwoks 3, 535–549.

Widrow, B., Lehr, M.A. (1990). 30 Years of Adaptive Neural Networks: Perception, Madaline, and Backprpagation. Proceedings of the IEEE Neural Networks I: Theory & Modeling 78, 1415–1442.

Winkler, R.L. (1989). Combining forecasts: A philosophical basis and some current issues. International Journal of Forecasting 5, 605–609.

Winters P.R. (1960). Forecasting sales by exponentially weighed moving averages. Management Science 6, 324–342.

Wolpert, D. (1992). Stacked generalization. Neural Networks 5, 241–259.

Wong, B.K., Bodnovich, T.A., Selvi, Y. (1995). A bibliography of neural network business applications research: 1988-1994. Expert Systems 12, 253–262.

Wong, B.K., Lai, V.S., Lam, J. (2000). A bibliography of neural network business applications research: 1994-1998. Computers & Operations Research 27, 1045–1076.

Wong, B.K., Selvi, Y. (1998). Neural network applications in finance: a review and analysis of literature (1990-1996). Information & Management, 34, 129–139.

Wu, B. (1995). Model-free forecasting for nonlinear time series (with application to exchange rates). Computational Statistics & Data Analysis, 19, 433–459.

Xu, L., Krzyzak, A., Suen, C.Y. (1992). Methods of combining multiple classifiers and their applications to handwriting recognition. IEEE Transactions on Systems, Man and Cybernetics 22, 418–435.

Yam, J.Y.F., Chow, W.S. (1997). Extended least squares based algorithm for training feedforward networks. IEEE Transactions on Neural Networks 8, 806-810.

Yan, X., Zhang, C., Zhang, S. (2003). Toward databases mining: pre-processing collected data. Applied Artificial Intelligence 17, 545–561.

Yang, G.Y., Tang, X.W., Ma, Y.K. (1996). The discussions of many problems about combined forecasting with non-negative weights. Journal of XiDian University 25(2), 210–215.

Yang, S., Browne, A. (2004). Neural network ensembles: combining multiple models for enhanced performance using a multistage approach. Expert Systems 21, 279–288.

Yang, Y. (2000). Regression with multiple candidate models: selecting or mixing? Working paper, Department of Statistics, Iowa State University.

Yang, Y. (2001). Combining time series models for forecasting. Working paper, Department of Statistics, Iowa State University.

Yao, J.T., Li, Y., Tan, C.L. (1997). Forecasting the exchange rates of CHF vs USD using neural networks. Journal of Computational Intelligence in Finance 15(2), 7–13.

Yao, J.T., Tan, C.L. (2000). A case study on using neural networks to perform technical forecasting of forex. Neurocomputing 34, 79–98.

Yaser, S.A.M., Atiya, A.F. (1996). Introduction to financial forecasting. Applied Intelligence 6, 205–213.

Yu, L., Wang, S.Y., Lai, K.K. (2003). Exchange rates rolling forecasting using neural network model with parameter adaptive control. In: Chen S. et al. (eds.). Financial Systems Engineering 2, 330–346.

Yu, L., Wang, S.Y., Lai, K.K. (2004). A novel nonlinear ensemble forecasting model incorporating GLAR and ANN for foreign exchange rates. Computers & Operations Research, forthcoming.

Yu, L., Wang, S.Y., Lai, K.K. (2005a). A novel adaptive learning algorithm for stock market prediction. Lecture Notes in Computer Science 3827, 443–452.

Yu, L., Lai, K.K., Wang, S.Y. (2005b). Double robustness analysis for determining optimal feedforward neural network architecture. Lecture Notes in Computer Science 3610, 382–386.

Yu, L. Wang, S.Y., Lai, K.K. (2005c). A novel nonlinear ensemble forecasting model incorporating GLAR and ANN for foreign exchange rates. Computers & Operations Research 32(10), 2523–2541.

Yu, L. Wang, S.Y., Lai, K.K. (2005d). Adaptive smoothing neural networks in foreign exchange rate forecasting. Lecture Notes in Computer Science 3516, 523–530.

Yu, L. Wang, S.Y., Huang, W., Nakamori, Y. (2005e). Are foreign exchange rates predictable? A survey from artificial neural network perspective. MADIS Working Paper, Chinese Academy of Sciences.

Yu, L., Wang, S.Y., Lai K.K. (2005f). Mining stock market tendency using GA-based support vector machines. Lecture Notes in Computer Science 3828, 336–345.

Yu, L., Wang, S.Y., Lai, K.K., Nakamori, Y. (2005g). Time series forecasting with multiple candidate models: selecting or combining? Journal of Systems Science and Complexity 18(1), 1–18.

Yu, L., Lai, K.K., Wang, S.Y. (2005h). Designing a hybrid AI system as a forex trading decision support tool. Proceedings of IEEE International Conference on Tools with Artificial Intelligence, 89–93.

Yu, L., Wang, S.Y., Lai, K.K. (2006a). An integrated data preparation scheme for neural network data analysis. IEEE Transactions on Knowledge and Data Engineering 18(2), 217–230.

Yu, L., Wang, S.Y., Lai, K.K. (2006b). An online learning algorithm with adaptive forgetting factors for feedforward neural networks in financial time series forecasting. Nonlinear Dynamics and Systems Theory 7(1), 51–66.

Yu, L., Wang, S.Y., Lai, K.K. (2006c). An adaptive BP algorithm with optimal learning rates and directional error correction for foreign exchange market trend prediction, Lecture Notes in Computer Science 3973, 498–503.

Yu, L., Wang, S.Y., Lai, K.K. (2006d). A novel SVM-based neural network ensemble model for foreign exchange rates forecasting. Journal of Computational Information System, forthcoming.

Yu, L., Lai, K.K., Wang, S.Y., Huang, W. (2006e). A bias-variance-complexity trade-off framework for complex system modeling. Lecture Notes in Computer Science 3980, 518–527.

Yu, L., Lai, K.K., Wang, S.Y., Huang, W. (2006f). Developing an intelligent forex rolling forecasting and trading decision support system for online E-service. International Journal of Intelligent Systems, forthcoming.

Yu, S.W. (1999). Forecasting and arbitrage of the Nikkei stock index futures: An application of back-propagation networks. Asia-Pacific Financial Markets 6, 341–354.

Yu, X.H. (1992). Can back-propagation error surface not have local minima? IEEE Transactions on Neural Networks 3, 1019–1021.

Yu, X.H., Chen, G.A., Cheng, S.X. (1995). Dynamic learning rate optimization of the back propagation algorithm. IEEE Transactions on Neural Networks 6(3), 669–677.

Zhang, G. P. (2004). Neural Networks in Business Forecasting. IRM Press.

Zhang, G.P. (2003). Neural Networks in Business Forecasting. Information Science Publishing.

Zhang, G.P. (2003). Time series forecasting using a hybrid ARIMA and neural network model. Neurocomputing 50, 159–175.

Zhang, G.P., Berardi, V.L. (2001). Time series forecasting with neural network ensembles: an application for exchange rates prediction. Journal of the Operational Research Society, 52, 652–664.

Zhang, G.P., Patuwo, B.E., Hu, M.Y. (1998). Forecasting with artificial neural networks: the state of art. International Journal of Forecasting, 14, 35–62.

Zhang, G.Q., Hu, M.Y. (1998). Neural network forecasting of the British Pound/US dollar exchange rate. OMEGA 26(4), 495–506.

Zhang, S., Zhang, C., Yang, Q, (2003). Data preparation for data mining. Applied Artificial Intelligence 17, 375–381.

Zhang, Y., Li, X.R. (1999). A fast U-D factorization-based learning algorithm with applications to nonlinear system modeling and identification. IEEE Transactions on Neural Networks 10, 930–938.

Zhou, Z.H., Wu, J.X., Tang, W. (2002). Ensembling neural networks: many could be better than all. Artificial Intelligence 137, 239–263.

Zhu, M.L., Fujita, M., Hashimoto, N. (1994). Application of neural networks to runoff prediction. In: Hipel, K.W., Mcleod, A.I., Panu, U.S., Singh, V.P. (Eds.). Stochastic and Statistical Methods in Hydrology and Environmental Engineering. Kluwer Academic, Dordrecht.

Subject Index

Early Titles in the
INTERNATIONAL SERIES IN
OPERATIONS RESEARCH & MANAGEMENT SCIENCE
Frederick S. Hillier, Series Editor, *Stanford University*

Saigal/ *A MODERN APPROACH TO LINEAR PROGRAMMING*
Nagurney/ *PROJECTED DYNAMICAL SYSTEMS & VARIATIONAL INEQUALITIES WITH APPLICATIONS*
Padberg & Rijal/ *LOCATION, SCHEDULING, DESIGN AND INTEGER PROGRAMMING*
Vanderbei/ *LINEAR PROGRAMMING*
Jaiswal/ *MILITARY OPERATIONS RESEARCH*
Gal & Greenberg/ *ADVANCES IN SENSITIVITY ANALYSIS & PARAMETRIC PROGRAMMING*
Prabhu/ *FOUNDATIONS OF QUEUEING THEORY*
Fang, Rajasekera & Tsao/ *ENTROPY OPTIMIZATION & MATHEMATICAL PROGRAMMING*
Yu/ *OR IN THE AIRLINE INDUSTRY*
Ho & Tang/ *PRODUCT VARIETY MANAGEMENT*
El-Taha & Stidham/ *SAMPLE-PATH ANALYSIS OF QUEUEING SYSTEMS*
Miettinen/ *NONLINEAR MULTIOBJECTIVE OPTIMIZATION*
Chao & Huntington/ *DESIGNING COMPETITIVE ELECTRICITY MARKETS*
Weglarz/ *PROJECT SCHEDULING: RECENT TRENDS & RESULTS*
Sahin & Polatoglu/ *QUALITY, WARRANTY AND PREVENTIVE MAINTENANCE*
Tavares/ *ADVANCES MODELS FOR PROJECT MANAGEMENT*
Tayur, Ganeshan & Magazine/ *QUANTITATIVE MODELS FOR SUPPLY CHAIN MANAGEMENT*
Weyant, J./ *ENERGY AND ENVIRONMENTAL POLICY MODELING*
Shanthikumar, J.G. & Sumita, U./ *APPLIED PROBABILITY AND STOCHASTIC PROCESSES*
Liu, B. & Esogbue, A.O./ *DECISION CRITERIA AND OPTIMAL INVENTORY PROCESSES*
Gal, T., Stewart, T.J., Hanne, T. / *MULTICRITERIA DECISION MAKING: Advances in MCDM Models, Algorithms, Theory, and Applications*
Fox, B.L. / *STRATEGIES FOR QUASI-MONTE CARLO*
Hall, R.W. / *HANDBOOK OF TRANSPORTATION SCIENCE*
Grassman, W.K. / *COMPUTATIONAL PROBABILITY*
Pomerol, J-C. & Barba-Romero, S. / *MULTICRITERION DECISION IN MANAGEMENT*
Axsäter, S. / *INVENTORY CONTROL*
Wolkowicz, H., Saigal, R., & Vandenberghe, L. / *HANDBOOK OF SEMI-DEFINITE PROGRAMMING: Theory, Algorithms, and Applications*
Hobbs, B.F. & Meier, P. / *ENERGY DECISIONS AND THE ENVIRONMENT: A Guide to the Use of Multicriteria Methods*
Dar-El, E. / *HUMAN LEARNING: From Learning Curves to Learning Organizations*
Armstrong, J.S. / *PRINCIPLES OF FORECASTING: A Handbook for Researchers and Practitioners*
Balsamo, S., Personé, V., & Onvural, R./ *ANALYSIS OF QUEUEING NETWORKS WITH BLOCKING*
Bouyssou, D. et al. / *EVALUATION AND DECISION MODELS: A Critical Perspective*
Hanne, T. / *INTELLIGENT STRATEGIES FOR META MULTIPLE CRITERIA DECISION MAKING*
Saaty, T. & Vargas, L. / *MODELS, METHODS, CONCEPTS and APPLICATIONS OF THE ANALYTIC HIERARCHY PROCESS*
Chatterjee, K. & Samuelson, W. / *GAME THEORY AND BUSINESS APPLICATIONS*
Hobbs, B. et al. / *THE NEXT GENERATION OF ELECTRIC POWER UNIT COMMITMENT MODELS*
Vanderbei, R.J. / *LINEAR PROGRAMMING: Foundations and Extensions, 2nd Ed.*
Kimms, A. / *MATHEMATICAL PROGRAMMING AND FINANCIAL OBJECTIVES FOR SCHEDULING PROJECTS*
Baptiste, P., Le Pape, C. & Nuijten, W. / *CONSTRAINT-BASED SCHEDULING*
Feinberg, E. & Shwartz, A. / *HANDBOOK OF MARKOV DECISION PROCESSES: Methods and Applications*
Ramík, J. & Vlach, M. / *GENERALIZED CONCAVITY IN FUZZY OPTIMIZATION AND DECISION ANALYSIS*
Song, J. & Yao, D. / *SUPPLY CHAIN STRUCTURES: Coordination, Information and Optimization*
Kozan, E. & Ohuchi, A. / *OPERATIONS RESEARCH/ MANAGEMENT SCIENCE AT WORK*
Bouyssou et al. / *AIDING DECISIONS WITH MULTIPLE CRITERIA: Essays in Honor of Bernard Roy*

Early Titles in the
INTERNATIONAL SERIES IN OPERATIONS RESEARCH & MANAGEMENT SCIENCE
(Continued)

Cox, Louis Anthony, Jr. / *RISK ANALYSIS: Foundations, Models and Methods*
Dror, M., L'Ecuyer, P. & Szidarovszky, F. / *MODELING UNCERTAINTY: An Examination of Stochastic Theory, Methods, and Applications*
Dokuchaev, N. / *DYNAMIC PORTFOLIO STRATEGIES: Quantitative Methods and Empirical Rules for Incomplete Information*
Sarker, R., Mohammadian, M. & Yao, X. / *EVOLUTIONARY OPTIMIZATION*
Demeulemeester, R. & Herroelen, W. / *PROJECT SCHEDULING: A Research Handbook*
Gazis, D.C. / *TRAFFIC THEORY*
Zhu/ *QUANTITATIVE MODELS FOR PERFORMANCE EVALUATION AND BENCHMARKING*
Ehrgott & Gandibleux/ *MULTIPLE CRITERIA OPTIMIZATION: State of the Art Annotated Bibliographical Surveys*
Bienstock/ *Potential Function Methods for Approx. Solving Linear Programming Problems*
Matsatsinis & Siskos/ *INTELLIGENT SUPPORT SYSTEMS FOR MARKETING DECISIONS*
Alpern & Gal/ *THE THEORY OF SEARCH GAMES AND RENDEZVOUS*
Hall/*HANDBOOK OF TRANSPORTATION SCIENCE - 2nd Ed.*
Glover & Kochenberger/ *HANDBOOK OF METAHEURISTICS*
Graves & Ringuest/ *MODELS AND METHODS FOR PROJECT SELECTION: Concepts from Management Science, Finance and Information Technology*
Hassin & Haviv/ *TO QUEUE OR NOT TO QUEUE: Equilibrium Behavior in Queueing Systems*
Gershwin et al/ *ANALYSIS & MODELING OF MANUFACTURING SYSTEMS*
Maros/ *COMPUTATIONAL TECHNIQUES OF THE SIMPLEX METHOD*
Harrison, Lee & Neale/ *THE PRACTICE OF SUPPLY CHAIN MANAGEMENT: Where Theory and Application Converge*
Shanthikumar, Yao & Zijm/ *STOCHASTIC MODELING AND OPTIMIZATION OF MANUFACTURING SYSTEMS AND SUPPLY CHAINS*
Nabrzyski, Schopf & Węglarz/ *GRID RESOURCE MANAGEMENT: State of the Art and Future Trends*
Thissen & Herder/ *CRITICAL INFRASTRUCTURES: State of the Art in Research and Application*
Carlsson, Fedrizzi, & Fullér/ *FUZZY LOGIC IN MANAGEMENT*
Soyer, Mazzuchi & Singpurwalla/ *MATHEMATICAL RELIABILITY: An Expository Perspective*
Chakravarty & Eliashberg/ *MANAGING BUSINESS INTERFACES: Marketing, Engineering, and Manufacturing Perspectives*
Talluri & van Ryzin/ *THE THEORY AND PRACTICE OF REVENUE MANAGEMENT*
Kavadias & Loch/*PROJECT SELECTION UNDER UNCERTAINTY: Dynamically Allocating Resources to Maximize Value*
Brandeau, Sainfort & Pierskalla/ *OPERATIONS RESEARCH AND HEALTH CARE: A Handbook of Methods and Applications*
Cooper, Seiford & Zhu/ *HANDBOOK OF DATA ENVELOPMENT ANALYSIS: Models and Methods*
Luenberger/ *LINEAR AND NONLINEAR PROGRAMMING, 2nd Ed.*
Sherbrooke/ *OPTIMAL INVENTORY MODELING OF SYSTEMS: Multi-Echelon Techniques, Second Edition*
Chu, Leung, Hui & Cheung/ *4th PARTY CYBER LOGISTICS FOR AIR CARGO*
Simchi-Levi, Wu & Shen/ *HANDBOOK OF QUANTITATIVE SUPPLY CHAIN ANALYSIS: Modeling in the E-Business Era*

**** A list of the more recent publications in the series is at the front of the book ****

Printed in the United States of America